Professor Alice Roberts

Editor-in-Chief

Humans

The evolution of a species

Contents

Editor-in-Chief

ALICE ROBERTS
Alice is a biological anthropologist, anatomist, author, and broadcaster known for making science accessible through TV series including *Origins of Us*. Her TV work and books bridge human evolution, archaeology, and medicine, and her research spans palaeopathology, human biology, and human origins. With this breadth of expertise, Alice has guided the concept, content, and development of this book throughout.

Writers and contributors

EMILY TARREGA-SAUNDERS
Lecturer at Royal Holloway, University of London, using anatomy and behaviour to study the evolution of locomotion in humans and other apes.

MARK MASLIN
Professor at University College London and United Nations University, and a world-leading expert on human evolution, past climate changes, and the causes of the Anthropocene.

SALLY REYNOLDS
Associate Professor at Bournemouth University working on African hominin ecology and ancient footprint sites, such as White Sands, New Mexico.

ERIKA MATTHEWS
Palaeontologist working at the University of Arkansas, studying Miocene apes, particularly canine tooth morphology and sexual dimorphism.

LUCY TIMBRELL
Postdoctoral Research Fellow at the University of Liverpool and the Max Planck Institute of Geoanthropology, studying cultural variability in past and present hunter-gatherer societies.

DEREK HARVEY
Naturalist with a special interest in evolutionary biology, Derek has taught a generation of biologists and now concentrates on writing science and natural history books.

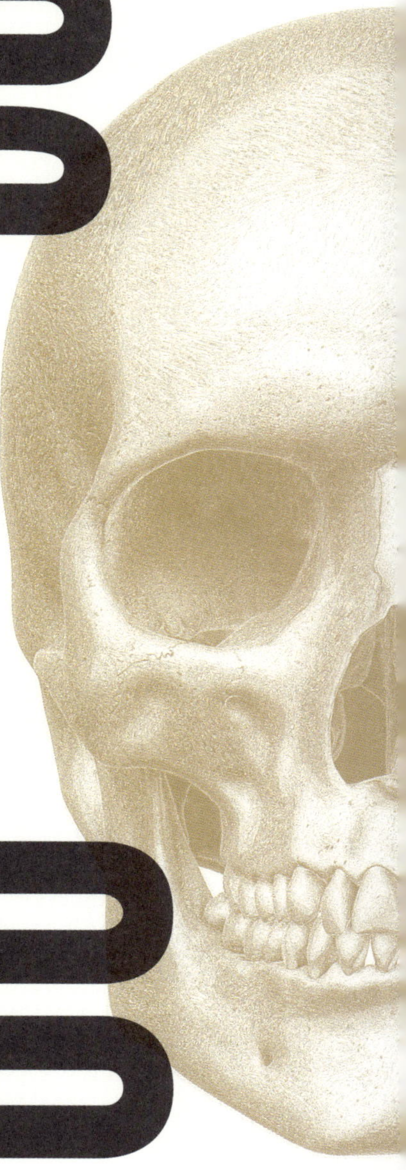

EMILY JEFFRIES
Anthropologist working at Durham University on the potential coevolution between storytelling and cooperation, in which storytelling may act to stabilize prosocial behaviour.

DUNCAN STIBBARD-HAWKES
Evolutionary anthropologist at Baylor University, Texas, studying what hunter-gatherer material culture can (and cannot) tell us about the deep past.

MICHAEL C WESTAWAY
Biological anthropologist and archaeologist at the University of Queensland, currently investigating socio-economic change in Australia and Papua New Guinea in the Holocene.

MORTEN H CHRISTIANSEN
Morten is the William R Kenan, Jr, Professor of Psychology at Cornell University, studying the interaction of biological and environmental constraints in the evolution of language.

MAURICIO GONZÁLEZ-FORERO
Senior Postdoctoral Fellow at the Konrad Lorenz Institute for Evolution and Cognition Research, developing mathematical tools to investigate brain size evolution.

KATJA HEUER
Postdoctoral Researcher at Institut Pasteur, Paris, using histology and neuroimaging to study the evolution and development of the brain and strongly involved in interdisciplinary research.

SALLY STREET
Inter-disciplinary researcher at Durham University interested in understanding large-scale patterns and processes in the evolution of behaviour, cognition, and culture.

MICHAEL MARSHALL
Science writer focused on life sciences, health, and the environment, writing a monthly newsletter about human evolution, *Our Human Story*, for *New Scientist*.

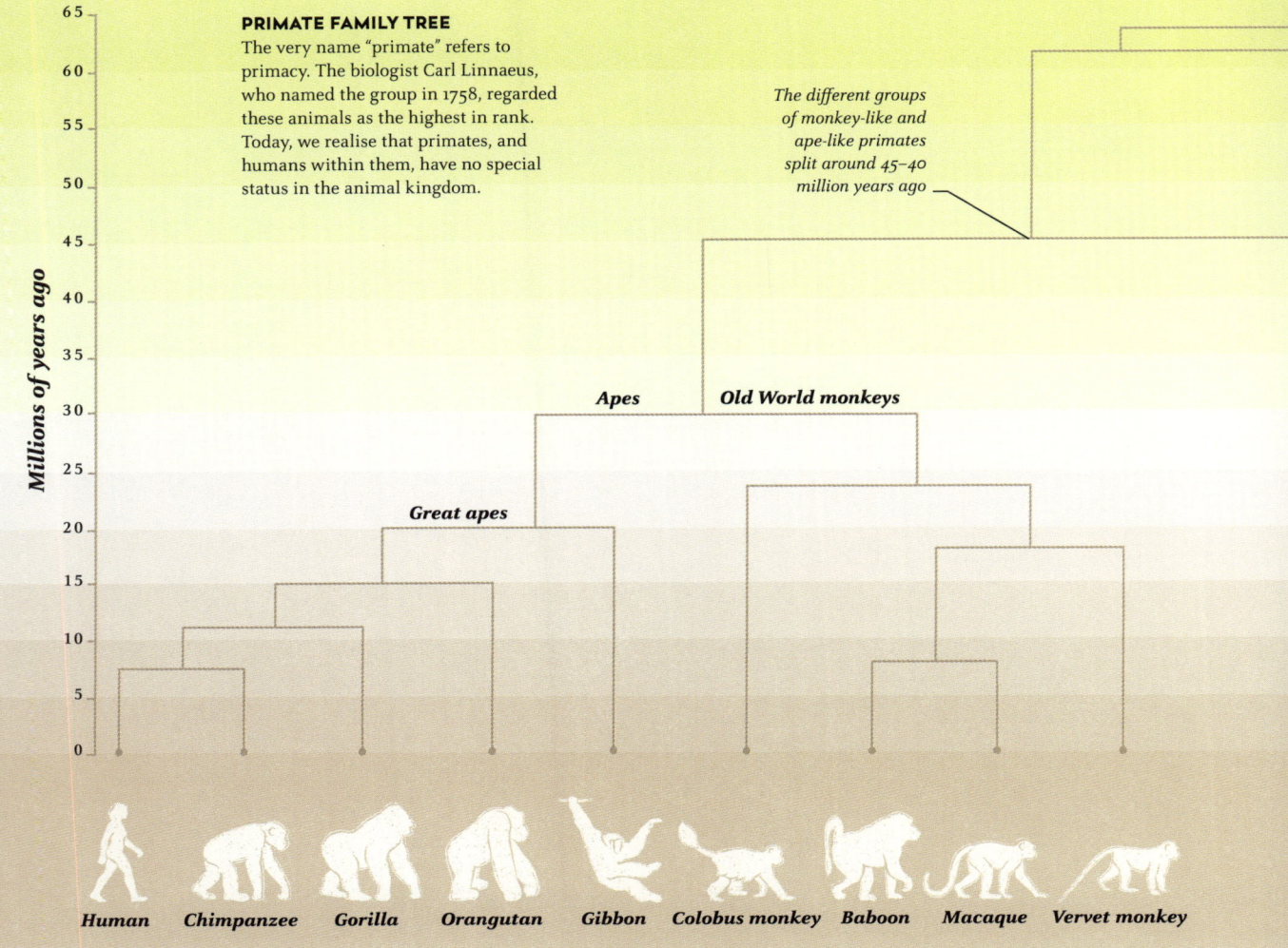

The very name "primate" refers to primacy. The biologist Carl Linnaeus, who named the group in 1758, regarded these animals as the highest in rank. Today, we realise that primates, and humans within them, have no special status in the animal kingdom.

The different groups of monkey-like and ape-like primates split around 45–40 million years ago

Millions of years ago

65
60
55
50
45
40
35
30
25
20
15
10
5
0

Apes

Old World monkeys

Great apes

Human   *Chimpanzee*   *Gorilla*   *Orangutan*   *Gibbon*   *Colobus monkey*   *Baboon*   *Macaque*   *Vervet monkey*

Introduction

Alice Roberts, Editor-in-Chief

It's very easy to think of humans as being so utterly exceptional in every way that we must be some sort of special creation – something that stands apart from the usual rules and laws governing the natural world. And yet, just like any other species on the planet, we're a product of our evolutionary heritage: the environments our ancestors survived in, the ways in which their bodies adapted, and the social and cultural solutions to challenges that they invented or stumbled upon. The best way to understand ourselves – to understand what it truly is to be human – is to compare ourselves with other animals. We can see where there are similarities, when evolution has produced a solution which works so well it tends to be conserved – and where there are differences, shining a light on the aspects of our bodies, brains and behaviour that really are quite special to us.

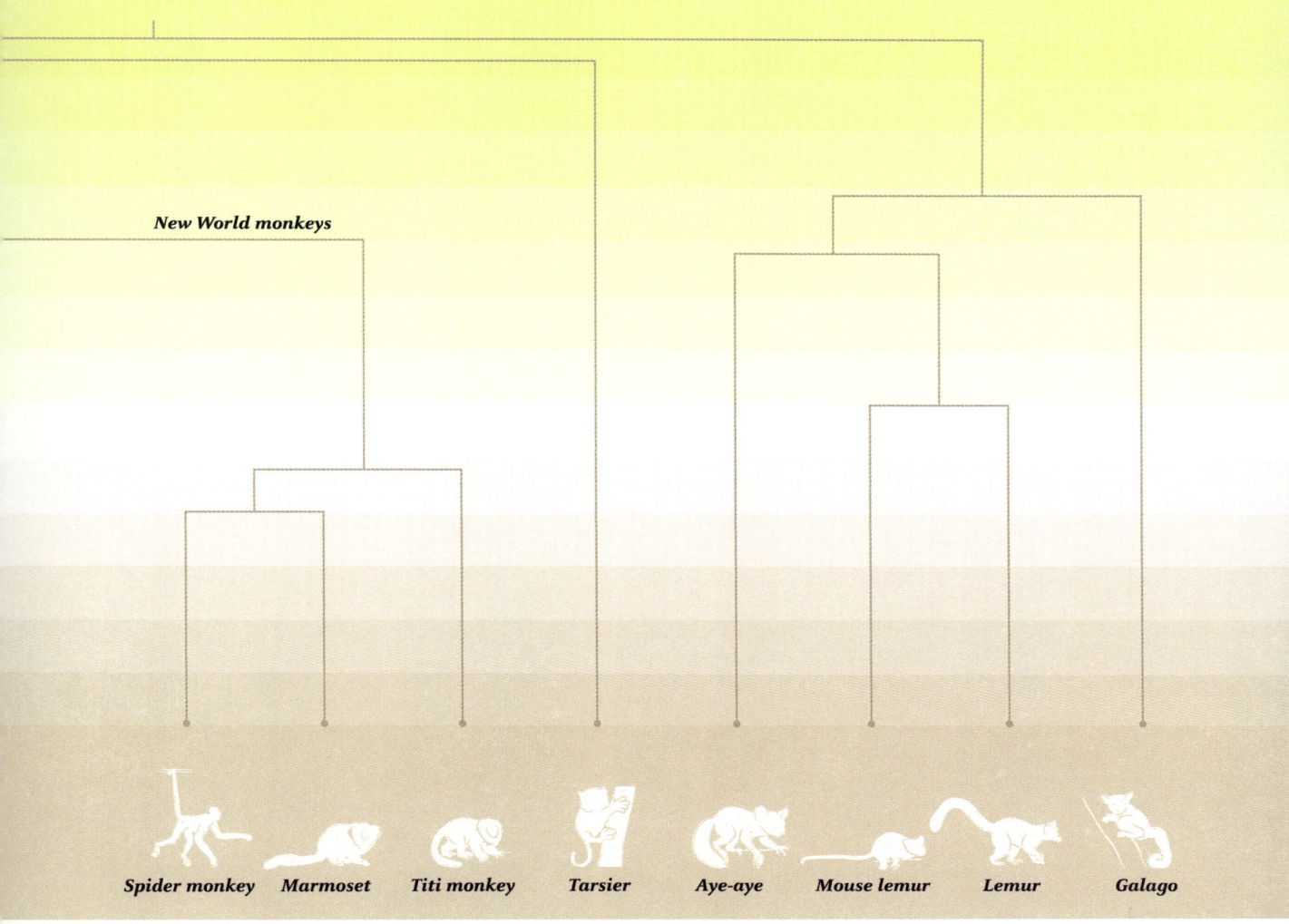

New World monkeys

Spider monkey   **Marmoset**   **Titi monkey**   **Tarsier**   **Aye-aye**   **Mouse lemur**   **Lemur**   **Galago**

"As buds give rise by growth to fresh buds, and these, if vigorous, branch out and overtop on all sides many a feebler branch, so by generation I believe it has been with the great Tree of Life, which fills with its dead and broken branches the crust of the Earth, and covers the surface with its ever branching and beautiful ramifications."

Charles Darwin, 1859

This comparative approach helps us to understand ourselves, but it also leads to a deeper comprehension and grasp of biology itself. For a long time, philosophers and scientists realised that there were some quite compelling similarities between humans and other animals. For an anatomist like Galen, surgeon to gladiators and physician to the emperor Marcus Aurelius in the first century CE, this meant that he could hope to understand human anatomy better by dissecting other mammals; there was enough similarity in those different bodies.

Over the centuries, other scholars tried to make sense of these similarities. They suggested that there was an underlying structure to life on Earth: that living organisms could be ranked in a linear fashion according to degrees of complexity – and perfection.

5

As early as the fourth century BCE, Aristotle was ranking animals in this way, developing a concept which would become known as the "Great Chain of Being" or *scala naturae*. This fundamental idea, always with humans at the top of the scale, of course, would persist right up until the 19th century.

In 1859, Charles Darwin published *On the Origin of Species by Means of Natural Selection*, and the entire rich tapestry of biology was re-hung on a new frame. At the point when Darwin was thinking and writing, there was plenty of evidence of the fact of evolution – from fossils as well as the comparative study of living animals and their embryos. Darwin's real breakthrough – and a concept which had also been proposed by Alfred Russell Wallace – was to explain how evolution happened, through the process of natural selection. But Darwin also realised that this mechanism helped to explain how all life on Earth was linked – not in a ladderlike *scala naturae*, but in a great, branching tree. His idea of "descent through modification" described how species would change over time, evolving away from each other, in a treelike pattern – an idea he had begun to sketch in his notebooks 20 years previously. Some of Darwin's contemporaries went on to unify all life-forms then known into a "tree of life". Ernst Haeckel, like others of his time, placed "man" centrally at the summit of all this evolutionary progress, as he saw it, with lower life-forms, including primitive worms, ranked below.

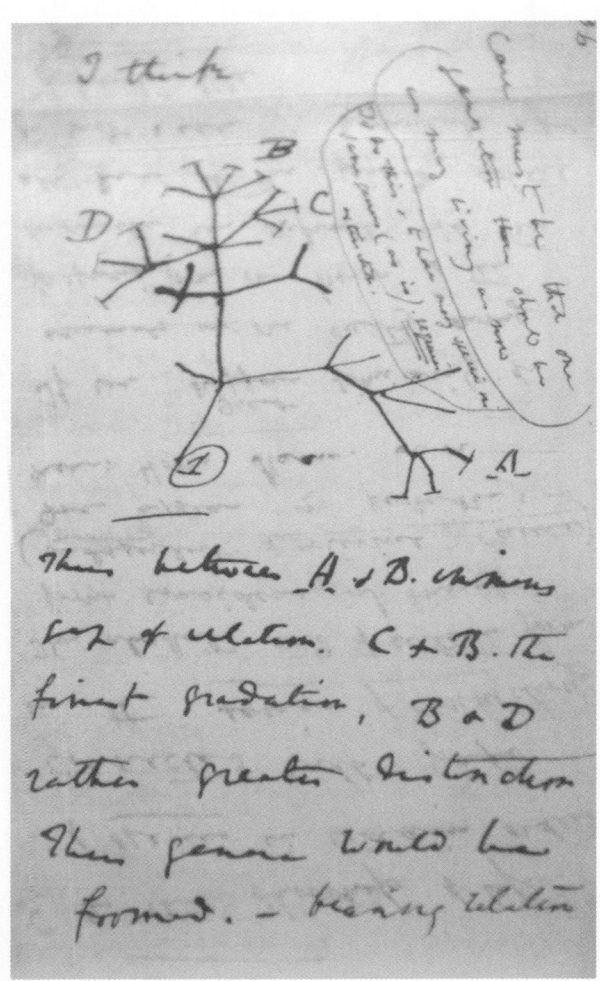

Darwin's notebook
1837

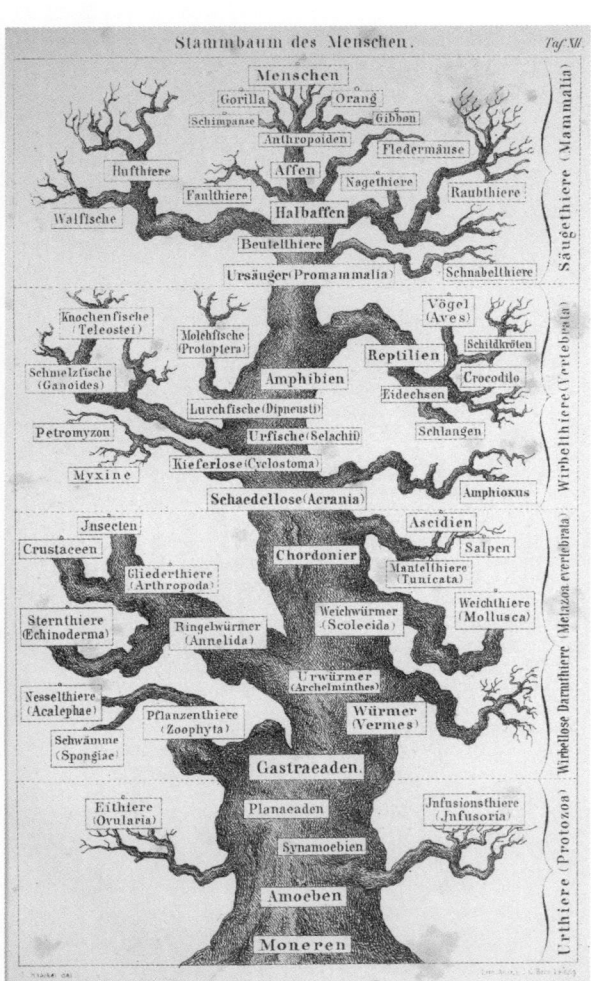

Haeckel's tree of life
1879

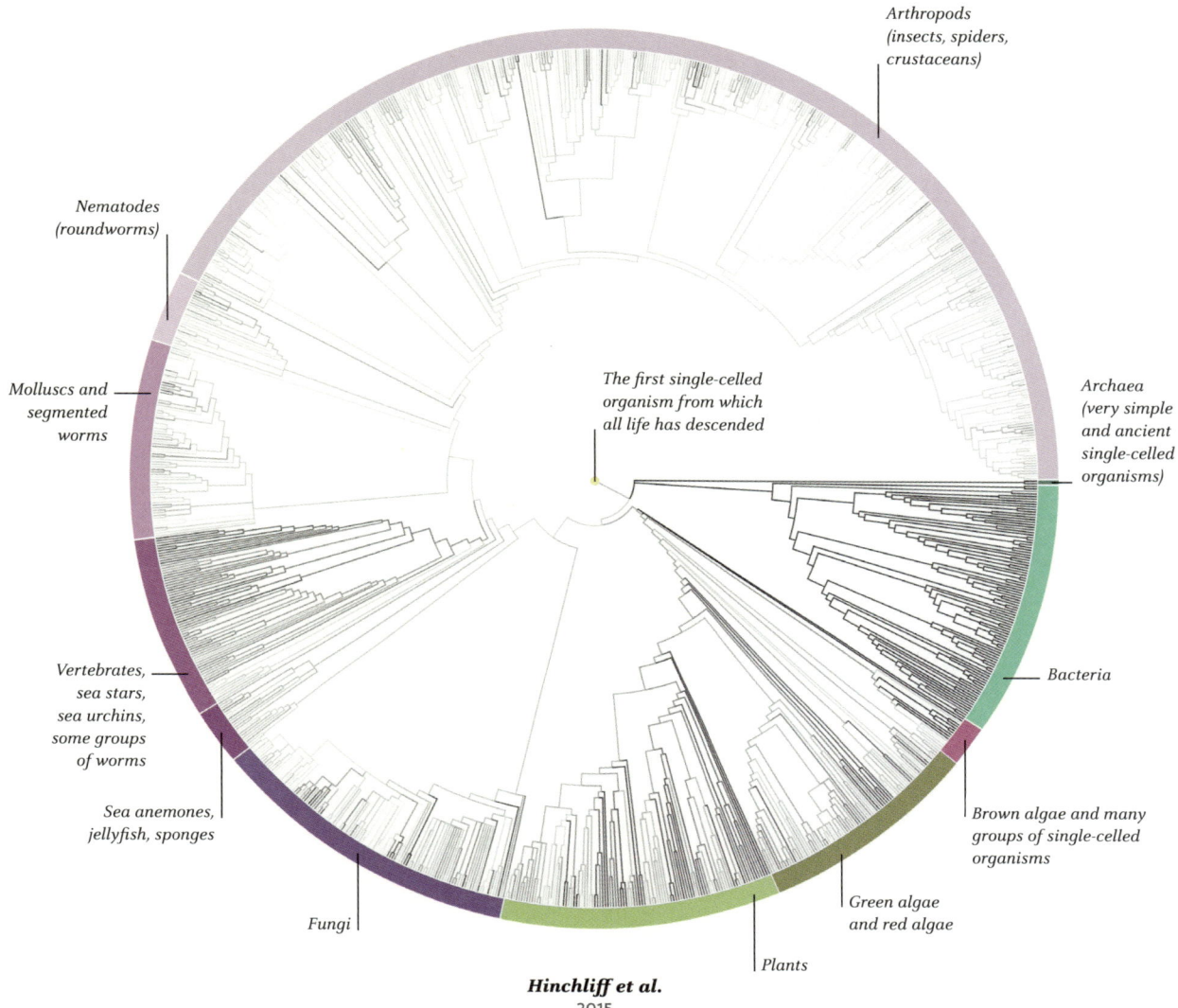

Arthropods
(insects, spiders,
crustaceans)

Nematodes
(roundworms)

Molluscs and
segmented
worms

The first single-celled
organism from which
all life has descended

Archaea
(very simple
and ancient
single-celled
organisms)

Bacteria

Vertebrates,
sea stars,
sea urchins,
some groups
of worms

Brown algae and many
groups of single-celled
organisms

Sea anemones,
jellyfish, sponges

Green algae
and red algae

Fungi

Plants

Hinchliff et al.
2015

When scientists try to visualize the tree of life today, they tend to represent living species as "twigs" in an egalitarian way. The tree is rooted in the centre – then extant species are all branches around the edge, like millions of knights sitting at the famous (though apocryphal) round table of King Arthur.

Where does this leave us, as humans? We're not the pinnacle of evolution, clearly, as there is no pinnacle. And even the bit of the tree of life that includes multicellular organisms – fungi, plants and animals – is actually quite a small, and recent sprig, compared with all the single-celled life-forms living on Earth today.

When we put ourselves in this context, it's quite humbling. In a very real way, we are one, tiny twig on the gloriously branching tree of life. But this means we are part of nature, not separate from it. And it means we can come to understand ourselves better by using this context – comparing ourselves with other animals. By doing that, we can find out what really makes us special.

Alice Roberts

7

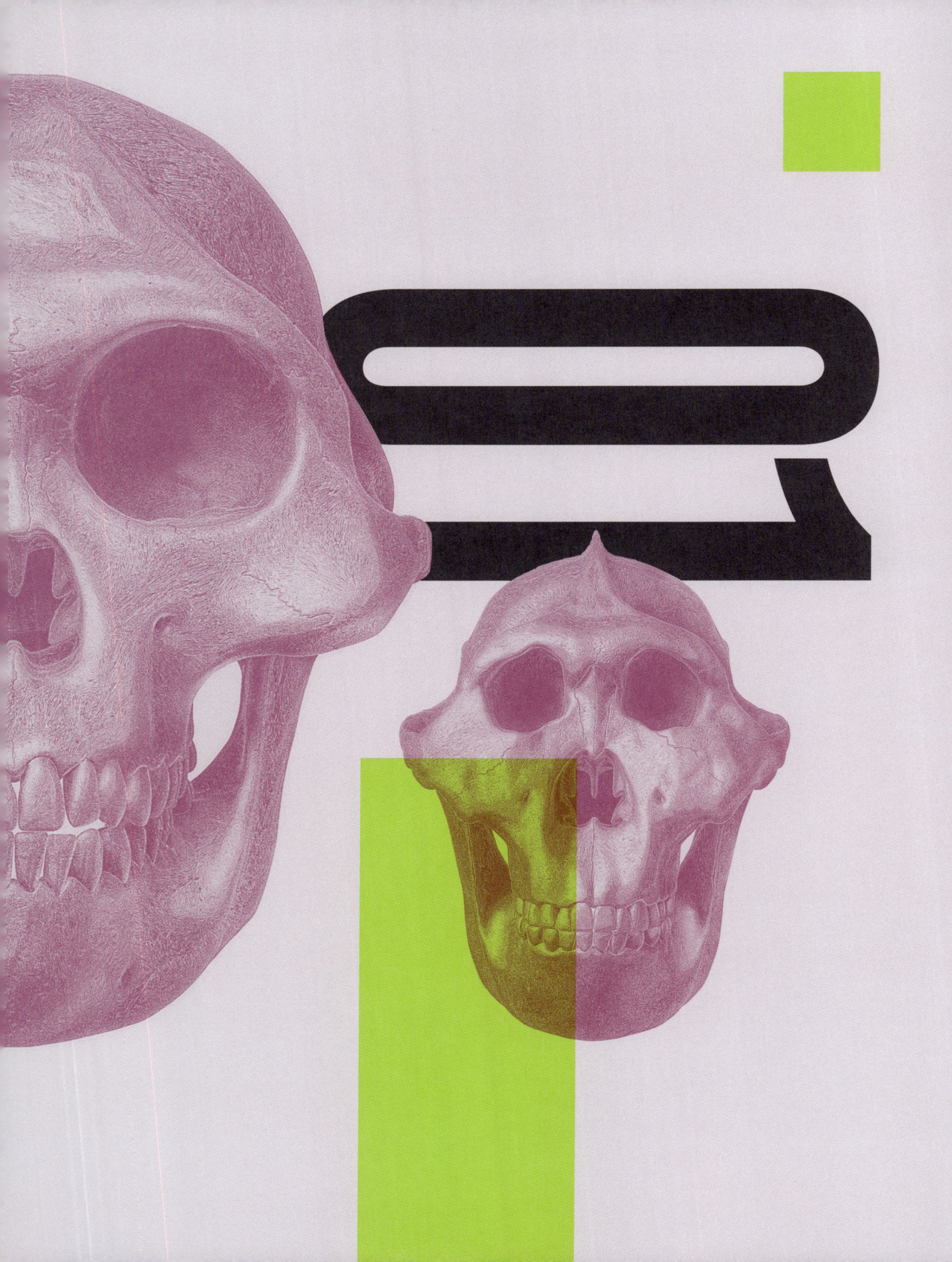

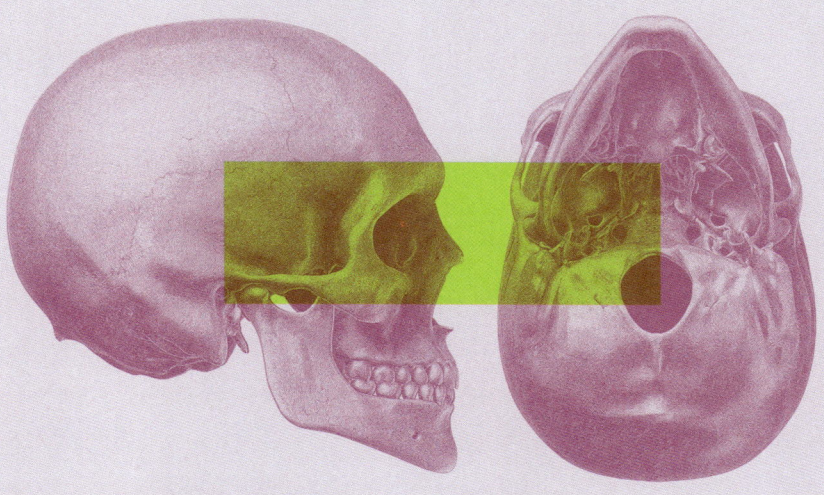

Cast of characters

The story of human evolution is not just the history of Homo sapiens – *it is far older than that. The most recent acts have been played out by a cast of many species of early humans and other more remote ape relations.*

The period spanning 23–5.3 million years ago (MYA) is often called the "Real Planet of the Apes", because of the large number of apes that lived during this time. The apes form a superfamily of primate species that includes humans.

Our ape ancestors

Palaeontologists sometimes call the Miocene epoch the "Planet of the Apes". During this period, countless species of ape existed around the world. Apes form a taxonomic superfamily, taxonomy being the classification of organisms based on shared characteristics and evolutionary relationships. Apes first evolved in Africa, with many fossils being found in Kenya, Ethiopia, and Uganda. Some early apes, such as *Proconsul* and *Ekembo*, had a long, flexible spine and walked along branches on four legs, in a not too dissimilar way to modern monkeys. In contrast, *Morotopithecus* had a shorter, more stable lower back that was better adapted to climbing vertically (see p.58). It also had limbs capable of arm-hanging, similar to modern apes. None of these apes had a tail, which differentiated them from most monkey species.

SUBTROPICAL EUROPE

This period saw increasing warmth and humidity, conditions that allowed vast subtropical forests to spread northward into Europe and Asia, bringing apes with them – fossils of *Griphopithecus,* from 17–16 MYA, can be found as far north as Germany.

RISING UP

By the late Miocene, many species possessed characteristics that define all apes today – no tail, large brains for their body size, flexible shoulders

By following the spread of forests, apes migrated from Africa into Europe and Asia

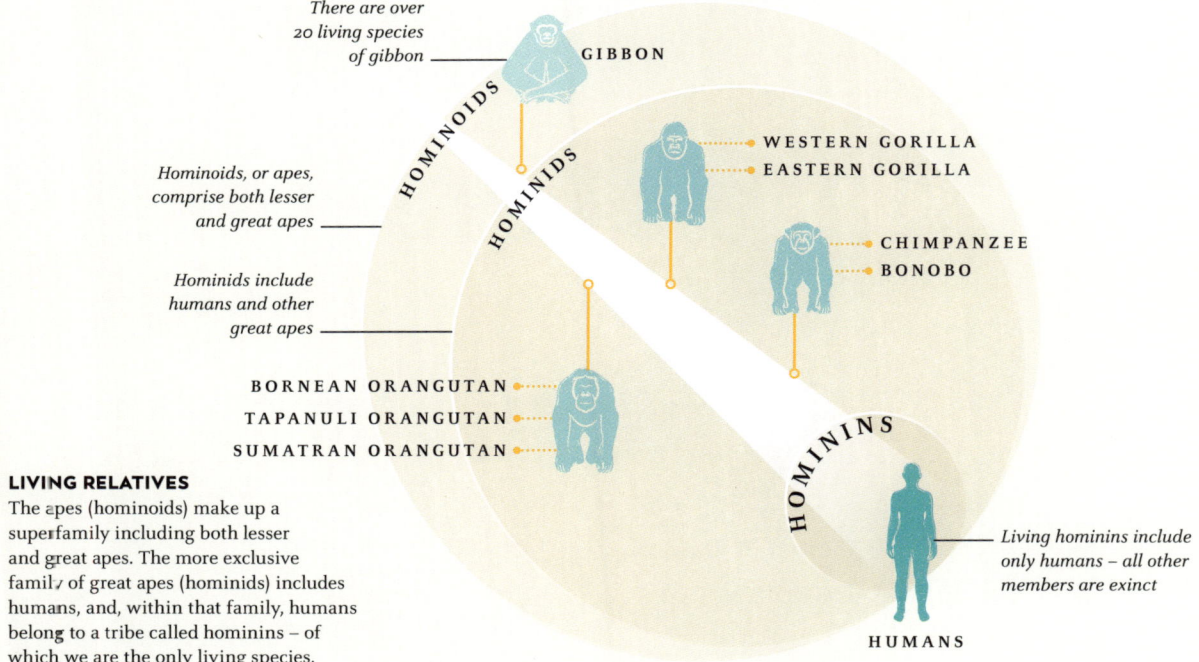

There are over 20 living species of gibbon ———— GIBBON

HOMINOIDS

HOMINIDS

WESTERN GORILLA
EASTERN GORILLA

CHIMPANZEE
BONOBO

Hominoids, or apes, comprise both lesser and great apes ————

Hominids include humans and other great apes ————

BORNEAN ORANGUTAN
TAPANULI ORANGUTAN
SUMATRAN ORANGUTAN

HOMININS

HUMANS

Living hominins include only humans – all other members are exinct

LIVING RELATIVES

The apes (hominoids) make up a superfamily including both lesser and great apes. The more exclusive family of great apes (hominids) includes humans, and, within that family, humans belong to a tribe called hominins – of which we are the only living species.

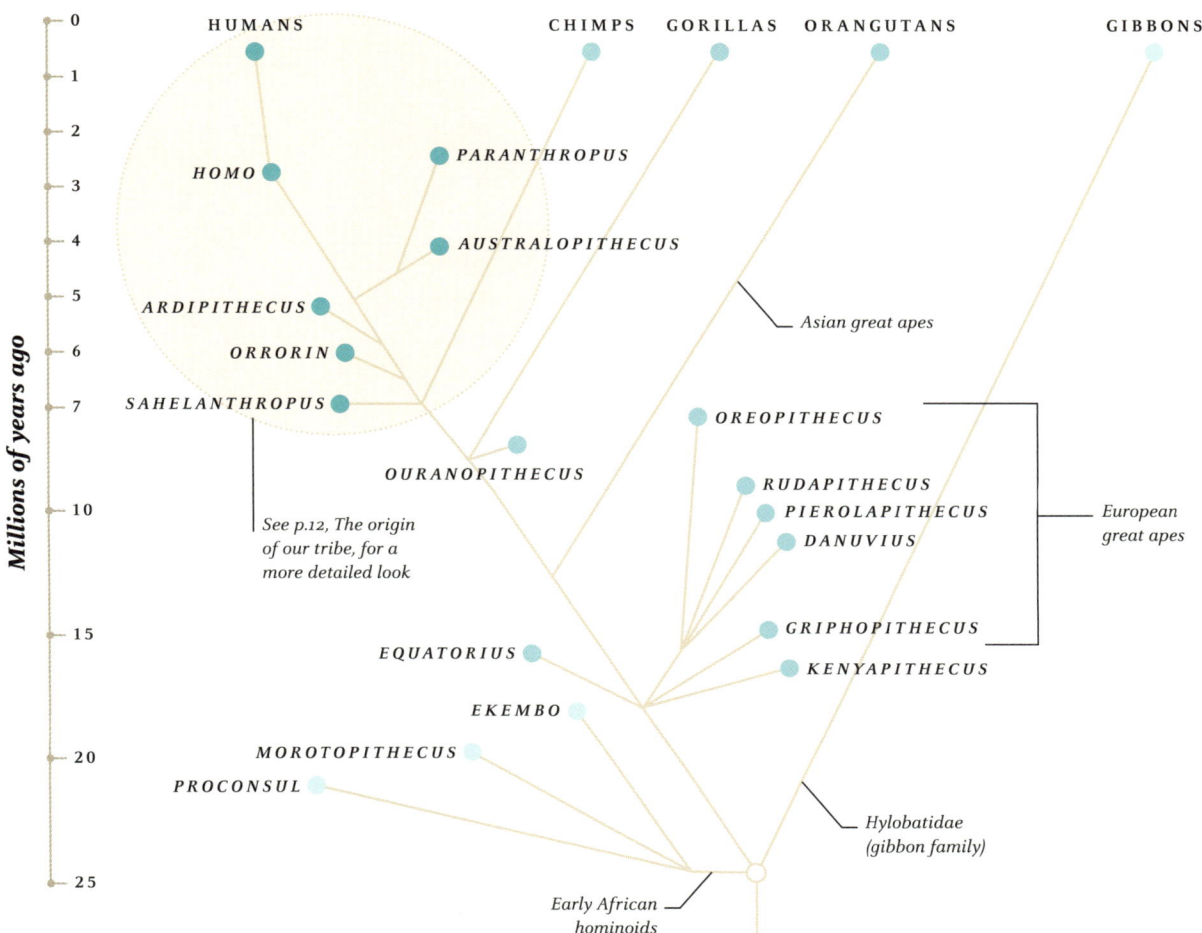

Millions of years ago

HUMANS   CHIMPS   GORILLAS   ORANGUTANS   GIBBONS

HOMO
PARANTHROPUS
AUSTRALOPITHECUS
ARDIPITHECUS
ORRORIN
SAHELANTHROPUS
OURANOPITHECUS

Asian great apes

OREOPITHECUS
RUDAPITHECUS
PIEROLAPITHECUS
DANUVIUS

European great apes

GRIPHOPITHECUS
KENYAPITHECUS

EQUATORIUS

EKEMBO

MOROTOPITHECUS
PROCONSUL

Hylobatidae (gibbon family)

Early African hominoids

See p.12, The origin of our tribe, for a more detailed look

HOMINOIDS
HOMINIDS
HOMININS

EXTINCT RELATIVES

The apes are a diverse and widespread group that includes species that once existed across Africa, Europe, and Asia. Hominins, including humans, are both apes (hominoids) and great apes (hominids).

are known as hominins. By studying the hominins, we can see the gradual emergence of traits that differentiate humans from other apes – small, blunt canine teeth; upright, two-legged walking; enormous brains for our body size; and much more (see pp.12–13).

The evolution of apes over the last 23 million years has resulted, so far, in the emergence of not only our own species, but also two species of gorilla, three species of orangutan, two species of chimpanzee, and around 20 species of gibbon. Alongside humans, all of these apes were equally successful at finding their unique ecological niche and adapting superbly to it. It is only much more recently, when *Homo sapiens* took control of habitats, that our species began distinguishing itself as unusual.

Apes typically had larger brains, flexible shoulders and wrists, broad chests, and shorter lower backs than monkeys

and wrists, broad chests, and short lower backs. Despite these similarities, experts argue that they practiced widely different types of locomotion. In the trees, apes such as *Rudapithecus*, *Danuvius*, *Oreopithecus*, and *Pierolapithecus* may have clambered below branches or walked upright on top of them. Other apes, such as *Ouranopithecus*, may have walked on the ground on all fours.

SPLITTING LINEAGES

The Miocene also saw the split of the human and chimpanzee lineages around 7 MYA. Members of our lineage, both living and extinct,

Pages 12–13

- 🟡 *SAHELANTHROPUS*
- 🟠 *ORRORIN*
- 🔴 *ARDIPITHECUS*
- 🟢 *KENYANTHROPUS*
- ⚪ *PAN*

Pages 14–15

- 🟣 *AUSTRALOPITHECUS*
- 🔵 *HOMO*
- 🟢 *PARANTHROPUS*

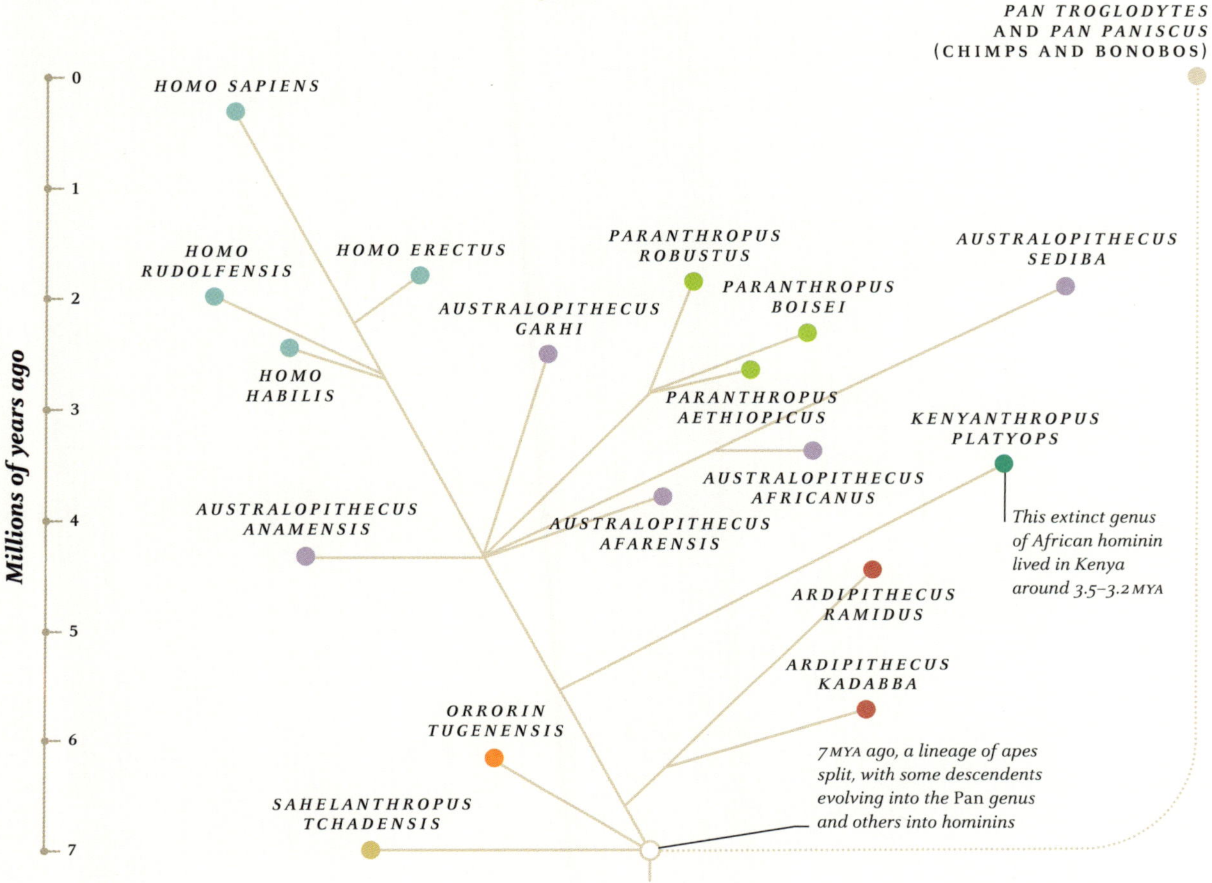

PAN TROGLODYTES
AND *PAN PANISCUS*
(CHIMPS AND BONOBOS)

Millions of years ago

0

HOMO SAPIENS

1

HOMO RUDOLFENSIS

HOMO ERECTUS

PARANTHROPUS ROBUSTUS

AUSTRALOPITHECUS SEDIBA

2

PARANTHROPUS BOISEI

AUSTRALOPITHECUS GARHI

HOMO HABILIS

3

PARANTHROPUS AETHIOPICUS

KENYANTHROPUS PLATYOPS

AUSTRALOPITHECUS AFRICANUS

4

AUSTRALOPITHECUS ANAMENSIS

AUSTRALOPITHECUS AFARENSIS

This extinct genus of African hominin lived in Kenya around 3.5–3.2 MYA

ARDIPITHECUS RAMIDUS

5

ARDIPITHECUS KADABBA

6

ORRORIN TUGENENSIS

7 MYA ago, a lineage of apes split, with some descendents evolving into the Pan genus and others into hominins

SAHELANTHROPUS TCHADENSIS

7

The origin of our tribe

After splitting from a lineage of other apes 7 million years ago, human ancestors formed a distinct taxonomic tribe – the hominins. This tribe includes humans and other apes with the distinctive feature of walking on two legs.

The first hominins appeared around 7 million years ago (MYA). At the time, they would have looked remarkably similar to other ape species in Africa, with only a few traits to set them apart. Most researchers agree that the earliest of these unique characteristics to emerge included smaller canine teeth and features linked to walking on two legs. Small canine teeth may have evolved due to changes in competition between males (see p.186) or changes in diet (see p.138), however two-legged walking was almost certainly influenced in part by climatic change.

One of the first species to display these traits was *Sahelanthropus tchadensis*, emerging in Africa around 7 MYA. Although represented by only a handful of fossils, some experts have

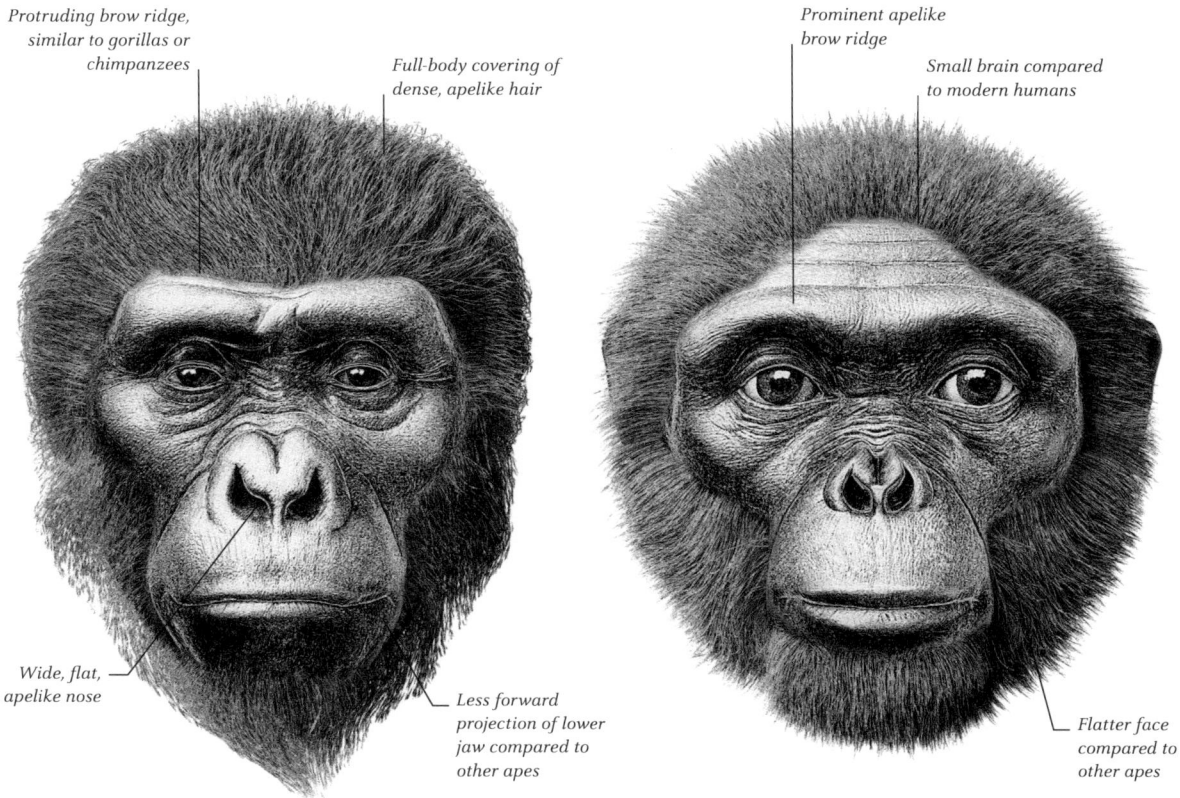

Protruding brow ridge,
similar to gorillas or
chimpanzees

Full-body covering of
dense, apelike hair

Wide, flat,
apelike nose

Less forward
projection of lower
jaw compared to
other apes

Sahelanthropus

Prominent apelike
brow ridge

Small brain compared
to modern humans

Flatter face
compared to
other apes

Ardipithecus

claimed that *Sahelanthropus* possessed bipedal capabilities, evidenced by a centrally located foramen magnum (where the spinal cord exits the skull) that was better positioned to support the head in a vertical posture. Quadrupeds generally have a foramen magnum positioned towards the back of the skull. *Sahelanthropus* also had smaller canine teeth. However, its forearm bones were similar to other apes, alluding to a tree-dwelling lifestyle, and skull remains indicate that it had a small brain, also characteristic of apes. Together, this evidence has made experts reluctant to include *Sahelanthropus* within the hominin tribe.

Fossils of another species, *Orrorin tugenensis*, also display reduced teeth and a thick thigh bone adapted to upright walking. However, like *Sahelanthropus*, long, weight-bearing arms and curved finger bones generate debate as to whether *Orrorin* can be fully classified as a hominin.

ONE STEP CLOSER

By 4.4 MYA, two species emerged that are now more widely accepted as early members of the hominin tribe – *Ardipithecus kadabba* and

CLOSE RESEMBLANCE

Both *Sahelanthropus* and *Ardipithecus* had relatively apelike features, such as a prominent brow, flatter nose, and small brain. However, unlike other apes, both genera exhibited reduced prognathism (forward projection) of the lower jaw.

Ardipithecus ramidus. Like their earlier relatives, these *Ardipithecus* species possessed features compatible with a tree-dwelling lifestyle, such as long arms and a grasping big toe engineered for climbing. However, they also had a centrally placed foramen magnum under their skull and reduced canines, like later hominins. Moreover, *Ardipithecus* possessed a "transitional" pelvis with a short, humanlike hip bone compatible with an S-shaped, humanlike spine. This anatomy suggests that these hominins were as much at home walking upright on the ground as they were clambering in the trees.

Ardipithecus *bridged the gap between tree-dwelling* Sahelanthropus *and land-dwelling hominins*

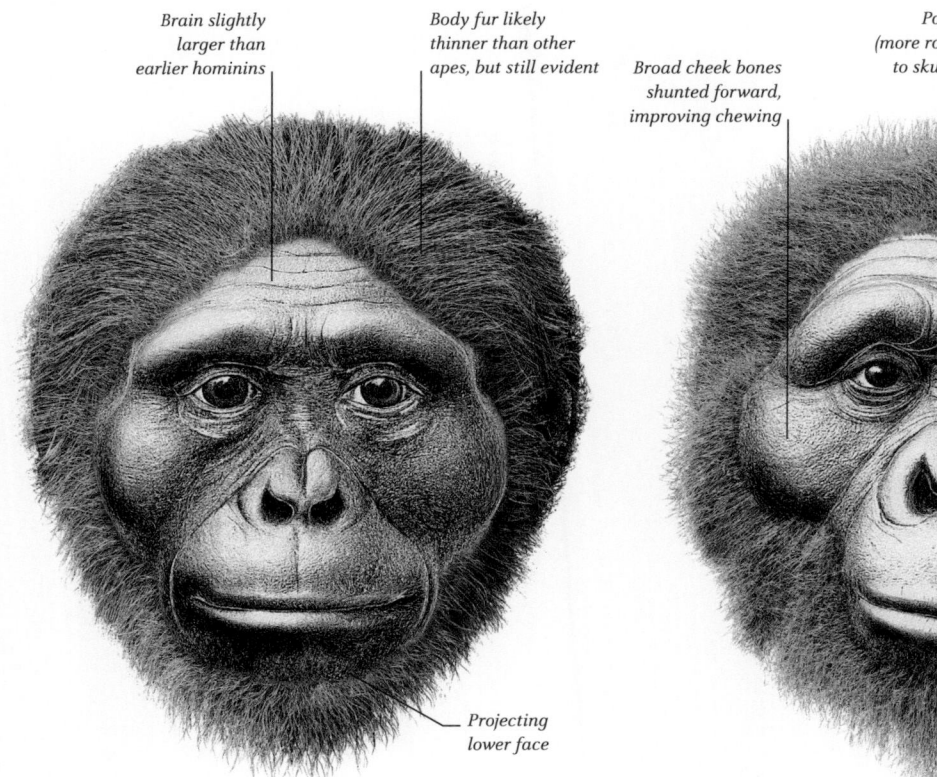

Brain slightly larger than earlier hominins

Body fur likely thinner than other apes, but still evident

Broad cheek bones shunted forward, improving chewing

Powerful chewing muscles (more robust in males) anchored to skull by sagittal crest (bone along midline of skull)

Projecting lower face

Lower face projects outward

AUSTRALOPITHECUS 4.2–2 MYA
Among the australopithecines were the ancestors of early humans. Members of this genus blended ape and humanlike features, with long arms and a spine, hips, and legs adapted to walking on two legs.

PARANTHROPUS 2.7–2.3 MYA
Paranthropus had an upright posture and apelike, cone-shaped torso. Its large, robust teeth and strong chewing muscles reflected an anatomy suited to processing a tough, fibrous diet.

Homo emerges

Around 4 million years ago, Australopithecus appeared in east Africa. It was a competent biped, possibly travelling great distances on two legs. From the australopithecines arose Homo, the genus from which modern humans evolved.

By 4 million years ago, hominins were highly adept walkers

As hominins committed to the ground, they evolved to be even more efficient bipeds. *Australopithecus* was an enormously successful group. Early species, such as *A. anamensis* and *A. afarensis,* were competent upright walkers, and it is believed that later species, such as *A. africanus* and *A. sediba,* may have even been able to traverse long distances on two legs. Although these hominins retained some tree-dwelling features, such as curved fingers and long arms similar to orangutans, they also had the full suite of bipedal characteristics, from a humanlike pelvis to inline big toes. Large molar teeth with thick enamel allowed australopithecines to exploit tough foods, such as leaves and nuts. However, slightly larger brains and dexterous hands may have enabled some of them to use basic tools capable of slicing meat from bone – two 3.4-million-year-old fossilized mammal bones from Ethiopia display cut marks and are attributed to *A. afarensis*. It is unlikely that these small-bodied hominins were hunting large prey at this time, instead relying on chasing off predators and stealing prey.

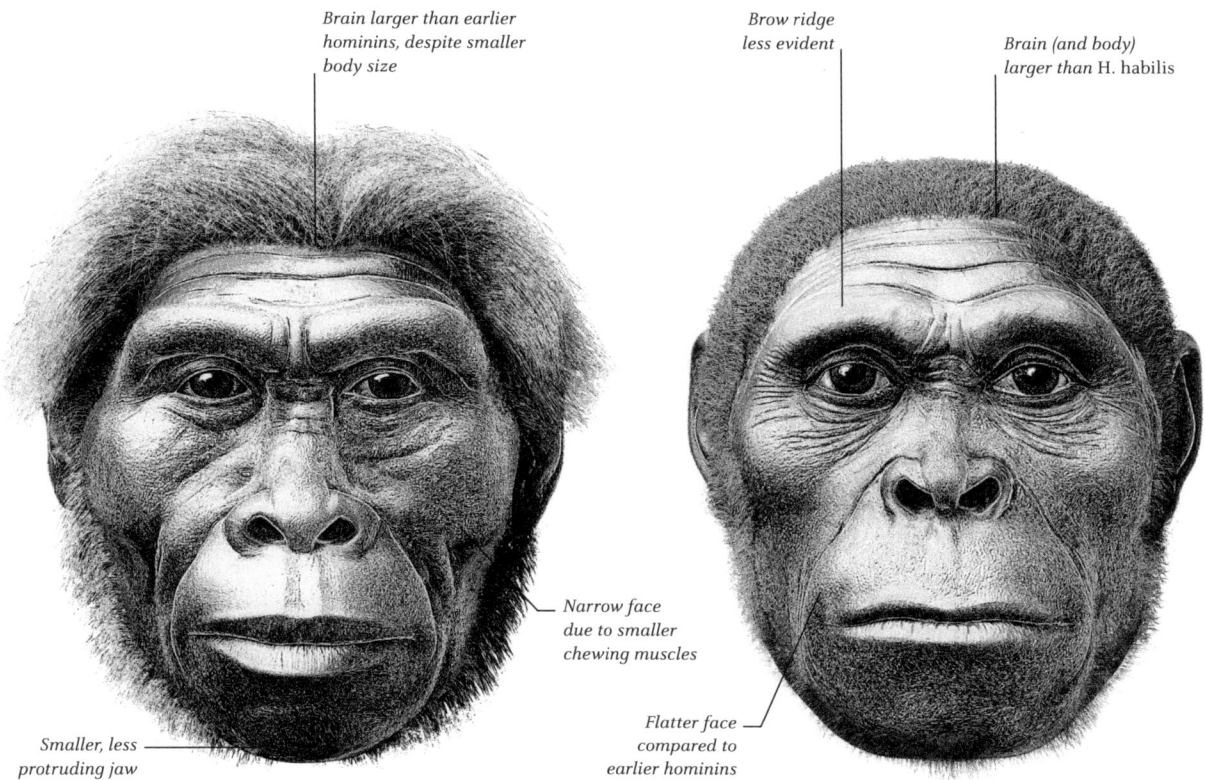

Brain larger than earlier hominins, despite smaller body size

Brow ridge less evident

Brain (and body) larger than H. habilis

Narrow face due to smaller chewing muscles

Smaller, less protruding jaw

Flatter face compared to earlier hominins

HOMO HABILIS 2.4–1.4 MYA
This human ancestor was the first formally named *Homo* species. *H. habilis* was relatively short, between 100–135 cm (39–53 in), and had long, apelike arms and an upright stance.

HOMO RUDOLFENSIS 1.9–1.8 MYA
Although fossil evidence is limited, experts hypothesize that this species was larger and more robust than *Homo habilis*, with an average height of 148–175 cm (58–69 in).

DISTANT RELATIVE

The descendants of the australopithecines branched off in two very different directions. One branch led to our own genus, *Homo*, while the other evolved into *Paranthropus*. Members of this genus developed the australopithecines' powerful chewing abilities, evolving enormous molars to break down low-quality, abrasive foods, such as grass. Small-brained and bipedal, some species may have used tools to process the occasional scavenged carcass, suggesting that while their anatomy was adapted for herbivory, they were more than capable of omnivory, too.

EARLY HUMANS

The first members of *Homo,* or early humans, emerged 2.8 MYA in eastern Africa, but fossils are too fragmentary to assign them to any particular species. Experts disagree on how to group the first named *Homo* species, *H. habilis* and *H. rudolfensis*, which retained many australopithecine traits. Small-bodied with long arms and relatively short legs, these early humans may have still spent plenty of time in the trees. However, they possessed smaller teeth, larger brains, and started to regularly make and use stone tools. It is thought that larger brains and toolmaking (see pp.34–35) created a feedback loop in which bigger brains innovated better tools, enabling a broader diet to support the growth of even larger brains. Like their ancestors, the first early humans were confined to Africa. But, evidence suggests that by 2 MYA they were on the move.

Early humans had small bodies with long, apelike arms, but also had larger brains, a flatter face, and a broader toolkit

15

Around 300,000 years ago, up to four species of Homo *coexisted in Africa, with further species in Europe and Asia. By 40,000 years ago, only* Homo sapiens *remained – spread across five continents.*

Humans overlap through space and time

HOMO HABILIS

HOMO RUDOLFENSIS

2.6 2.4 2.2 2 1.8 1.6 1.4

Millions of years ago

Early humans – hominins belonging to the *Homo* genus – continued to evolve in Africa, where they first emerged, but some populations expanded into Eurasia, probably several times to varying degrees of success. Some of the earliest fossils outside Africa were found in Georgia and date to around 1.8 million years ago.

Some researchers propose that most of these specimens, inside and out of Africa, belong to a single highly variable and long-lived species called *Homo erectus*. Others suggest that only the Asian specimens should be considered *Homo erectus*, with the Georgian fossils belonging to a separate species called *Homo georgicus* and the African specimens belonging to *Homo ergaster*.

Regardless of their classification, hominins established themselves across Eurasia, evolving as they went. Most are united by certain trends, such as bigger brains and more advanced tools, however these elements did not define our entire genus. Enormous brains, some larger than modern averages, could be found in later members of *Homo erectus* and *Homo longi,* while diminutive brains characterized other species such as *Homo floresiensis* and *Homo naledi*. Advanced and basic tools alike helped *Homo* species get a foothold in a variety of habitats, from frigid mountain plateaus to remote tropical islands.

These hominins overlapped in space, time, and strategy. Throughout the last 2 million years they cohabited the Earth, making encounters with one another a virtual guarantee.

HOMO GEORGICUS

● AFRICA
● ASIA
● EUROPE
● NORTH AMERICA
● AUSTRALIA

AND THEN THERE WAS ONE
Coexistence allowed species to exchange everything, from tools and technology to genes through interbreeding. But, by 40,000 years ago, each species had gradually winked out of existence – leaving only one.

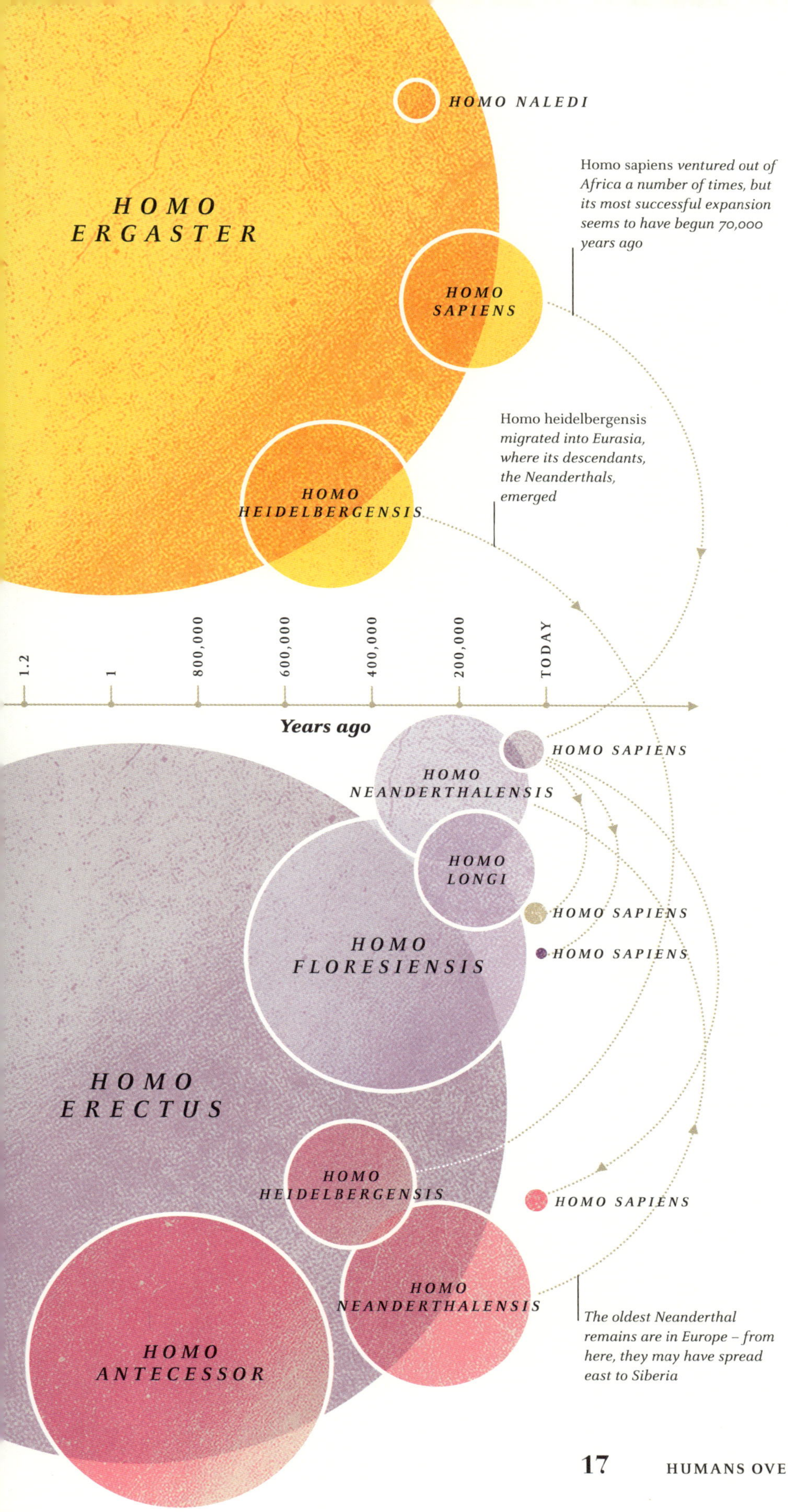

HOMO ERGASTER

HOMO NALEDI

Homo sapiens *ventured out of Africa a number of times, but its most successful expansion seems to have begun 70,000 years ago*

HOMO SAPIENS

Homo heidelbergensis *migrated into Eurasia, where its descendants, the Neanderthals, emerged*

HOMO HEIDELBERGENSIS

The *Homo* genus encompasses a wide variation in braincase sizes. The smallest cranium (*H. habilis*) is the earliest, while the largest craniums belong to the most recent species, *H. neanderthalensis* and *H. sapiens*.

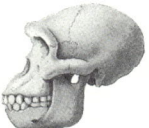

Homo habilis

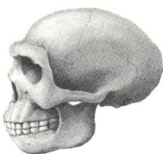

Homo erectus

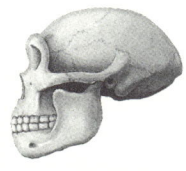

Homo heidelbergensis

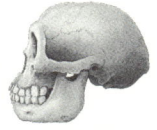

Homo floresiensis

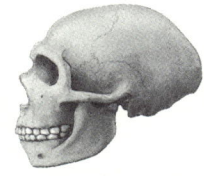

Homo neanderthalensis

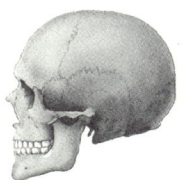

Homo sapiens

| 1.2 | 1 | 800,000 | 600,000 | 400,000 | 200,000 | TODAY |

Years ago

HOMO NEANDERTHALENSIS

HOMO SAPIENS

HOMO LONGI

HOMO SAPIENS

HOMO FLORESIENSIS

HOMO SAPIENS

HOMO ERECTUS

HOMO HEIDELBERGENSIS

HOMO SAPIENS

HOMO NEANDERTHALENSIS

The oldest Neanderthal remains are in Europe – from here, they may have spread east to Siberia

HOMO ANTECESSOR

The birth of our species

Homo sapiens first emerged 300,000 years ago in Africa to a world replete with other hominins. When we evolved, at least five other species of early human walked the Earth. By 40,000 years ago, only one remained.

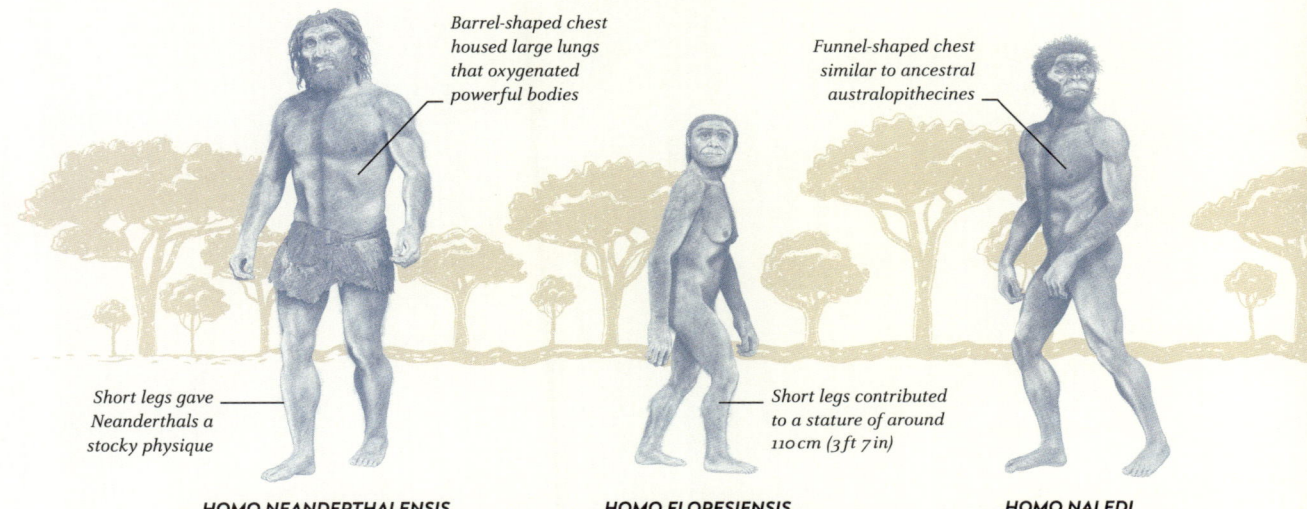

Barrel-shaped chest housed large lungs that oxygenated powerful bodies

Funnel-shaped chest similar to ancestral australopithecines

Short legs gave Neanderthals a stocky physique

Short legs contributed to a stature of around 110 cm (3 ft 7 in)

HOMO NEANDERTHALENSIS
Neanderthals mostly lived in Europe and western Asia, dying out around 40,000 years ago.

HOMO FLORESIENSIS
Adapting to scarce resources may have led to the diminutive stature of H. floresiensis.

HOMO NALEDI
Small-brained H. naledi possessed a combination of modern and archaic features.

In many ways, our species, *Homo sapiens*, was not vastly different from other *Homo* species, or early humans, present at the time. By 300,000 years ago, most early humans shared certain traits, such as a big brain and flatter face. However, *H. sapiens* were lankier and taller than our other human relatives, which experts suggest was an adaptation to our African habitat (taller, more slender builds facilitate heat loss).

Despite belonging to the same species, early *H. sapiens* did not look quite like modern humans. They did not possess the globular skull, defined chin, and smaller teeth that identify humans today. At their inception, *H. sapiens* presented a mixture of archaic and modern features – their skull was more elongated, like Neanderthals, and they retained a suggestion of the prominent brow

> The first H. sapiens did not look like modern humans

ridge that was more pronounced in earlier *Homo* species, such as *Homo erectus*. Although possessing a retracted face and the precursor of a chin, early *H. sapiens* still had larger teeth and jawbones, possibly reflective of a diet that still contained tough, fibrous foods.

GRADUAL SHIFTS

Fossils reveal that between 100,000–35,000 years ago, our ancestors' skulls evolved into the more globular shape we possess today, followed by a noticeable reduction in the size of teeth and jaws by around 30,000 years ago. While the latter is linked to increasing tool use, cooking, and food processing (see pp.148–53), which introduced softer foods into our diet, they were also the product of a long-term evolutionary trend towards

HOMO SAPIENS
Appearing in Africa at least 300,000 years ago, H. sapiens is the last surviving Homo species, spread across the world.

Shorter arm length relative to leg likely inherited from earlier Homo species

Only Homo sapiens and Neanderthals have left evidence of making clothing – it is unknown whether other species were clothed

LIVING TOGETHER
When *H. sapiens* first emerged, they shared the planet with at least five other *Homo* species – and experts believe it was likely more. Although belonging to the same genus, different early human species had adaptations to their local environment that set them apart.

Humanlike hands capable of a powerful grip

Large nasal opening possibly an adaptation to colder weather

HOMO LONGI
Also known as Denisovans, H. longi is believed to have had a powerful build – an adaptation to the cold.

HOMO ERECTUS
Robust H. erectus had longer legs and shorter arms with modern humanlike proportions.

smaller teeth and jaws, one that continues. Today, some individuals have jaws that are so small they cannot accommodate wisdom teeth (see p.139) – a result of this continuing evolutionary process.

AND THEN THERE WAS ONE
By 40,000 years ago, we were the sole remaining human species settled across Africa, Europe, and Asia. Debates rage as to why our species remains the last hominin standing. *H. sapiens* may have interbred with other humans and assimilated them into their own groups. Perhaps we out-competed them in the face of a changing climate due to a more adaptable and innovative nature. It may have just been luck. In the end, it is likely that many factors played a role in the persistence and ubiquity of *H. sapiens*.

LAST SURVIVORS
In the last few tens of thousands of years, *H. sapiens* domesticated dogs, innovated more advanced tools, and created new forms of art. By 10,000 years ago, some human groups had established the first permanent settlements. Foragers began to experiment with farming and a sedentary lifestyle, leading to the emergence of towns and cities. Settlements traded goods, religions, and ideas at a pace unimaginable to earlier human groups. Today, we are a global species wielding incredible technologies – our bodies, minds, and creative abilities a testament to a vast and deep hominin history.

By 10,000 years ago, humans had domesticated dogs and created permanent settlements

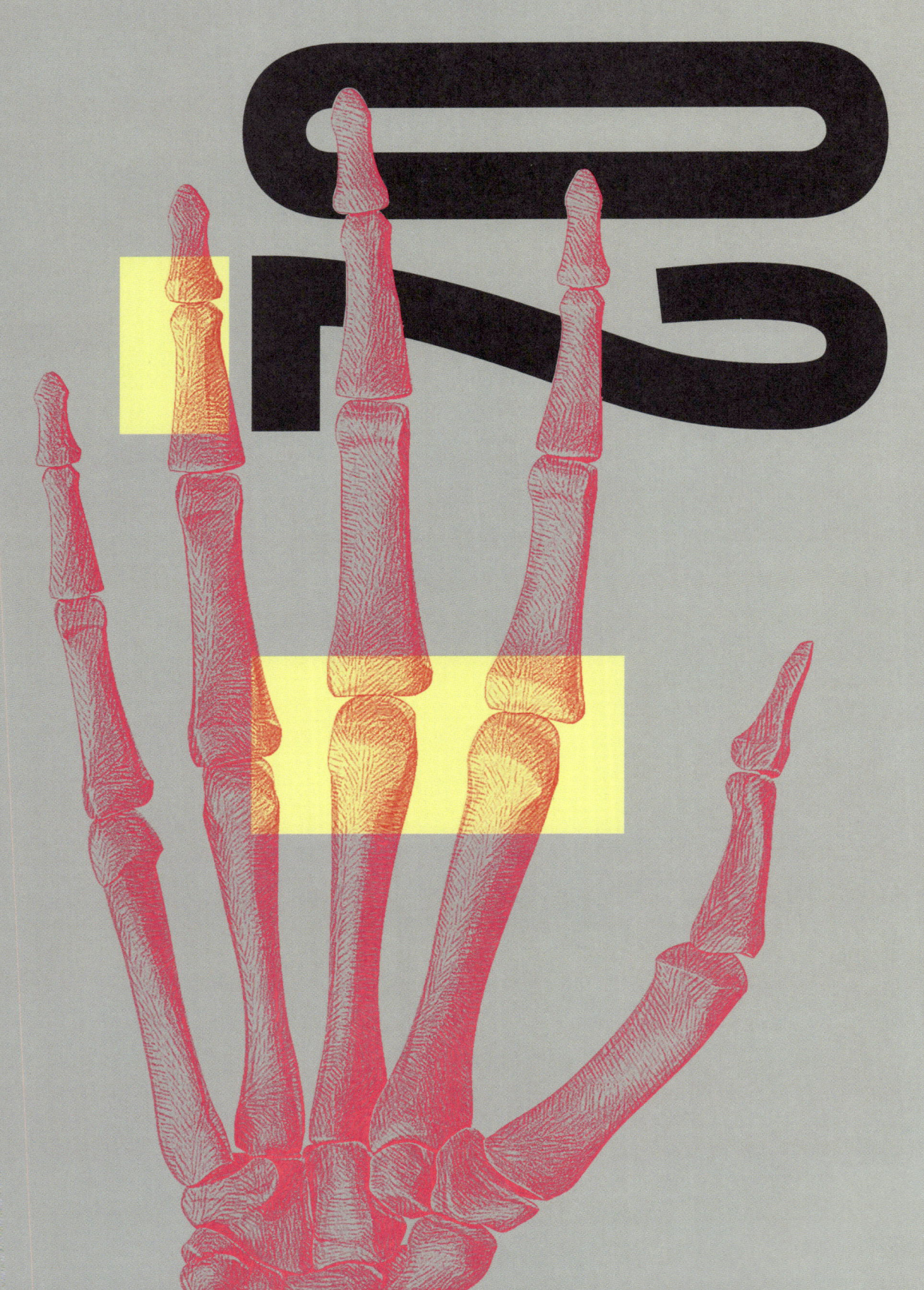

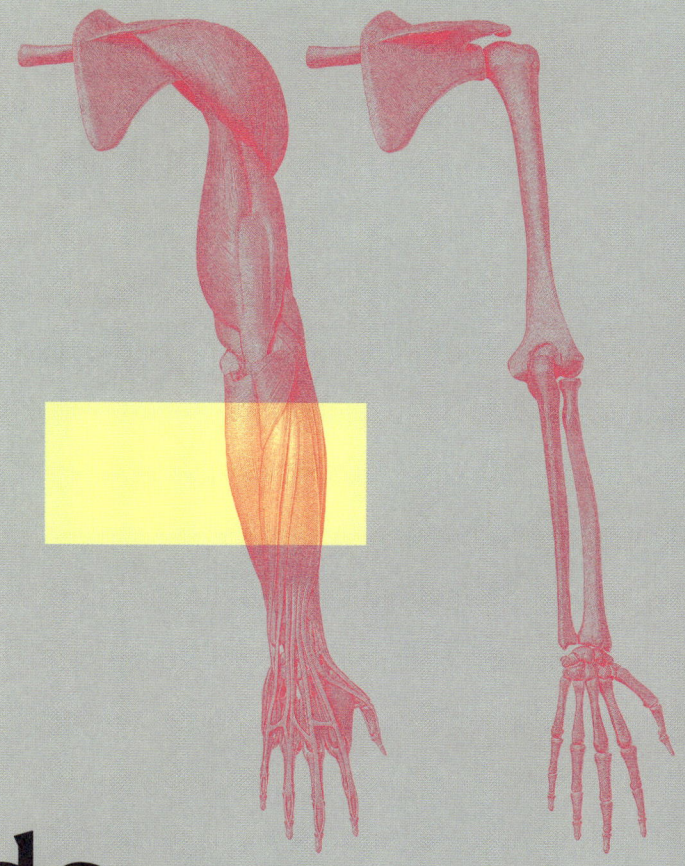

Hands and arms

The human knack of manipulating things with power and precision is rooted in our treeliving ancestors, but early humans developed it further, to knap stones, throw spears, and gesture ideas. Our dexterity is one our most distinctive traits.

Trace your family tree back roughly 375 million years, and your ancestor did not have hands or limbs, but fins. The earliest tetrapods (animals with four limbs) evolved forelimbs first, from pectoral fins. These primitive arms often had more than five digits and were probably used for interacting with the ground, perhaps for helping movement through swamps, but not for walking on the ground out of water. Even today, mammal forelimbs share skeletal similarities with fins, and the genes that control limb development in a human embryo, known as Homeobox genes (Hox for short), also control fin development in fish.

NOT JUST FOR WALKING

Once animals started walking on all four limbs, forelimbs began to vary in function. Mammals have perhaps the greatest diversity of forelimb shapes (compare human hands to cat paws and bat wings), and their colonization of different environments may be due to their wide range of forelimb functions, which includes foraging and grooming as well as movement. Primates developed hands with five separate digits and nails, which were useful for grasping branches and foraging with greater dexterity. This paved the way for using tools, which occurs in many primates but was honed by hominins, once they stopped relying on their forelimbs to move around and evolved hands specialized for dexterous and powerful object manipulation.

A ground-based, bipedal lifestyle freed hominin forelimbs to become adapted to dexterous tasks

Embracing change

Modern human hands bear little resemblance to the fins of fish, yet that is where their origin lies. Our forelimbs show vestiges of adaptations that evolved in our ancestors, from walking on all fours and swinging from trees to throwing a spear.

Early tetrapods had more than five digits – Acanthostega, for example, had eight

LAND VERTEBRATES

375–55 MYA

DEVELOPING DIGITS
The first animals to develop forelimbs were aquatic. They evolved digits that helped them move across the beds of swamps or lagoons, but their limbs could not bear their weight on land.

Fin bones of the lobe-finned fish Elpistostege share a common origin with mammal forelimb bones

ORIGIN OF LIMBS
The ancestors of tetrapods belonged to a group called the lobe-finned fish. Their pectoral fins were the evolutionary precursors to forelimbs.

Similarities in embryonic fins and forelimbs reveal their shared evolutionary history

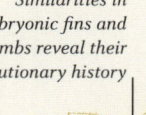

REGULATOR GENES
A mammalian embryo's developing limb buds have growth that is organized by the same group of Hox genes as fish appendages.

Forelimb diversification started in early mammals, such as Megaconus mammaliaformis

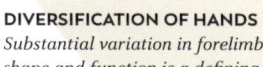

DIVERSIFICATION OF HANDS
Substantial variation in forelimb shape and function is a defining feature of mammals.

GRASPING HANDS

Primates developed long, curved fingers with nails and opposable thumbs. These hands had a powerful grip and enabled safe movement around the treetops.

Grasping hands and opposable thumbs were an early primate adaptation to life in the trees

FROM FINS TO HANDS

From the pectoral fins of ancient fish, our ancestors' forelimbs underwent a myriad of adaptations to each habitat and use. Once hominins no longer needed to use their forelimbs for moving around the trees, their arms and hands gradually became honed for both precision and power.

Apes have short thumbs, but their long fingers can hold objects in a variety of grips

DEXTERITY

The shape of primates' hands allowed them to manipulate objects with more control and begin using them as tools.

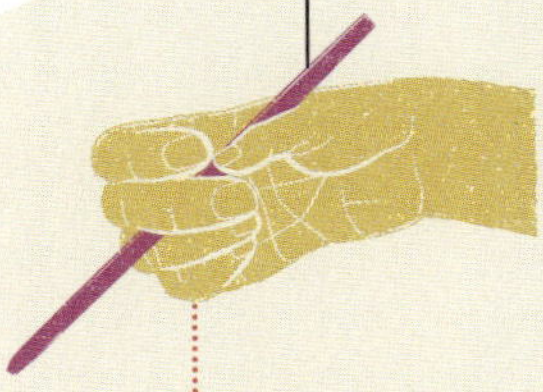

Stretching back the mobile shoulder builds elastic energy for throwing

THROWING

Later hominins developed wide, sideways-facing shoulders, which, together with torso and hip rotations, enabled them to throw with power and accuracy.

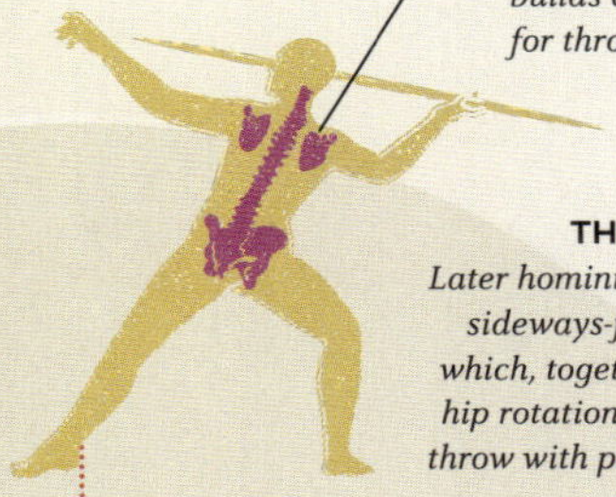

PRIMATES

55–7 MYA

HOMININS

7 MYA–PRESENT

Chimpanzee shoulder bones are angled upwards, to enable hanging beneath branches and climbing

Our long thumbs can provide powerful grips to oppose the other fingers

Short forearms can be held up by a flexed elbow during running

MOBILE ARMS

Large-bodied apes developed mobile shoulders and wrists suited to moving between branches and hanging underneath them.

PRECISION GRIP

Long thumbs with unique musculature allowed hominins to hone a precision grip and carry out tasks – such as making tools – that required great dexterity.

SHORT, SWINGING ARMS

Human running relies on arms that swing back and forth, which can only happen with wide shoulders and relatively short arms.

23

If you pinch the tips of your thumb and any finger together, this is a '"precision grip" – something that only humans possess, and which was critical for the development of tool use in hominins.

Precision grip and pinch power

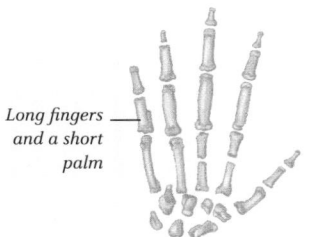

Long fingers and a short palm

Ardipithecus ramidus

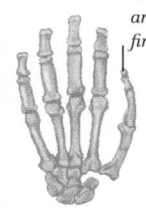

Longer thumb and shorter fingers

Australopithecus sediba

Over time, hominin hands have generally become shorter and wider as they were used less for moving around the trees, and more for manipulating objects. The earliest hominins had long, curved fingers for gripping branches, although their hands were still shorter than chimpanzee hands.

The discovery of the 4.4-million-year-old hominin *Ardipithecus ramidus*, nicknamed "Ardi", suggests that the hands of our last common ancestor with chimpanzees (an unknown species from approximately 7 million years ago) may have been more humanlike than researchers first thought. *Ardipithecus* had long fingers, short palms, and flexible wrists. Humans rely on short palms and flexible joints to achieve a precision grip, so even though Ardi probably didn't use tools, these features may have pre-adapted hominin hands for manual dexterity.

Ardipithecus hands show traits that may have laid the foundations of later dexterity

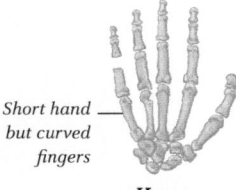

Short hand but curved fingers

Homo naledi

EVIDENCE OF A PRECISION GRIP

Around 2–3 million years ago, most Australopithecines retained curved fingers, but their fingers were shorter relative to their thumbs (although their thumbs were still small compared with ours). Cut marks on bones suggest that 3.3-million-year-old *Australopithecus afarensis* could cut with sharp-edged stones. Researchers have also found evidence of precision grip in *A. africanus* specimens by examining the bone structure within their metacarpals. Throughout our lifetime, the internal matrix of bone in our joints changes orientation to withstand the particular forces exerted on the joint – and the digits of *A. africanus* imply that they engaged in behaviours that required pressing the pads of their thumbs and fingers together.

The hand of *A. sediba* was even more like those of modern humans, with a longer and more robust thumb and shorter fingers that may have been producing stone tools. Early *Homo sapiens* and Neanderthals had more modern hands that could make and use complex tools, but other *Homo* species had different combinations of features. *Homo naledi*, for example, who lived

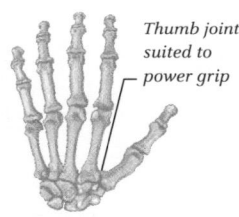

Thumb joint suited to power grip

Homo neanderthalensis

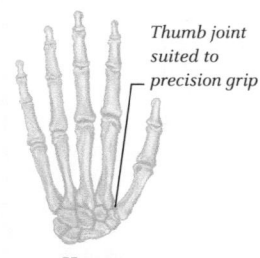

Thumb joint suited to precision grip

Homo sapiens

CHANGING HANDS
These hand bones of hominin species show how thumbs became longer and fingers became shorter as hand uses changed, but different species had a combination of adaptations.

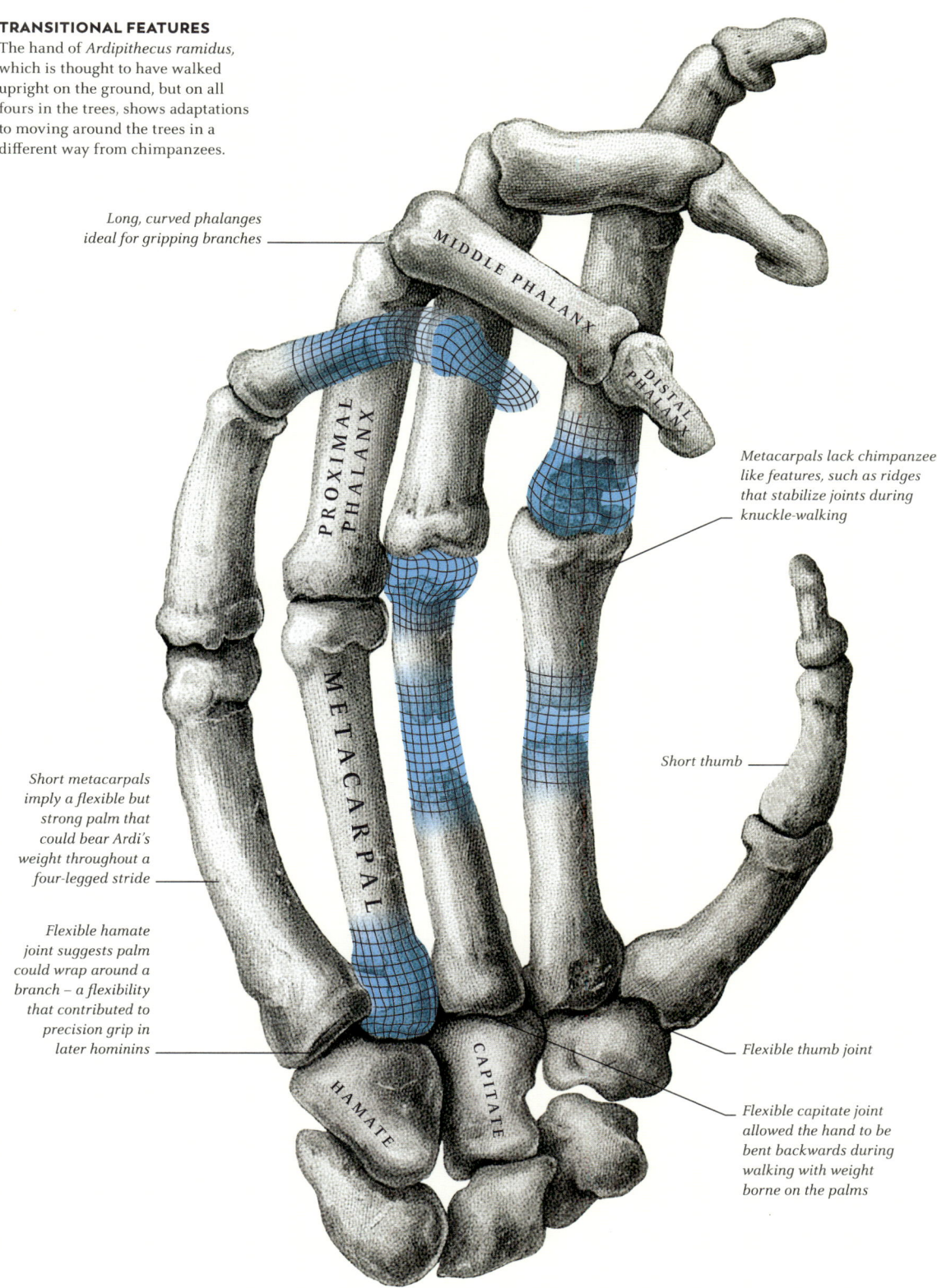

The hand of *Ardipithecus ramidus*, which is thought to have walked upright on the ground, but on all fours in the trees, shows adaptations to moving around the trees in a different way from chimpanzees.

Long, curved phalanges ideal for gripping branches

MIDDLE PHALANX

DISTAL PHALANX

PROXIMAL PHALANX

METACARPAL

Metacarpals lack chimpanzee-like features, such as ridges that stabilize joints during knuckle-walking

Short thumb

Short metacarpals imply a flexible but strong palm that could bear Ardi's weight throughout a four-legged stride

Flexible hamate joint suggests palm could wrap around a branch – a flexibility that contributed to precision grip in later hominins

HAMATE

CAPITATE

Flexible thumb joint

Flexible capitate joint allowed the hand to be bent backwards during walking with weight borne on the palms

Ardipithecus ramidus

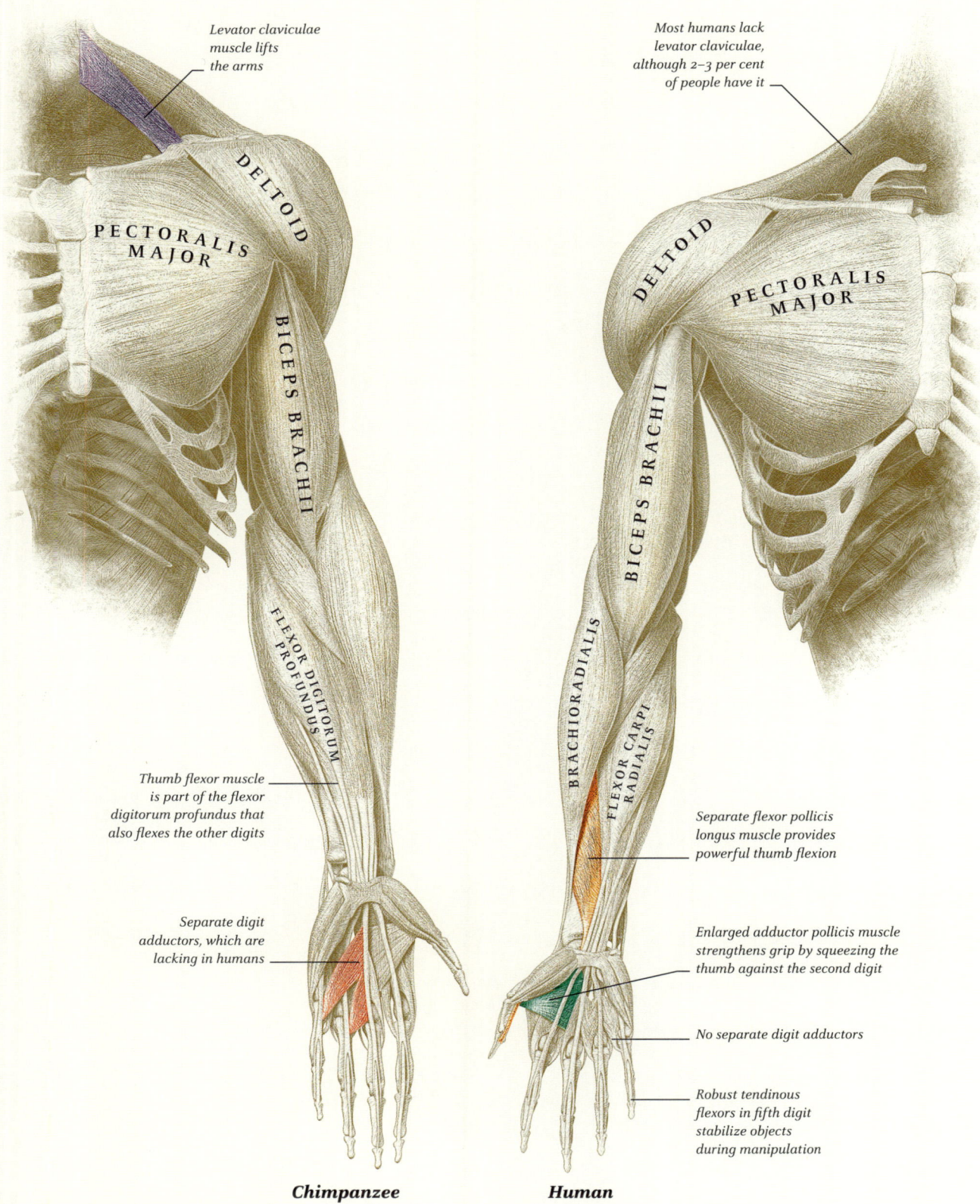

Levator claviculae
muscle lifts
the arms

Most humans lack
levator claviculae,
although 2–3 per cent
of people have it

DELTOID

PECTORALIS
MAJOR

PECTORALIS
MAJOR

DELTOID

BICEPS BRACHII

BICEPS BRACHII

FLEXOR DIGITORUM
PROFUNDUS

BRACHIORADIALIS

FLEXOR CARPI
RADIALIS

Thumb flexor muscle
is part of the flexor
digitorum profundus that
also flexes the other digits

Separate flexor pollicis
longus muscle provides
powerful thumb flexion

Separate digit
adductors, which are
lacking in humans

Enlarged adductor pollicis muscle
strengthens grip by squeezing the
thumb against the second digit

No separate digit adductors

Robust tendinous
flexors in fifth digit
stabilize objects
during manipulation

Chimpanzee

Human

MUSCULAR TOOLKITS

These illustrations show the arm and hand muscles in a chimpanzee and human, from a palm view. Unique muscles in the human hand facilitate manual dexterity, while chimpanzee hands and arms are adapted to moving around the trees.

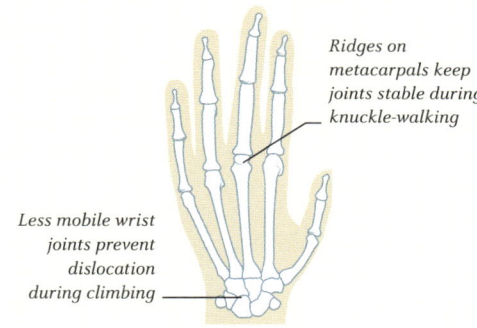

Ridges on metacarpals keep joints stable during knuckle-walking

Less mobile wrist joints prevent dislocation during climbing

CHIMPANZEE HAND
Long fingers can grip branches, while stable joints bear heavy loads during climbing and hanging

300,000 years ago, had a modern wrist and palm but extremely curved phalanges – potentially showing how climbing and tool use adaptations could co-exist.

A DIFFERENT DIRECTION

In contrast to our short, flexible hands with long thumbs, chimpanzees have long palms and fingers that aid in climbing and hanging underneath branches, and stiff wrist and hand joints that withstand carrying high loads during climbing. They also have ridges on their metacarpal bones, which stabilize joints during knuckle-walking on the ground. Early hominins lacked many of these features, suggesting they evolved more recently in chimpanzees and gorillas, rather than being ancestral to humans.

UNIQUELY HUMAN MUSCLES

Development of manual dexterity in hominins also resulted in muscular changes, and when you compare the muscles of a human and chimpanzee arm, they become more different as you move from the shoulder towards the fingers. In the forearm, we find the first of two muscles that are unique to humans: flexor pollicis longus. This runs across our palm to the end of our thumb, allowing the thumb to flex (bend towards the palm) independently of the other digits to achieve powerful and precision gripping. Other apes have this flexor as part of another muscle that flexes the other fingers as well, rather than as a separate unit. However, the way that it attaches to the bone is very similar, making it difficult to assess the muscular anatomy of fossil hominins. Researchers have argued that the thumbs of *Ardipithecus ramidus* and the 6-million-year-old hominin *Orrorin tugenensis* indicate separate flexor muscles.

The second unique thumb muscle, which was probably present in many hominin species, is extensor pollicis brevis that runs across the top of the hand and straightens the thumb away from the palm. The main muscle that brings the thumb in towards our fingers, adductor pollicis, is much larger in humans than in chimpanzees, and is helped by a deeper muscle, the accessory adductor pollicis, which was thought to be unique to humans but sometimes exists in bonobos.

All these peculiarities of our hand anatomy let us not only pinch our thumb and finger pads together, but also do it with considerable power. We can grip objects in a range of positions with enough strength to withstand a forceful impact, which is

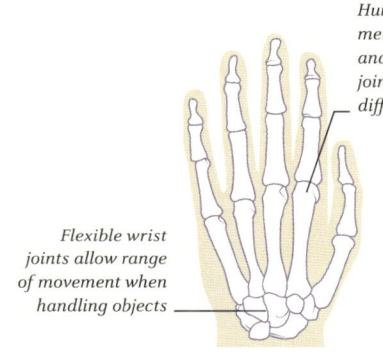

Humans lack metacarpal ridges and have flexible joints that permit different grips

Flexible wrist joints allow range of movement when handling objects

MODERN HUMAN HAND
A long thumb and flexible wrist and finger joints are important adaptations for tool use

ALL FINGERS AND THUMBS

Hand bones in a human and chimpanzee, viewed from the back. Chimpanzees have long fingers, short thumbs, and stiff joints for stability during locomotion. Humans have flexible joints and long thumbs for manipulating objects.

Unique human flexor muscles in the hand and arm allow us to pinch our thumb and fingers together powerfully

HUMAN HAND GRIPS

Human hands are capable of achieving a variety of grips, which would have been used in making and using many different tools and weapons.

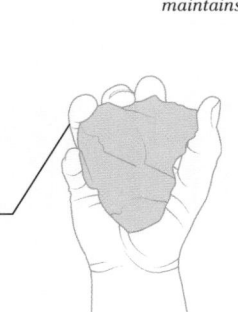

Powerful squeezing maintains grip

Relies on squeezing the pads of all digits together

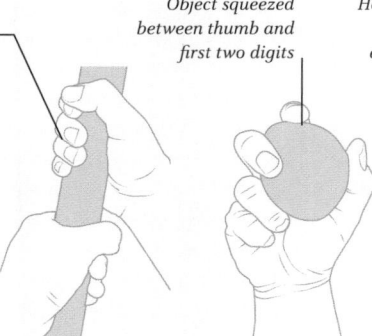

Object squeezed between thumb and first two digits

How Homo erectus may have pierced shells to eat the molluscs inside

FIVE-JAW GRIP
The powerful flexion of our thumb and fifth digit help to stabilize a roundish object.

CLUBBING GRIP
Our short palm and fingers, and mobile thumb, can squeeze a club powerfully.

THREE-JAW GRIP
This strong, stable precision grip was almost certainly used in stone tool production.

POWER GRIP
Our broad finger pads can pinch together to hold small objects powerfully.

Hand strength would have been equally as important for making tools and using weapons

important when thrusting a weapon or striking a flint to produce a sharp-edged tool. The combination of flexibility and strength also gives us "precision handling": holding an object using only the digits of that hand. The robust musculature in our thumb and fifth digit are particularly influential here, as they are often responsible for stabilizing an object. Precision handling would have benefitted our hominin ancestors as they honed their tool use skills – and while we cannot reliably estimate the exact precision handling abilities of many ancient hominins, it almost certainly existed in Neanderthals, who are associated with sophisticated tool use as well as cave paintings.

> Our hand anatomy not only lets us pinch our thumb and finger pads together, but also do it with considerable power.

THE STORY IN A HALF-A-MILLION-YEAR-OLD SHELL

Some of the best evidence we have of early humans' dexterity is from the objects that they interacted with. In 2015, researchers revealed a startling level of precision handling (and associated brain power) in *Homo erectus* – not from their hand anatomy, but from a collection of *Pseudodon* mussel shells.

Found in Trinil, a site in Java where *Homo erectus* remains were discovered, many of the shells had a hole punched through them, and some even had geometric patterns engraved on the surface. The holes were over the position of the adductor muscle that holds the shell closed, which – once damaged – would cause the shell to open without breaking, in a similar way to an oyster opened with a shucker. Experiments revealed that they were probably made using shark teeth (many of which were found at the site), as they are sharp enough to pierce the shell, but still require substantial

The Trinil shell
discovery pushes
manual dexterity
further back in
time than thought
previously

precision control. Before the shells' discovery, this level of manual control was only associated with *Homo sapiens* from the last 100,000 years, but the Trinil collection revealed how *Homo erectus* populations living 500,000 years ago were gathering mussel shells, expertly breaking them open to eat the molluscs inside, possibly using them as tools, and occasionally drawing patterns on them.

The reason the researchers can paint such a detailed picture of this foraging behaviour is by ruling out alternative explanations for the holes: they do not resemble holes made by other animals (such as birds) but are consistent with those made by modern humans to open the shell. The shells are also all adult-sized, but have different shapes, which means large mussels were selected from different locations along a river before being discarded together at the Trinil site, to be found 500,000 years later.

DRILLING FOR THE PRIZE
This 500,000-year-old fossil *Pseudodon* shell from Trinil, Java, demonstrates the precision abilities of *Homo erectus*. Numbers indicate how many shells were found with holes in each section; most are where the adductor muscle holds the shell closed.

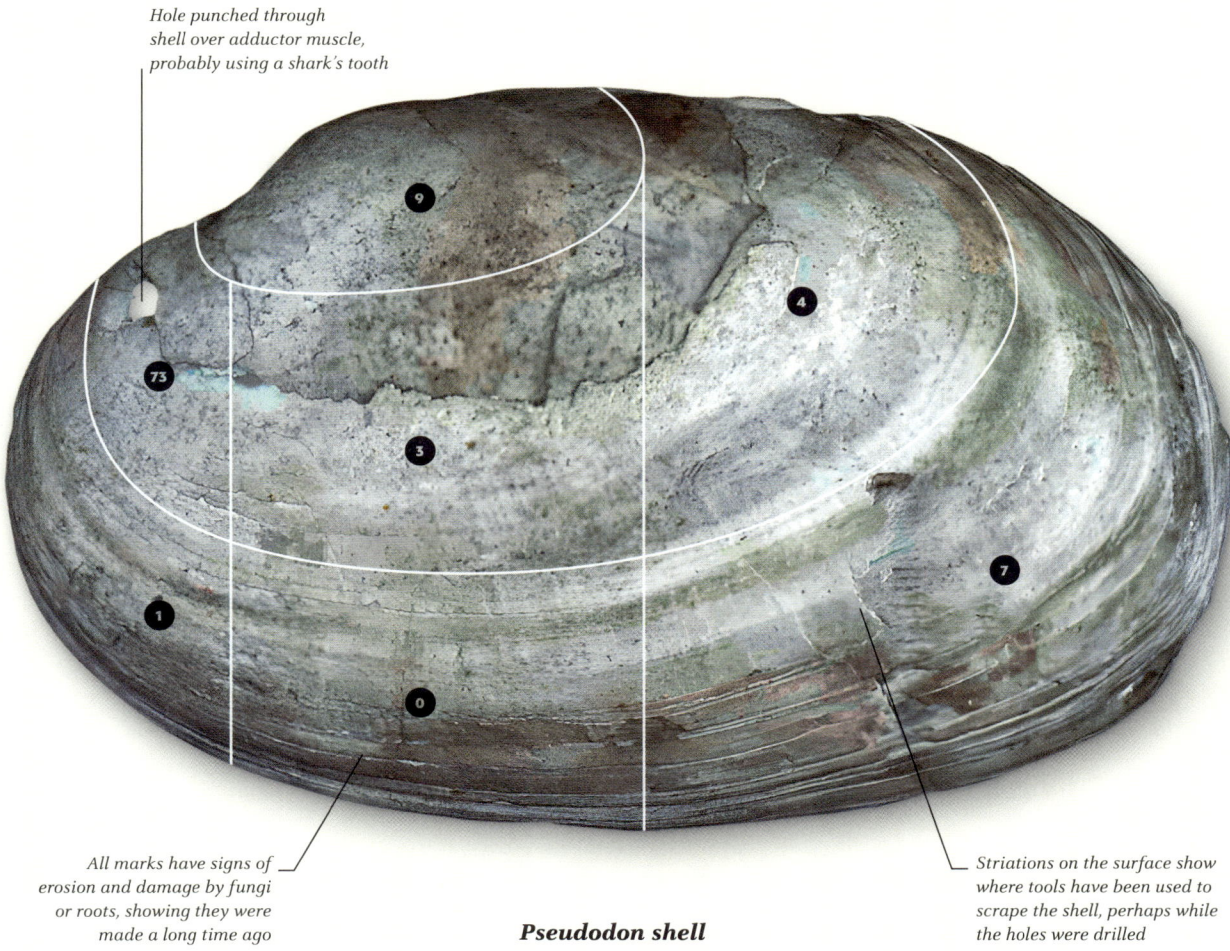

Hole punched through shell over adductor muscle, probably using a shark's tooth

All marks have signs of erosion and damage by fungi or roots, showing they were made a long time ago

Striations on the surface show where tools have been used to scrape the shell, perhaps while the holes were drilled

Pseudodon shell

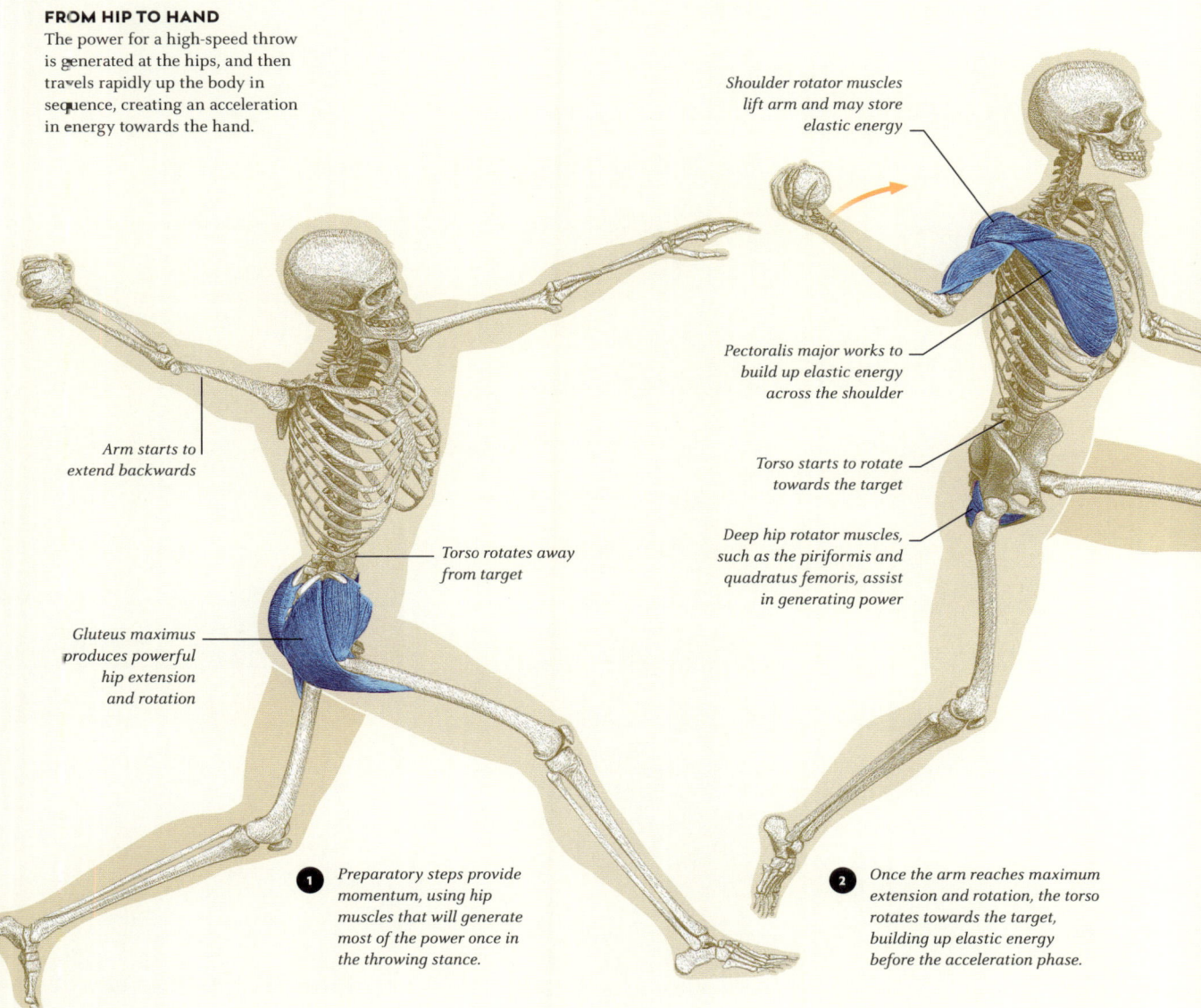

FROM HIP TO HAND
The power for a high-speed throw is generated at the hips, and then travels rapidly up the body in sequence, creating an acceleration in energy towards the hand.

Shoulder rotator muscles lift arm and may store elastic energy

Pectoralis major works to build up elastic energy across the shoulder

Arm starts to extend backwards

Torso starts to rotate towards the target

Torso rotates away from target

Deep hip rotator muscles, such as the piriformis and quadratus femoris, assist in generating power

Gluteus maximus produces powerful hip extension and rotation

1 *Preparatory steps provide momentum, using hip muscles that will generate most of the power once in the throwing stance.*

2 *Once the arm reaches maximum extension and rotation, the torso rotates towards the target, building up elastic energy before the acceleration phase.*

How humans throw

Being able to throw may have given Homo sapiens *a considerable advantage over other early humans. Throwing relies not just on strong arms, but on a coordinated system of skeletal and muscular adaptations across the whole body, some of which are used in upright walking and running.*

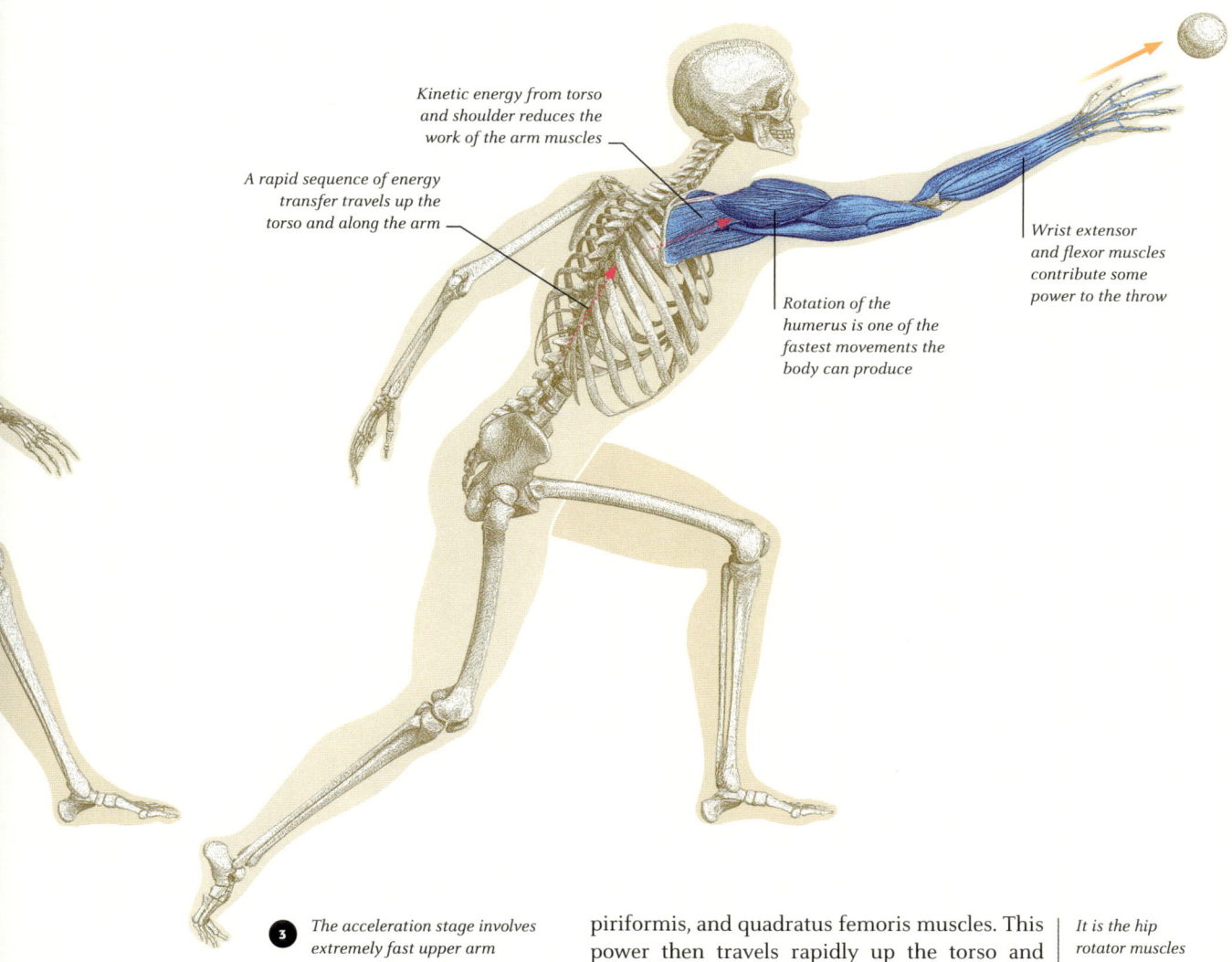

Kinetic energy from torso and shoulder reduces the work of the arm muscles

A rapid sequence of energy transfer travels up the torso and along the arm

Rotation of the humerus is one of the fastest movements the body can produce

Wrist extensor and flexor muscles contribute some power to the throw

3 *The acceleration stage involves extremely fast upper arm rotation, elbow extension, and wrist flexion, ending when the projectile is released.*

From the first projectiles thrown by our ancestors at prey, or in combat with adversaries, to javelins launched at 96 kph (60 mph) by elite athletes, high-speed, accurate throwing is a trademark of our species. Other apes do throw, but they lack the skeletal and muscular adaptations to throw with the same speed and precision as us, most of which developed over the last 2 million years.

The power for an overarm throw is generated in the hips and legs, with most being produced by hip rotators such as the gluteus maximus, piriformis, and quadratus femoris muscles. This power then travels rapidly up the torso and through the arm with whiplike acceleration, resulting in high-velocity projection from the hand. The arm-cocking phase (when the arm reaches back) also builds up elastic energy, contributing substantial power to the throw. During this phase, the torso rotates towards the target, but the throwing arm lags behind, its flexed elbow rotating the humerus (upper arm bone) even further away from the direction of the torso. This stretches tendons and ligaments across the shoulder, which then recoil to help power the forwards acceleration of the arm.

It is the hip rotator muscles that generate the majority of the force involved in overarm throwing

REFLECTED IN THE SKELETON

These fast rotations place high demands on the arm and shoulder, which are reflected in the skeleton: baseball pitchers develop a thicker humerus with a more backwards-facing shoulder joint to withstand the forces generated during

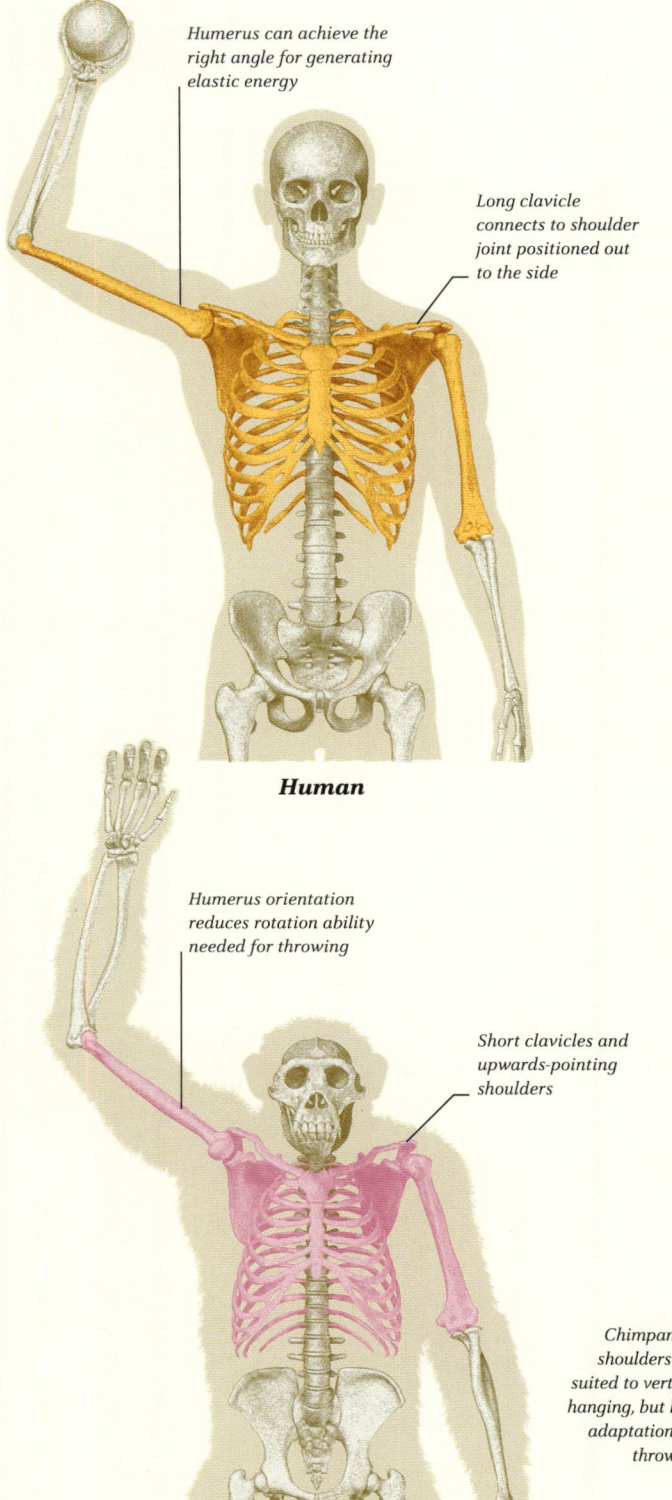

Humerus can achieve the right angle for generating elastic energy

Long clavicle connects to shoulder joint positioned out to the side

Human

Humerus orientation reduces rotation ability needed for throwing

Short clavicles and upwards-pointing shoulders

Chimpanzee

Chimpanzee shoulders are suited to vertical hanging, but lack adaptations to throwing

THROWING VERSUS HANGING

Both humans and chimpanzees have mobile shoulders, but differences in chimps' ribcage shape and shoulder position affect their throwing performance. Humans' low, more sideways-facing shoulders allow powerful throwing, but we struggle to bear our own weight from our arms when elevated.

throwing. The rapid transfer of energy through different parts of the body also requires extremely complex motor control, which our ape relatives lack. Chimps can throw objects at about one-third of the speed of an adult human's throw, despite having more muscular shoulders and arms.

UPRIGHT SKELETON

Although early hominins such as *Australopithecus* were not proficient throwers, their habitual upright walking enabled a vertical, flexible spine, a balanced upright posture, and pelvic muscles that could generate power in an upright position. These adaptations were crucial for the torso rotations that powered throwing in *Homo sapiens*. Upright walking also freed the hands, and the evolution of manual dexterity gave early humans the precision grips to throw stones and power grips to throw spears (see p.26).

BROAD-SHOULDERED BODIES

The most game-changing adaptations for hominin throwing were in the shoulders. Both humans and chimpanzees have mobile shoulders, but chimpanzee shoulders are oriented upwards due to their funnel-shaped ribcage. This helps them hang from their arms while moving through the trees, but reduces the arm's extension and rotation abilities, and the ability of the pectoralis major muscle to build up elastic energy

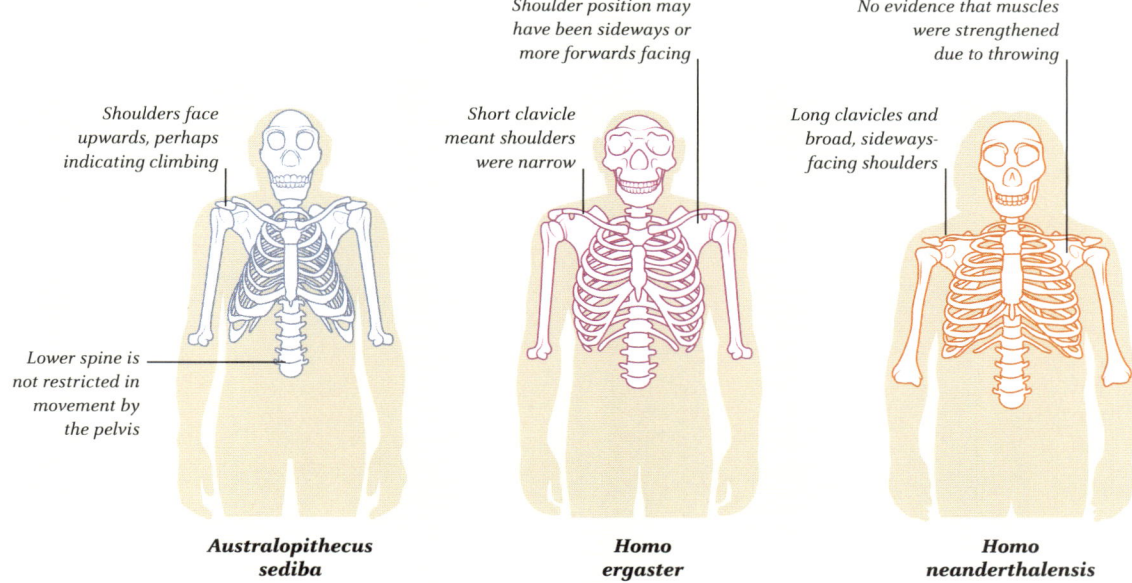

Shoulder position may
have been sideways or
more forwards facing

No evidence that muscles
were strengthened
due to throwing

Shoulders face
upwards, perhaps
indicating climbing

Short clavicle
meant shoulders
were narrow

Long clavicles and
broad, sideways-
facing shoulders

Lower spine is
not restricted in
movement by
the pelvis

**Australopithecus
sediba**

**Homo
ergaster**

**Homo
neanderthalensis**

*Hominin
evolution shows
a gradual shift
in ribcage and
shoulder anatomy
over time*

in the shoulder. Most australopithecines' shoulders still point slightly upwards, but in *Homo ergaster* and later hominins, we start to see the emergence of a more barrel-shaped ribcage with a longer, horizontally positioned clavicle (collarbone) that pushes the scapula (shoulder blade) round towards the back of the ribcage and the shoulders out to the side. This anatomy allowed their arms to produce power, and perhaps build elastic energy, in the right position for throwing. However, the orientation of the shoulder joint in Turkana Boy – a nearly complete *Homo ergaster* fossil from 1.5 million years ago, who still had shorter clavicles – is hotly debated, meaning it is unclear whether he could throw.

THE MYSTERY OF NEANDERTHALS

We must also consider the possibility that much of this "throwing anatomy" could instead be interpreted as adaptations to running, which requires arm swinging and torso rotation. Neanderthals, whose anatomy is relatively like that of modern humans, possessed sideways-facing shoulders, like us, but their bones do not show other signs of the demands of throwing, suggesting that the spears they made were thrusted, rather than thrown.

The earliest evidence of projectile weapons themselves dates to approximately 400,000 years ago, based on spears possibly crafted as

THE IMPETUS FOR CHANGE

These torso and shoulder recreations of three early hominins suggest that over time shoulders became lower and sideways-facing (possibly initially for running). The flexible spines that enabled bipedal walking would eventually also provide powerful rotation when throwing projectiles.

projectiles. However, additional evidence of 300,000-year-old stone spear tips containing fractures also indicate high-velocity impact from throwing. The makers of these tools are as yet unknown, as is their purpose, which could have been hunting, defence, or combat. The first projectiles (stones) used by early hominins were probably for combat and defence, given that other apes throw objects during antagonistic interactions. It has also been noted that hunter-gatherers today rarely use throwing during hunting, yet this may be due to the invention of complex projectiles such as the bow and arrow. Neanderthals were proficient tool-makers, but *Homo sapiens*' ability to throw projectiles may have contributed to our success and their eventual displacement.

33

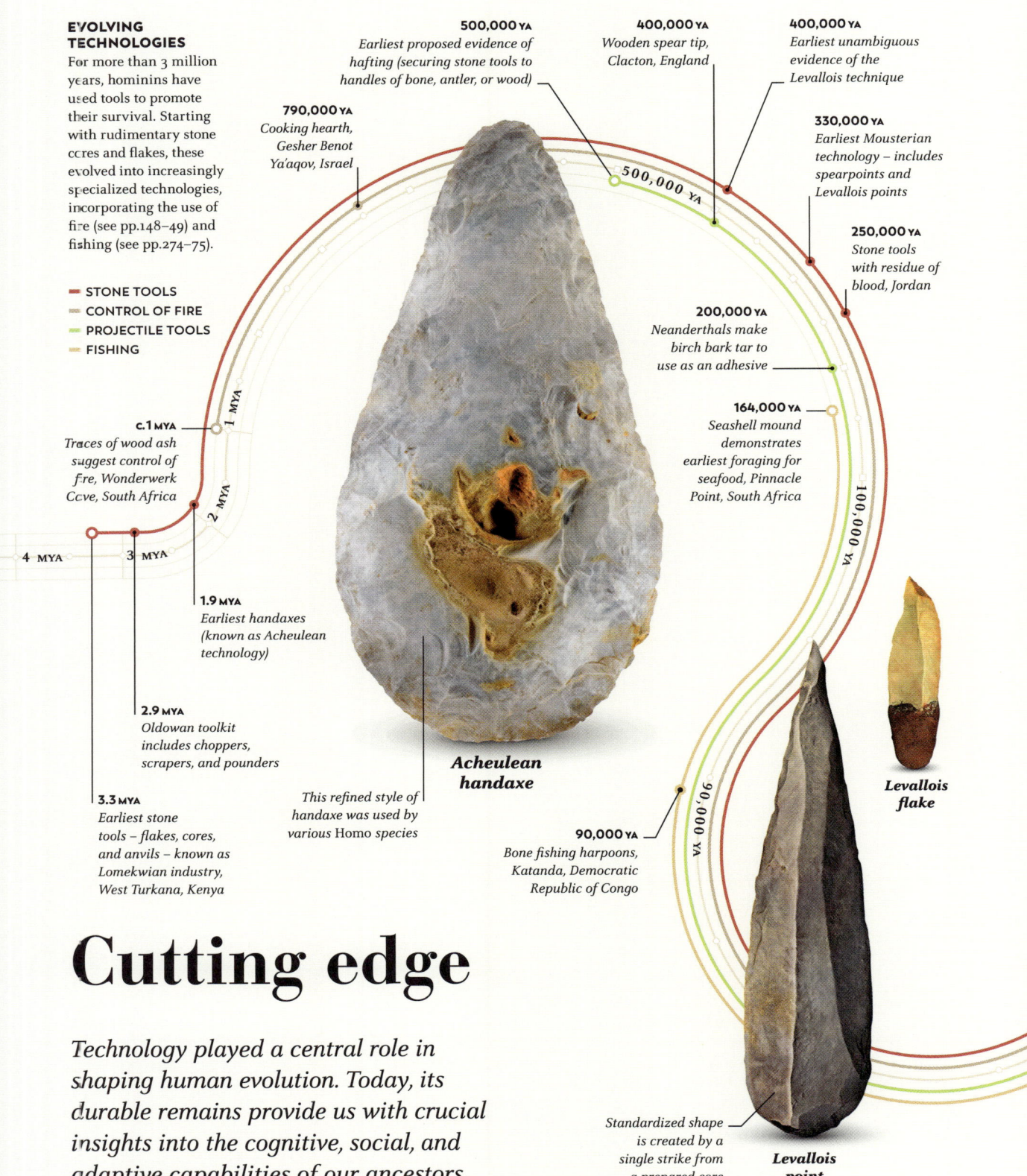

EVOLVING TECHNOLOGIES

For more than 3 million years, hominins have used tools to promote their survival. Starting with rudimentary stone cores and flakes, these evolved into increasingly specialized technologies, incorporating the use of fire (see pp.148–49) and fishing (see pp.274–75).

- ▬ STONE TOOLS
- ▬ CONTROL OF FIRE
- ▬ PROJECTILE TOOLS
- ▬ FISHING

790,000 YA
Cooking hearth, Gesher Benot Ya'aqov, Israel

500,000 YA
Earliest proposed evidence of hafting (securing stone tools to handles of bone, antler, or wood)

400,000 YA
Wooden spear tip, Clacton, England

400,000 YA
Earliest unambiguous evidence of the Levallois technique

330,000 YA
Earliest Mousterian technology – includes spearpoints and Levallois points

250,000 YA
Stone tools with residue of blood, Jordan

200,000 YA
Neanderthals make birch bark tar to use as an adhesive

164,000 YA
Seashell mound demonstrates earliest foraging for seafood, Pinnacle Point, South Africa

c.1 MYA
Traces of wood ash suggest control of fire, Wonderwerk Cave, South Africa

1.9 MYA
Earliest handaxes (known as Acheulean technology)

2.9 MYA
Oldowan toolkit includes choppers, scrapers, and pounders

3.3 MYA
Earliest stone tools – flakes, cores, and anvils – known as Lomekwian industry, West Turkana, Kenya

90,000 YA
Bone fishing harpoons, Katanda, Democratic Republic of Congo

Acheulean handaxe

This refined style of handaxe was used by various Homo species

Levallois flake

Levallois point

Standardized shape is created by a single strike from a prepared core

Cutting edge

Technology played a central role in shaping human evolution. Today, its durable remains provide us with crucial insights into the cognitive, social, and adaptive capabilities of our ancestors.

One of the most enduring forms of early technology are stone tools, which offer a tangible and datable record of technological behaviour over millions of years. These artefacts are especially valuable because they survive well in the archaeological record, unlike tools made from more perishable materials, such as wood. For this reason, experts cannot determine exactly when hominins (human ancestors) began using tools, however many believe that organic technologies likely did precede, or at least accompanied, stone tool use.

THE FIRST STRIKE

The earliest form of stone tool production used relatively simple techniques to strike flakes from pebble cores, creating sharp edges for cutting and scraping. The first instance of stone technology belongs to what is known as the Lomekwian industry, discovered in West Turkana, Kenya, and dating from around 3.3 million years ago (MYA). Lomekwian tools were made by anvil percussion, in which a large, heavy core is struck against a hard surface to detach flakes, yielding sharp edges on both flake and core. The discovery of Lomekwian technology provokes controversy as

The earliest stone tools were found in Kenya and are around 3.3 million years old

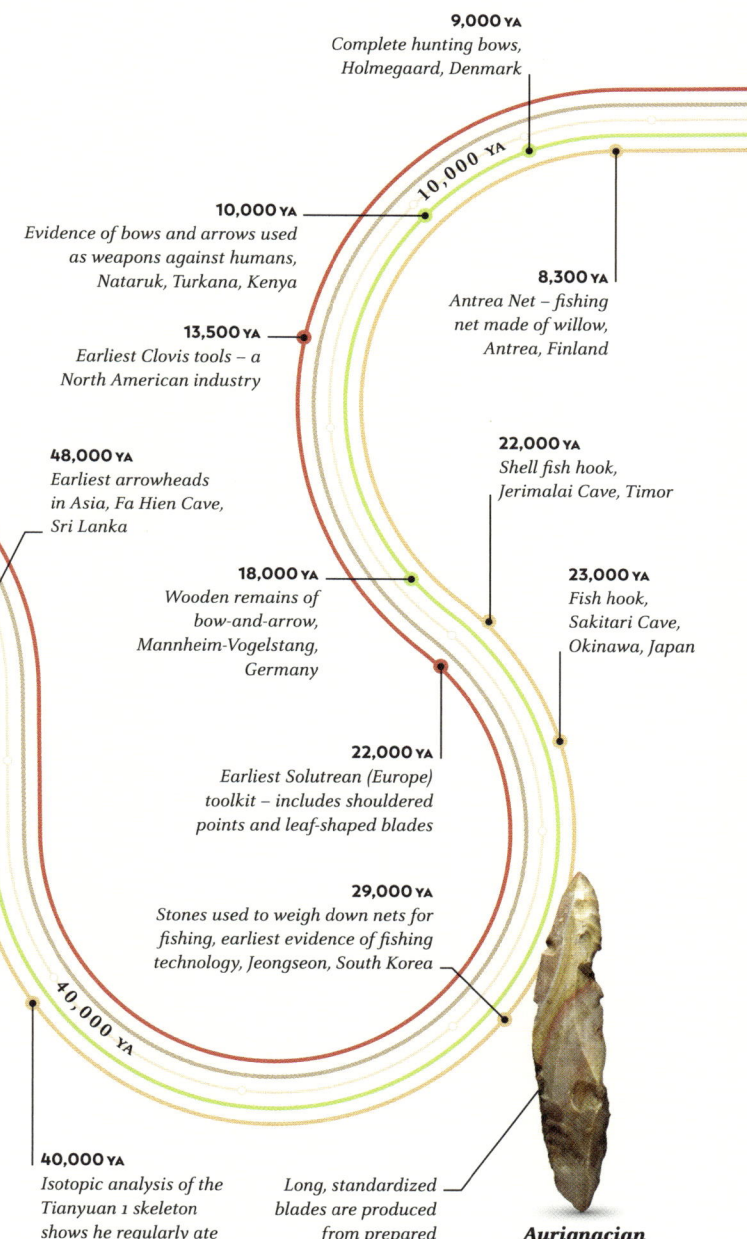

9,000 YA
Complete hunting bows, Holmegaard, Denmark

10,000 YA

10,000 YA
Evidence of bows and arrows used as weapons against humans, Nataruk, Turkana, Kenya

8,300 YA
Antrea Net – fishing net made of willow, Antrea, Finland

13,500 YA
Earliest Clovis tools – a North American industry

54,000 YA
Damaged stone points (possibly earliest arrowheads in Europe), Grotte Mandrin, France

60,000 YA

22,000 YA
Shell fish hook, Jerimalai Cave, Timor

48,000 YA
Earliest arrowheads in Asia, Fa Hien Cave, Sri Lanka

18,000 YA
Wooden remains of bow-and-arrow, Mannheim-Vogelstang, Germany

23,000 YA
Fish hook, Sakitari Cave, Okinawa, Japan

22,000 YA
Earliest Solutrean (Europe) toolkit – includes shouldered points and leaf-shaped blades

70,000 YA

42,000 YA
Tuna, shark, and ray bones indicate offshore fishing, Jerimalai Cave, Timor

29,000 YA
Stones used to weigh down nets for fishing, earliest evidence of fishing technology, Jeongseon, South Korea

40,000 YA

74,000 YA
Stone points believed to be arrowheads, Shinfa-Metema 1, Ethiopia

Harpoons like this are among the earliest evidence of fishing technology

40,000 YA
Isotopic analysis of the Tianyuan 1 skeleton shows he regularly ate freshwater fish, China

Long, standardized blades are produced from prepared prismatic cores

Bone harpoon

Aurignacian blade

HANDHELD PERCUSSION

Oldowan cores could be made by holding a core in one hand while striking it with a "hammerstone" in the other, which enabled good control over where flakes were removed, as well as their size. This method was ideal for shaping and preparing cores efficiently.

Core was typically made of hard stone, such as volcanic rock

Longer, opposable human thumb enables forceful pad-to-pad precision

Nonhuman apes have long fingers and a short thumb that prevents precise gripping of smaller objects

Flake scar shows point of detachment from the striking platform

Nonhuman apes prefer a "power grip" – clamping objects between fingers and palm

STRONG AND PRECISE

Compared to other apes, humans have shorter, straighter fingers with a more muscular thumb. This allows us to hold objects between our fingertips, achieving a precision grip with greater force and control.

Removal of flake produces sharp edge

Oldowan pebble core

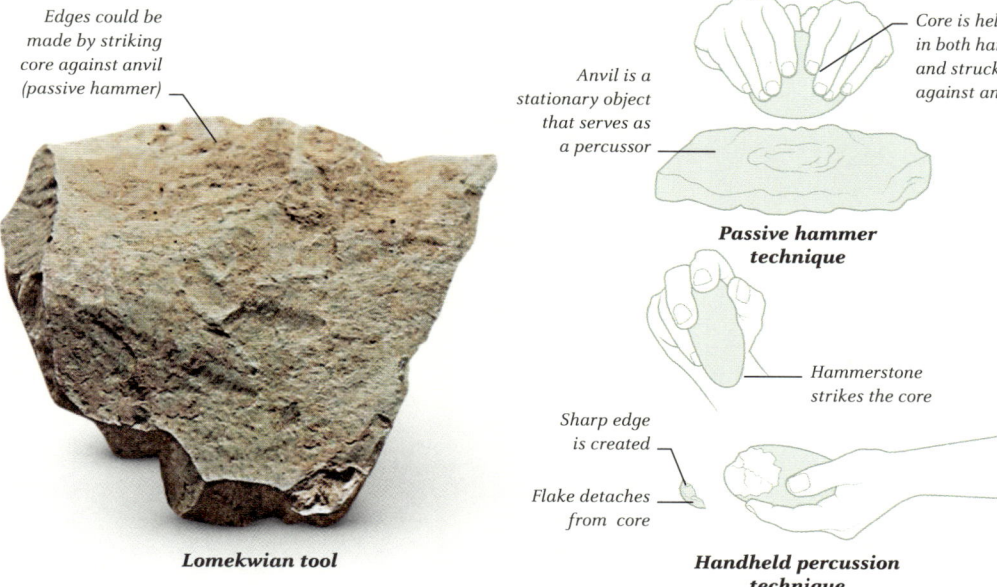

Edges could be made by striking core against anvil (passive hammer)

Lomekwian tool

Anvil is a stationary object that serves as a percussor

Core is held in both hands and struck against anvil

Passive hammer technique

Hammerstone strikes the core

Sharp edge is created

Flake detaches from core

Handheld percussion technique

PASSIVE STRIKING

Lomekwian tools were large and heavy and usually made by striking a stone core against a hammerstone on the ground. This approach required less dexterity than later handheld techniques.

TWO TECHNIQUES

The Oldowan industry transitioned away from the passive hammer technique and towards handheld percussion, which allowed for greater precision and control. Oldowan tools include both flakes and cores.

it predates the emergence of *Homo* by around 500,000 years. This challenges a belief long-held by some academics that toolmaking was unique to the lineage that gave rise to *Homo*, or early humans. Indeed, tools discovered in West Turkana occur within the same chronological and geographic range as *Kenyanthropus platyops* (see p.12), an earlier hominin that lived 3.5–3.2 MYA. Therefore, some experts have proposed that the Lomekwian represents a technological stage between a hypothetical, pounding-type stone tool used by earlier hominins, and the more intentional stone knapping (shaping through percussion) behaviour of later toolmakers.

WORKING HAND IN HAND

While initially associated with H. habilis, Oldowan technology may have been used by many different species of human

Uncovered in the 1930s at Oldupai (previously Olduvai) Gorge, Tanzania, Oldowan tools date from around 2.9–1.4 MYA. While initially attributed to *Homo habilis*, the diversity of Oldowan assemblages, along with the understanding that multiple hominin species coexisted in Africa, has led researchers to suggest that many species may have been responsible for Oldowan technology. Fossil and experimental evidence indicate that hominins at this time possessed hand anatomy well suited for precision gripping and toolmaking. Oldowan cores appear more refined than Lomekwian, and careful analysis has revealed evidence of flake removal occurring in a consistent orientation. This suggests that toolmakers possessed a degree of lateralized motor control (the ability for one side of the body to perform a task independent of the other). This skill worked hand in hand with the lateralization of brain functions (where certain cognitive abilities are dominant in one side of the brain) and ultimately the emergence of language capacity in early humans (see pp.46–51).

The development of Oldowan tools marked a significant evolutionary shift, providing hominins with a cutting edge that effectively extended their physical capabilities and greatly improved their ability to process food and other materials, such as wood.

A COGNITIVE LEAP

Emerging around 1.9 MYA and lasting until approximately 200,000 years ago, the Acheulean industry is linked with the emergence of *Homo*

CUTTING EDGE

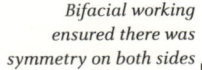

*Bifacial working
ensured there was
symmetry on both sides*

*Edges may have been shaped
by softer strikes to refine
symmetry and create a
balanced, "sexy" tool*

erectus and later species, such as *Homo heidelbergensis*. First identified at Saint-Acheul, France, Acheulean tools have been found across Africa, Europe, and Asia, and are typified by large cutting tools, such as handaxes, cleavers, and picks. Unlike the relatively simple flake production of earlier industries, Acheulean tools exhibit a deliberate and consistent bifacial shaping (shaped on both sides) of stone cores, producing teardrop- or ovate-shaped handaxes, some with a striking symmetry. This level of craftsmanship required not only more refined motor skills and greater control over flake removal, but also a mental template – an image of the desired end-product – reflecting a significant cognitive leap in abstract thought and planning ability.

Acheulean tools are thought to be highly versatile and likely served multiple purposes, including butchery, woodworking, and digging. Experimental archaeology (replicating ancient

SYMBOLIC GIFT
Beyond practical use, highly refined Acheulean handaxes, such as those found at Ma'ayan Barukh, could have served as social tokens or status symbols within early human communities.

**Acheulean handaxe,
Ma'ayan Barukh, Israel**

Blunt edge may have enabled better grip on the axe

Flakes were removed to create a somewhat rounded tip

This axe was made from quartzite, one of the hardest natural stones

ACHEULEAN CORES

These Acheulean handaxes were discovered in Aisne, France, and date from around 700,000–200,000 years ago. They are noted for their stone variety and fine workmanship, leading experts to speculate that they were valued more for their appearance rather than utility.

techniques in the present) has shown that these tools are especially effective at processing large animal carcasses, suggesting their use in scavenging or hunting. In Boxgrove, England, Acheulean handaxes from 500,000 years ago were found alongside butchered horse bones bearing cut marks. Elsewhere in England, in Clacton, researchers excavated the earliest known wooden spear tip, dating from around 400,000 years ago. Collectively, this provides compelling evidence that by this time early humans were technologically flexible – hunting large animals using a mix of organic and stone tools.

Acheulean technology appears in a range of environments, from the open savannahs of eastern Africa to the temperate woodlands of Europe. Experts suggest this demonstrates the versatility of Acheulean technology, which was suited to a variety of different locations.

Acheulean axes were discovered in a range of different environments

SEXY HANDAXE THEORY

The vast spread in time and space of Acheulean technology – over a million years and across three continents – seems to suggest one of two things. Perhaps isolated groups evolved separately in the same ways under the same pressures, leading to what is called convergent evolution. Alternatively, the Acheulean is a shared cultural tradition, transmitted by social learning including imitation, teaching, and possibly even rudimentary language.

Some researchers have proposed that Acheulean handaxes may have had symbolic or social functions as well. The "sexy handaxe" hypothesis proposes that the symmetry and refinement of some handaxes – far beyond what would have been necessary for practical tasks – could have served as displays of cognitive and motor skill in mate attraction, similar to how peacocks use their elaborate tails to signal genetic fitness. This idea is supported by the discovery of particularly large or finely made handaxes, such as those from Ma'ayan Barukh, Israel, which show little sign of actual use. While still debated, this hypothesis introduces the possibility that Acheulean tools represent not only a functional but also an aesthetic and communicative dimension in early human technology.

Taken together, the Acheulean industry signals a profound transformation in how early humans interacted with their world. With the emergence of these tools, we see not only increased efficiency and versatility in tool use, but also the first clear evidence of technological standardization and, possibly, cultural transmission.

PREPARED CORE

The emergence of prepared core technology marked another major evolutionary milestone in early human toolmaking. This technique, best exemplified by the Levallois method, involves carefully shaping a core by systematically removing flakes to prepare a striking platform. From this platform, toolmakers then strike off large, predetermined flakes with considerable control over their size, shape, and thickness. Unlike the more generalized products of earlier periods, these flakes demonstrate a significant increase in specialization and were used to make a wide range of flake tools, including scrapers, notches, and points – forms rarely seen in Oldowan and Acheulean industries. These innovations allowed early humans to tailor their tools more precisely to specific tasks, from hide working and wood processing to hunting and butchery.

Prepared core technology allowed early humans to tailor their tools to specific tasks

Prepared core technology tends to be associated with *Homo sapiens* and Neanderthals, but the archaeological contexts vary by region. For example, tools from 300,000–30,000 years ago are linked with early *H. sapiens*, with specific cultural industries in different regions, such as the Aterian of northern Africa, Lupemban in Central Africa, and Still Bay in southern Africa, all with unique tool styles and manufacturing techniques. As well as showing environmental adaptations, these regional differences have been interpreted as evidence of complex social structures, with materials being exchanged over long distances between culturally distinct groups. Artefacts like the finely crafted Still Bay points of Blombos Cave, South Africa; elaborate bone harpoons in Katanda, Democratic Republic of the Congo; and shell beads from sites across Africa (see p.104), often far from the coast, have been associated with the emergence of symbolic behaviours.

STRUNG TOGETHER

Another key innovation was the hafting of stone tools – attaching them to handles or shafts of wood, bone, or antler using adhesives, such as plant-based glue, and bindings, such as animal sinew or cord. Hafted tools, particularly points and scrapers, were more effective and versatile, essentially transforming simple stone flakes into composite tools, such as spears

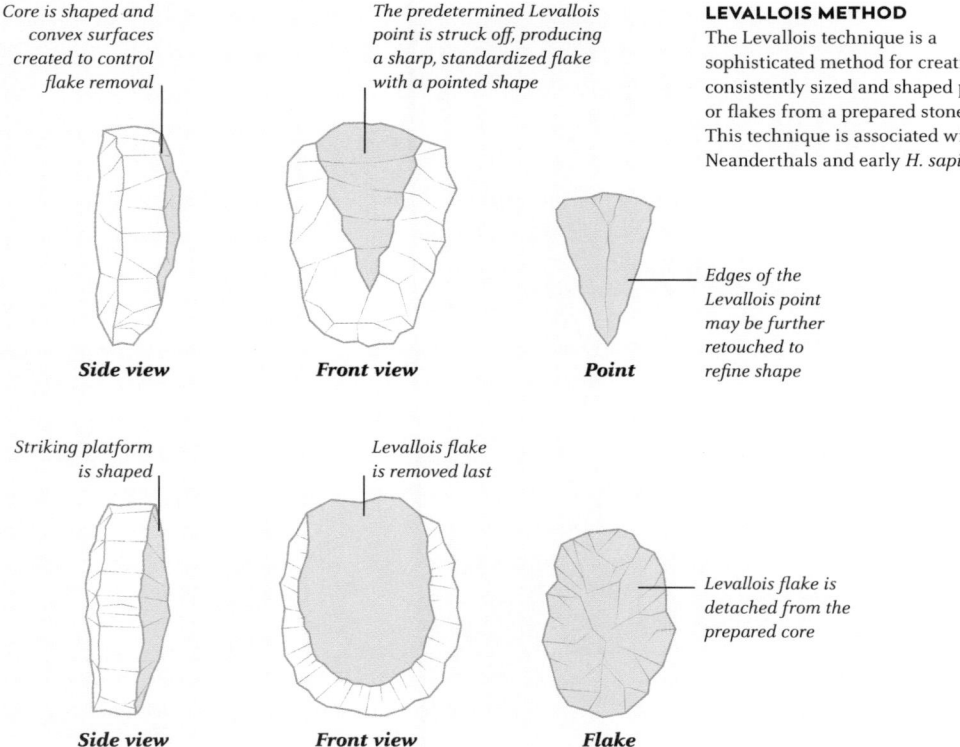

Core is shaped and convex surfaces created to control flake removal

Side view

The predetermined Levallois point is struck off, producing a sharp, standardized flake with a pointed shape

Front view

Point

Edges of the Levallois point may be further retouched to refine shape

LEVALLOIS METHOD
The Levallois technique is a sophisticated method for creating consistently sized and shaped points or flakes from a prepared stone core. This technique is associated with both Neanderthals and early *H. sapiens*.

Striking platform is shaped

Side view

Levallois flake is removed last

Front view

Flake

Levallois flake is detached from the prepared core

or knives. Archaeological evidence from Sibudu Cave in South Africa shows the use of compound adhesives as early as 70,000 years ago, which required complex knowledge of different material's properties and multistep planning. Additionally, tools from this period demonstrate heat treatment to improve stone flaking quality, suggesting increasingly sophisticated technological strategies.

The Mousterian tradition of Europe and southwest Asia is also dominated by prepared core technology, which occurred around 200,000–40,000 years ago and is closely linked with Neanderthals. Mousterian tools also employed the Levallois technique and included a similarly wide array of flake tools. Neanderthal technology was often adapted to diverse environmental conditions, such as the steppe-tundra landscapes of Molodova, Ukraine, where mammoth-bone semi-subterranean shelters were discovered (see p.259). Similar Mousterian sites also show evidence of hafting and fire use, challenging earlier assumptions that such complex behaviours were unique to *Homo sapiens*.

HAFTED TECHNOLOGY

Hafting is the process of combining a stone tool (such as a blade or projectile) and a handle to make a composite tool. Materials can be hafted together by tying using animal or plant matter, using adhesive, or fitting the stone tool into a slot or socket. These examples have been recreated by experimental archaeologists.

Tie made from raw (unsoftened) hide

Pointed arrowhead made from stone

String usually made from sinew or plant fibres twisted together

Wooden shaft and arrowhead held together with string and adhesive made from birch-bark tar

Polished Neolithic axe

Neolithic arrow

CUTTING EDGE

Prepared core technology is thought to reflect a profound increase in technological complexity, planning depth, and cultural expression. These tools enabled hominins to engage with their environments in more specialized and efficient ways. More than just functional advancements, these tools signal the emergence of distinct cultural traditions, intergroup interaction, and the roots of symbolic thinking – defining characteristics of modern human behaviour.

PRISMATIC BLADES

The emergence of blade technologies marked a further jump in the technical abilities of early humans. Blades would have also been created using a prepared core, however, the core would have been cylindrical in shape to generate long, parallel-sided flakes, or blades, when struck. Blades provided significantly more cutting edge per unit of raw material than earlier flake-based technologies. These blades were often retouched into specialized tools, such as backed blades, engravers, and endscrapers. Backed blades were modified along one edge to create a dulled surface that enabled hafting onto handles or shafts. The precision of blade technologies also reflects a growing reliance on soft-hammer production, using materials such as antler, bone, or soft stone that better controlled flake detachment and reduced breakage – both necessary for delicate retouching. Blade technology is exemplified in Europe by the Aurignacian industry, dating from around 43,000–26,000 years ago and associated with the first widespread arrival of *H. sapiens* in Europe following dispersal from Africa. Aurignacian tools were highly standardized, with finely retouched blades and bone and

PORTABLE TOOLKIT
The production of long, thin, sharp blades from stone cores emerged in Africa around 45,000 years ago. This method, here recreated by experimental archaeology, allowed humans to create a versatile range of tools, like a Swiss Army knife, from a single portable core.

antler points, and coincided with a rise in non-utilitarian objects, such as personal ornaments, carved figurines, and elaborate cave art.

The sophistication of Aurignacian tools reflects not only cognitive advancements, but also the adaptability of early humans to mould their technologies to suit their habitats and hunting strategies. Sites in Europe show that Aurignacian groups moved with the seasons, likely tracking migratory prey, such as reindeer, across open, steppe landscapes. The portability of blade-based tools would have been especially advantageous in these conditions, allowing groups to maintain effective toolkits while on the move.

ROOTS OF REFINEMENT
Microliths are small, standardized blades or flakes often produced using pressure flaking techniques, which use controlled pressure rather than striking to shape a stone. These tiny tools were typically hafted into composite implements, such as barbed arrows and spears, cutting tools, or woodworking implements. Microliths reflect a highly efficient and refined approach to stone tool production, further maximizing the cutting edge per unit of raw material. They emerged in Africa around 71,000 years ago, during a time when there was a growing reliance on bone implements, refined hunting strategies, and increasingly complex social organization.

WORKING SMARTER, NOT HARDER
The sequence shows, at each step, how more cutting edge can be produced from the same amount of raw material. This was important to early humans, since the raw material is heavy, and there might be many miles between localities providing suitable stone.

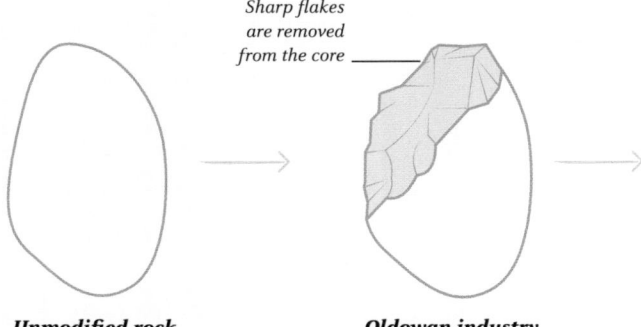

Sharp flakes are removed from the core

Unmodified rock   **Oldowan industry**

Blades are at least twice as long as they are wide

Prismatic core pre-shaped for blade removal

BLADE

PIERCER

KNIFE

SPEAR POINT

SCRAPER

BURIN

SAW

Also known as an awl; sharp tip is designed for puncturing material, such as hide

Long, continuous sharp edge for slicing or cutting materials

Combined with a shaft to create a composite weapon

Thick and sturdy, with retouched rounded end for scraping to remove flesh or smooth wood

Small flake removed from the edge to produce a chisel-like point used for engraving

Modified with notches to create a serrated edge for sawing tough materials

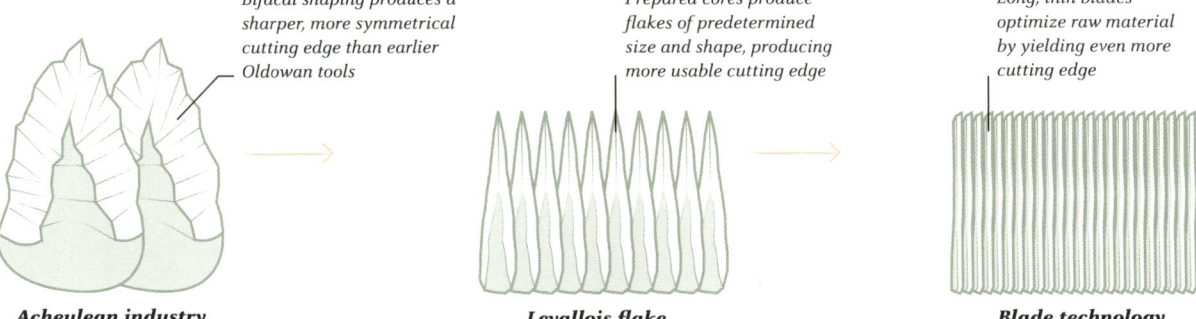

Bifacal shaping produces a sharper, more symmetrical cutting edge than earlier Oldowan tools

Prepared cores produce flakes of predetermined size and shape, producing more usable cutting edge

Long, thin blades optimize raw material by yielding even more cutting edge

Acheulean industry

Levallois flake

Blade technology

Like some previous technologies, microlithic tools have been found in widely separated parts of the world. In south Asia, microlith production dates to around 48,000–45,000 years ago. In Australia, microliths appear around 15,000 years ago and were long believed to have arrived with the dingo, although evidence now shows microliths predate the animal's introduction by around 4,000 years.

This widespread appearance of microliths has led some researchers to view them as a product of convergent evolution – parallel innovations arising in response to similar stresses rather than direct cultural transmission.

BUILT FOR STRENGTH

With the emergence of agriculture around 12,000 years ago, human societies began to develop increasingly specialized and durable tools to meet new challenges, such as tree felling and working the soil. The polished axe, crafted from hard stone, such as flint or jadeite, was used for tasks including forest clearance, construction, and early farming. Grindstones (see p.152) also became widespread during this time to process plant foods, marking a shift towards a diet more heavily reliant on cultivated crops. These tools reflect not only changes in human activity, but also growing social complexity; for instance, finely polished jadeite axes – sometimes found unused in burials – are thought to have served as prestige items, suggesting that by that time some societies had become stratified to include the concept of high or low status (see pp.296–99).

Wooden haft shaped and smoothed by stone scrapers

Ridges channel force of impact

Sharp and symmetrical edge

Robust axe head provides durable cutting edge capable of withstanding heavy force

Flakes would be struck off to sharpen the axe head

Mesolithic axe

A NEW AGE

The flaked stone head of a replica Mesolithic axe is relatively durable. However, ridges created by flaking focus the force of impact and risk shattering the stone after prolonged use. In contrast, a ground Neolithic-style axehead does not have ridges, can better withstand impact, and, therefore, is less prone to shattering.

Backward-pointing barb secures arrow in flesh of prey

BARBS

Small, retouched stone flakes are hafted onto arrowheads to create barbs.

Tip pierces flesh

Tool held together with hide string or adhesive

ARROWHEADS

Small points carefully retouched into special shapes ideal for hafting.

Opposite edge is sharp and used for cutting

CRESCENTS

Backed microliths that feature one blunted edge for secure hafting.

MINIATURIZED COMPONENTS

Microliths are small, standardized stone tools, often hafted onto wooden shafts to form composite implements. Microlithic technology reflects advances in planning, efficiency, and adaptability.

Smooth surface achieved by many hours of grinding – the huge investment making this a very valuable tool

Cutting edge could be resharpened by a little further grinding

Neolithic axe

A substantial part of the way we communicate with one another is through hand gestures. Gestures can convey a wealth of meaningful information, either on their own or to accompany speech, and it is thought that spoken language evolved in early humans from gestural communication. Since then, speech and gestures have continued to evolve together as part of the same integrated system.

All apes use gestures – some more than vocalizations – but they are most developed in humans. We are the only species able to use a gesture to indicate an object to another individual (by pointing, for example), and most human infants can do this long before they learn to speak. Research has also shown how influential hand gestures can be, to the extent that using them during questioning can affect a responder's recall of past events.

APING ACTIONS

Gestural communication in other apes is limited to attention-seeking or requesting an action from another individual. Chimpanzees have most, with up to 66 different gestures using different parts of the body, that they employ more intentionally than their vocalizations, which tend to be more reflexive.

Chimps have a gestural language of 66 gestures, which they use more deliberately than sounds

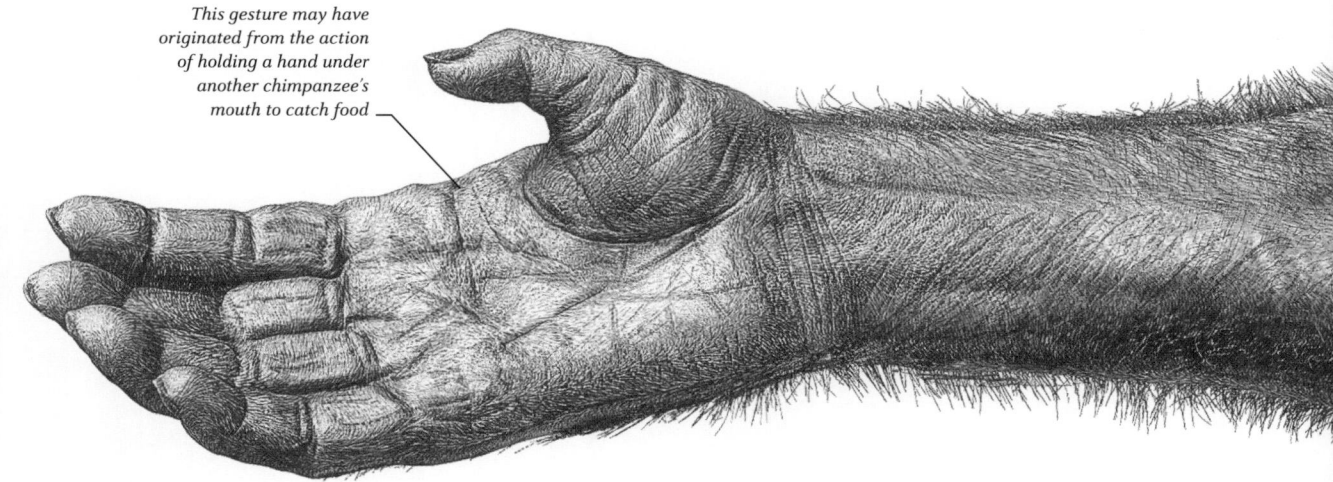

This gesture may have originated from the action of holding a hand under another chimpanzee's mouth to catch food

Gestures and communication

CHIMPANZEE GESTURES
A chimp extends its arm with the palm turned upwards to beg for food. This gesture may stem from a non-communicative action to gain food, implying a possible evolutionary origin for gestures in primates.

Almost all humans use their hands when they speak, and many populations have created sign languages using only gestures. Other apes gesture too, and it is thought that speech developed from gestural communication in our early human ancestors.

Humans are also able to learn and integrate new gestures, and while some gestures are now recognized across most of the world (such as a "thumbs up"), many gestures are specific to certain cultures. There is debate around other apes' ability to learn gestures from each other, even though some chimpanzees and gorillas have been taught sign language by humans, as cases of population-specific gestures among wild apes are rare. This implies that primate gestural communication may not have evolved via imitative learning (one individual copying another), but may stem from their use of other, non-communicative, actions in the same context.

HUMAN GESTURES

Humans use many hand gestures, most of which are culturally specific. Some gestures can be understood without speech, while others are co-speech gestures, adding emphasis or visual information to words.

POINTING

Humans can communicate about an object without touching it – babies can do this before speaking.

HAND TO MOUTH

Covering the mouth is a common gesture used to convey shock or surprise without using speech.

THUMBS UP

This gesture is used to emphasize speech, or in its absence. Its meaning can be culturally specific.

HANDS APART

Mostly used as a co-speech gesture to emphasize or provide a visual estimation of size.

GESTURES AND COMMUNICATION

Gestures are not just limited to the hands; both humans and chimpanzees use their faces to communicate, and in sign languages, facial expressions form the equivalent of tone and emphasis in speech. In early humans, facial gestures combined with vocalizations may have given rise to speech – perhaps once the hands were engaged in more long-term tasks.

Human facial expressions in combination with hand gestures may have given rise to speech

MOTOR CONTROL

Several areas in the brain are responsible for the coordination of hand movements, which rely on "sensorimotor feedback" loops. These involve the brain comparing what the eyes see, or nerves feel, with its instructions for movement.

Other evidence for a gestural origin of speech comes from the parts of the brain that process these skills. The control of hand movements involves many areas of the brain, and honing manipulation skills requires constant feedback between these areas and our sensory and motor systems to perfect a hand action. This feedback would have been a critical process for early humans as they perfected tool use behaviours. One of these brain regions – known as Broca's area – is involved in planning voluntary hand movements, like gestures, and is also one of the areas involved in language processing. Some researchers have suggested that a genetic mutation resulted in the integration of spoken language into Broca's area in early humans, although different parts of language (for

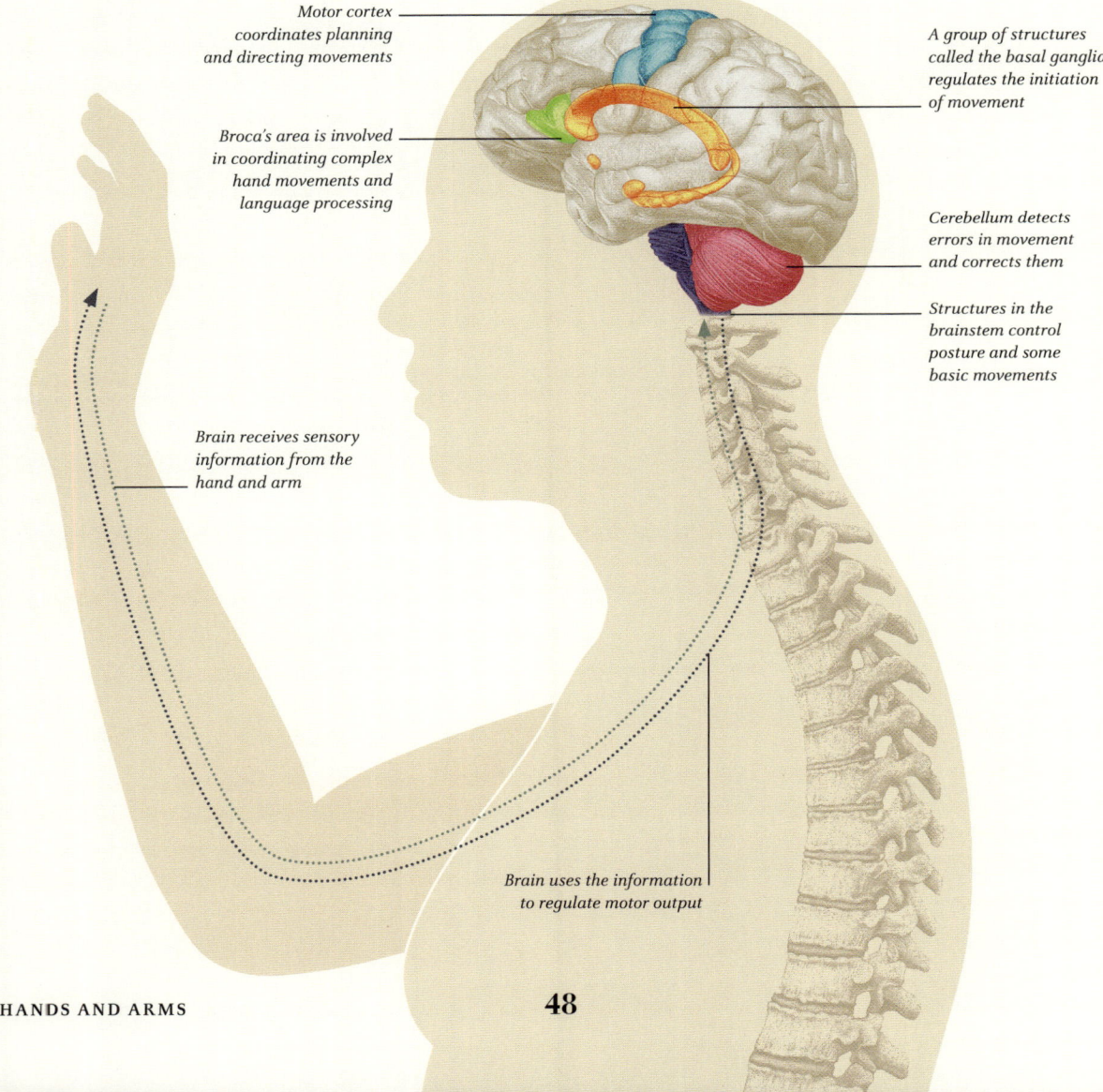

Motor cortex coordinates planning and directing movements

Broca's area is involved in coordinating complex hand movements and language processing

A group of structures called the basal ganglia regulates the initiation of movement

Cerebellum detects errors in movement and corrects them

Structures in the brainstem control posture and some basic movements

Brain receives sensory information from the hand and arm

Brain uses the information to regulate motor output

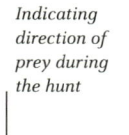

Used at the same time as crouching down to indicate prey is nearby and may spot them

Crouch down and be quiet!

Indicating direction of prey during the hunt

There they are/it is; keep quiet!

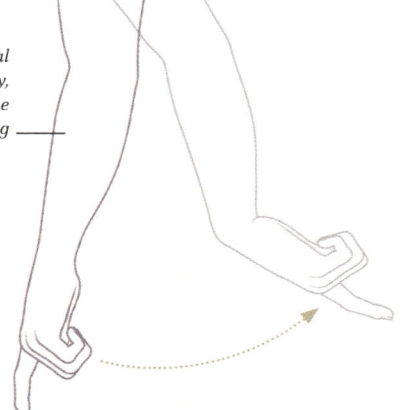

Indicating potential danger nearby, often at the same time as crouching

Get back!; do not come closer!

HUNTING SIGNALS
Above are three gestures used by modern Ju/'hoansi hunters in Botswana. Hunting requires cooperation for success and safety, and the need to stay quiet often favours gestures over speech.

instance, grammatical tenses and nouns) are associated with different systems in the brain, and likely required several evolutionary steps.

COMPLEX BRAIN CELLS
The development of gestural communication may have been helped by the presence of mirror neurons – a type of brain cell first discovered in monkeys and originally thought to be unique to primates, but which have also been found in birds. These neurons fire when an individual performs an action, such as reaching for food, as well as when they observe (or hear) another individual performing the same action. In humans, these cells form complex "mirror systems" in the brain that develop through sensorimotor learning (feedback loops involving senses such at sight, hearing, and touch, and the coordination of movement by the brain).

Reconstructing the degree of gestural and speech communication in our hominin and human ancestors is largely beyond current

Mirror neurons fire while performing an action, as well as when seeing or hearing others perform the same action

palaeontological methods, apart from guessing their hearing and speech capabilities from skeletal anatomy. This type of analysis has suggested similar speech abilities to humans in Neanderthals, but a more chimpanzee-like communication capacity in early hominins such as australopithecines. However, communication is a vital part of cooperative behaviours, such as hunting: failure to communicate about the type or location of prey, for example, could reduce hunting success or heighten danger to others. It is therefore likely that more complex gestural communication, evolved alongside these types of cooperative behaviour, and were associated with more open environments on the ground, rather than the cluttered treetop canopy where gestures cannot be easily seen.

THE UPPER HAND
Our dominant hand is the one we use most for manipulating objects, and no other primate shows handedness (the presence of a dominant hand) as strongly as humans: up to 90 per cent of humans worldwide are right-handed. Development of "laterality" (sidedness) in our limbs is linked to the development of brain laterality. The left brain controls movement of the right hand, and is also responsible for language processing, which suggests that differentiation of the brain hemispheres played a key role in the evolution of communication and speech. The evolution of such pronounced

49

The marks on Neanderthal front teeth show that they were frequently using stone tools held in their right hands to scrape hides gripped in their jaws.

In most humans, the left and right limbs are controlled by the opposite hemisphere of the brain. Tasks requiring the limbs to move differently, like flint knapping, make one side of the brain work differently, leading to lateralization in brain and body.

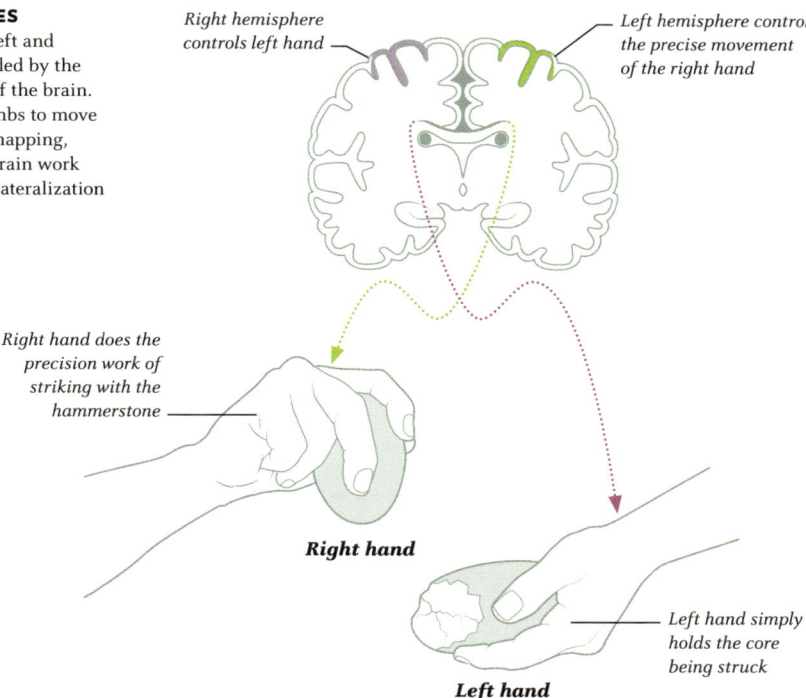

Right hemisphere controls left hand

Left hemisphere controls the precise movement of the right hand

Right hand does the precision work of striking with the hammerstone

Right hand

Left hand simply holds the core being struck

Left hand

handedness in humans may stem from early hominins who became ground-dwelling and used their hands less for climbing, meaning they no longer needed to retain maximum strength in both hands. Today, many humans have a stronger grip by up to 10 per cent in their dominant hand, although this is less pronounced in left-handed people.

Hominin behaviours that reinforced hand preferences may have included complex tool production (which requires the hands to perform different actions at the same time), the use of gestural communication, and throwing. The last action requires not only each hand, but each side of the body, to carry out different movements at the same time (see pp.30–33).

ON THE OTHER HAND

Chimpanzees and gorillas also show right-handed preferences, although not as strongly as humans, while orangutans tend to show a left-handed preference – perhaps because their treetop habitat requires a stronger right hand to grip branches while the left hand performs other tasks. In chimpanzees, handedness is context-specific: they often use their right hand for tasks such as poking a stick into a termite

Choice of which hand to use is context-specific in chimpanzees, and research suggests humans may be the same

mound to forage for insects, but choose their left for social settings, such as play. Research suggests that human handedness may also be more context-specific than originally thought, implying it may have co-evolved alongside factors such as more complex social structures.

RIGHT-HANDED HOMININS

Assessing handedness in early hominins requires bones from both hands of the same specimen, which is rare in the fossil record. However, a similar proportion to modern humans of 90 per cent right-handers has been suggested for Neanderthals – not from hand anatomy, but from their teeth. In modern hunter-gatherer populations, the action of processing material by holding it between the front teeth and one hand, and scraping it with a tool held in the other hand, leaves a characteristic pattern of wear on the incisors and canines. These same marks on Neanderthal teeth show that they were performing this action frequently, and that they were holding the processing tool with (in most cases) their right hand.

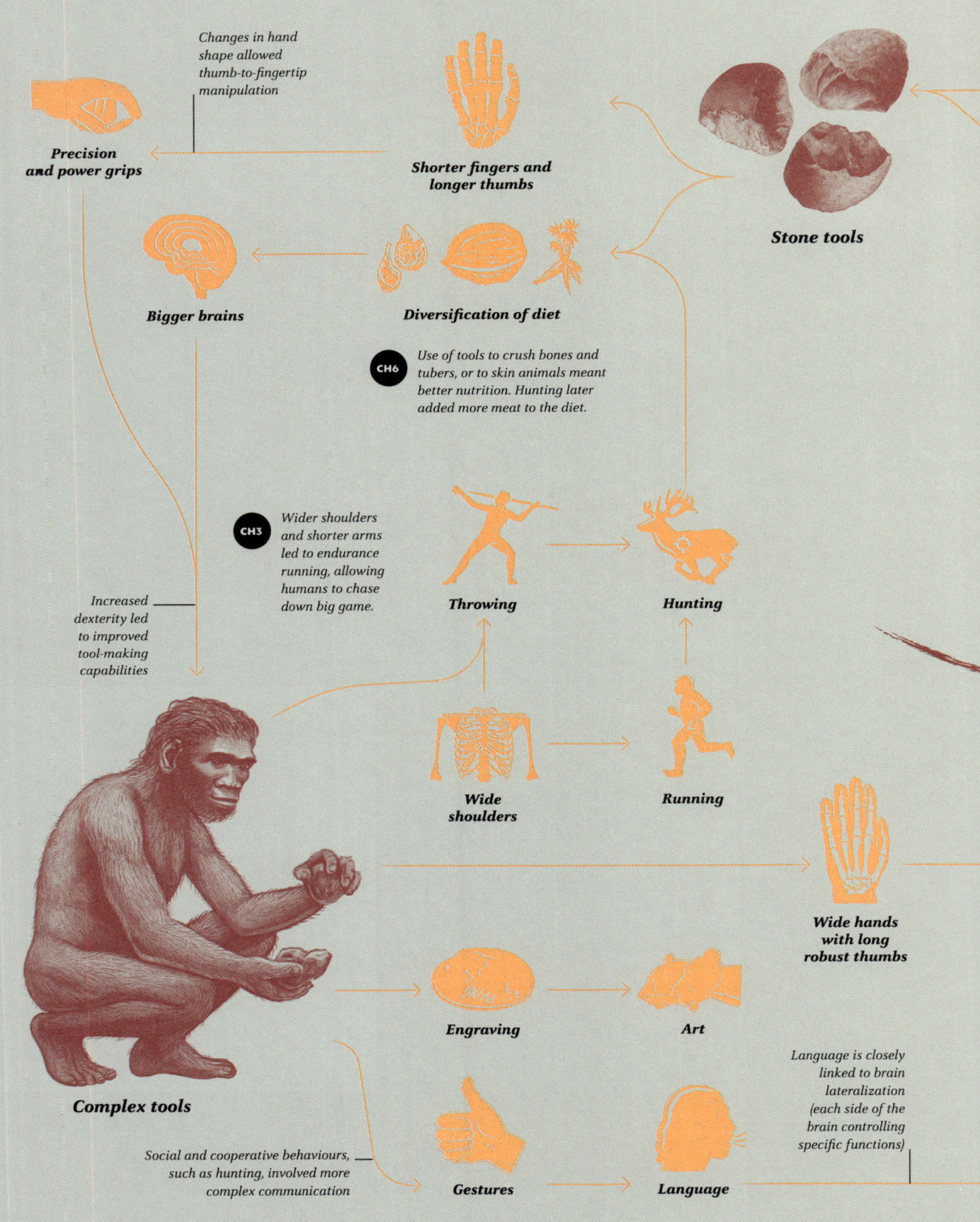

Changes in hand shape allowed thumb-to-fingertip manipulation

Precision and power grips

Shorter fingers and longer thumbs

Stone tools

Bigger brains

Diversification of diet

CH6 Use of tools to crush bones and tubers, or to skin animals meant better nutrition. Hunting later added more meat to the diet.

Increased dexterity led to improved tool-making capabilities

CH3 Wider shoulders and shorter arms led to endurance running, allowing humans to chase down big game.

Throwing

Hunting

Wide shoulders

Running

Complex tools

Engraving

Art

Wide hands with long robust thumbs

Language is closely linked to brain lateralization (each side of the brain controlling specific functions)

Social and cooperative behaviours, such as hunting, involved more complex communication

Gestures

Language

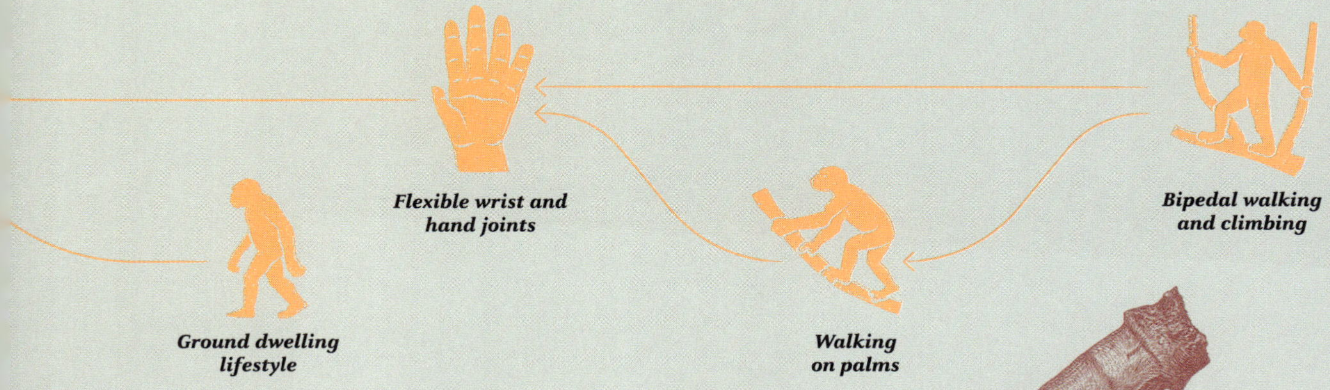

Flexible wrist and hand joints

Ground dwelling lifestyle

CH3 *Hands were freed up for other uses once we no longer needed them for grasping branches.*

Bipedal walking and climbing

Walking on palms

DEXTERITY CONNECTIONS

Our ape ancestors used their hands for moving through the trees, which required very different muscles and digit proportions to those we have in our hands today. The transition from one to the other was a slow process, with many evolutionary pressures.

Complex motor control

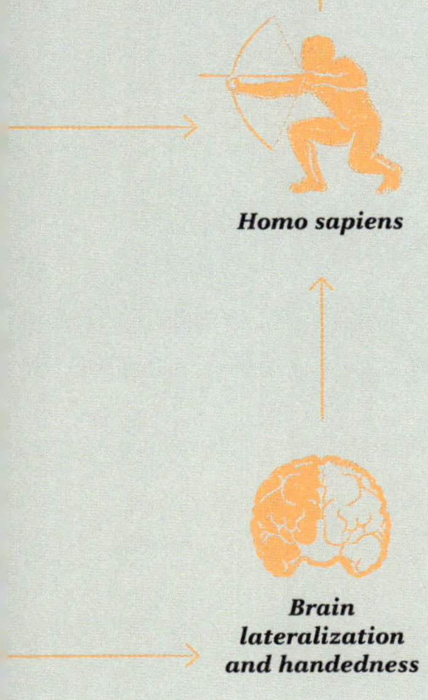

Homo sapiens

Brain lateralization and handedness

The making of our hands

The evolution of human dexterity and grip power is tightly linked to brain development, and went hand in hand with our ancestors' descent from the trees, stone tool production, and use of gestures to communicate.

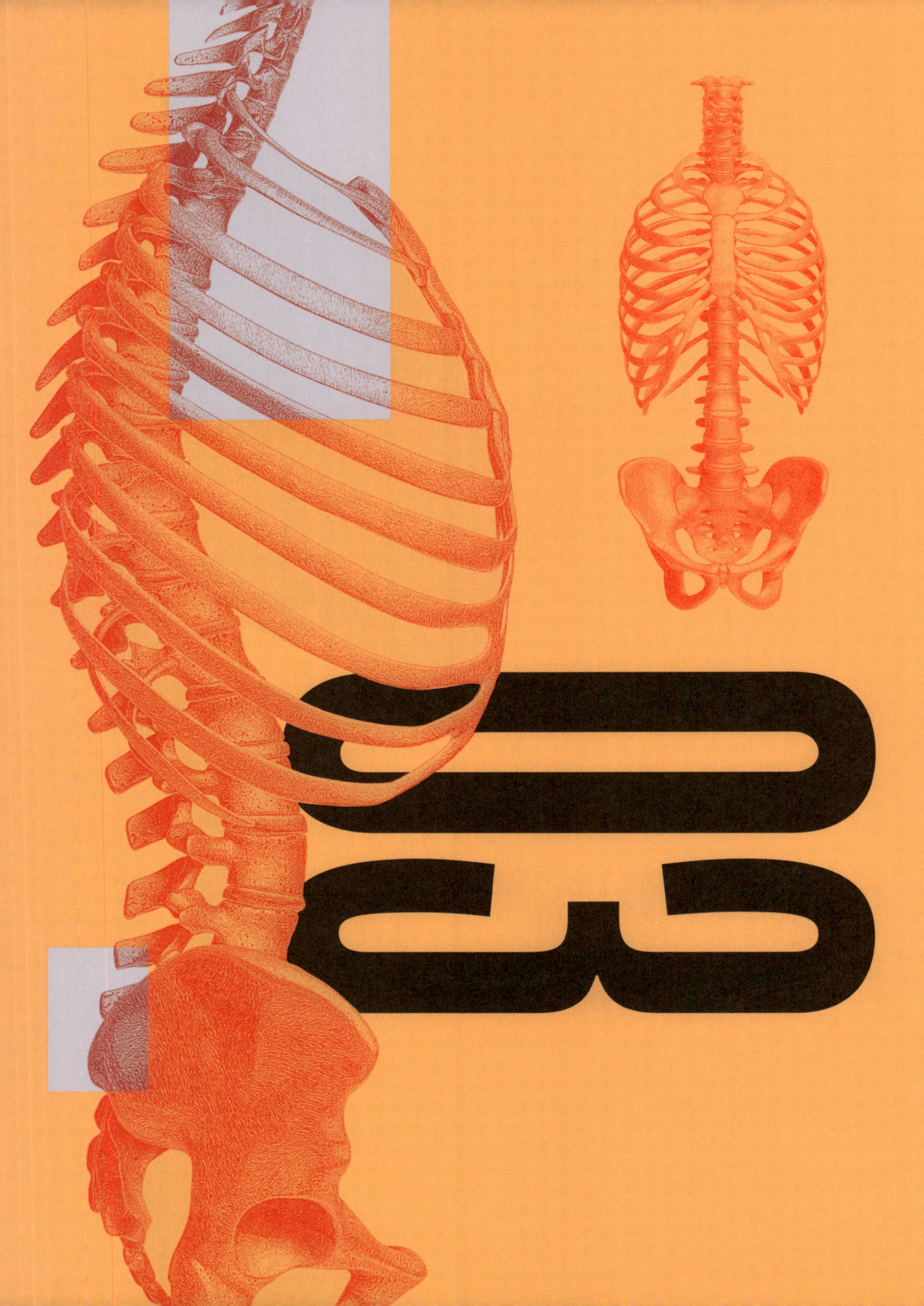

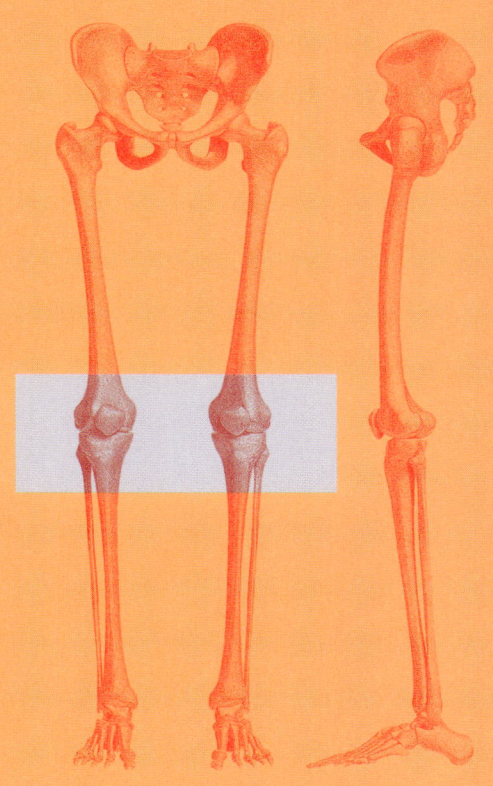

Spine, hips, and legs

Humans and other upright apes *balance their trunk and legs in a column using a suite of unusual skeletal features. The makeover of their spine, hips, legs, and feet turned tree climbers into efficient two-legged walkers and runners.*

The long walk to humankind

We owe our abilities to hold ourselves upright and walk on two legs to a series of evolutionary milestones stretching back 600 million years. During this time, the spine and limbs were reshaped by the forces of natural selection.

The earliest animals did not have a skeleton or limbs, but we can trace the evolution of our spine back to a rodlike structure called the notochord that developed in one group of early animals known as chordates. The notochord provided strength along the longitudinal axis of the body. It is still a feature of vertebrate embryos today, although it gets replaced in the developing embryo by the spine. The first spines formed in fish as a series of articulated segments, or vertebrae, strengthening the axis of the body and providing muscle attachment. When early amphibians moved out of water onto land, the spine adapted for weight-bearing. Pectoral fins became forelimbs, and pelvic fins became hind limbs. Fossils of early limbed vertebrates show how the hind limbs became firmly attached to the spine via the pelvic girdle, while the forelimbs moved away from the spine at the shoulder.

HAVING THE SPINE TO DIVERSIFY

Fish and amphibians have very uniform spines, whereas reptiles and mammals possess distinct regions – changes that seem to have been driven by the evolving nature of the limbs. Those distinct spinal regions could move and adapt independently of one another. This, together with diversification in types of feet, allowed mammals to move in many different ways and colonize a range of habitats. Hominins inherited primate anatomy shaped by life in the trees, which slowly adapted to suit life on the ground.

The recent history of our spine and legs involved adaptations to climbing trees

HIPS AND HIND LEGS

As the fins of certain fish slowly started evolving into limbs around 374 MYA, the hind pair remained attached to the spine by the pelvic (hip) girdle.

Hind limbs of the early tetrapod Ichthyostega *perhaps allowed some weight-bearing*

A SPINE MADE OF BONE

Spines developed first in fish to provide structural support and aid movement, but were unable to support weight on land.

Early fish vertebrae were made of unfused bones and cartilage

STIFF BUT FLEXIBLE ROD

The notochord provided a stiff, but flexible, axis along which an animal could flex its body to swim. Lancelets are among the few adult animals that have them today.

Notochord in an early animal similar to today's lancelet

VERTEBRATES WITH LIMBS

800–375 MYA

ANIMALS

Myotome muscle block contracts to flex fish's body

MOVING WITH THE BACKBONE

Before the development of limbs, vertebrates moved by flexing their spine from side to side with segments of muscle, called myotomes, along the backbone.

FEET THAT GRASP

Primates developed feet that could grasp branches, allowing them to specialize in effective treetop movement, often using their hands and feet simultaneously.

Most primates have five long, curved digits and opposable big toes

ARCHED SUSPENSION

Hominins developed stiff, arched feet with in-line toes, sacrificing grasping tree-climbing abilities in favour of efficient movement along the ground.

Foot arches provide shock absorption and save energy when walking and running

Backwards curvature of the lower spine brings our torso upright

7 MYA–PRESENT

HOMININS

55–7 MYA

PRIMATES

375–55 MYA

S-SHAPED SPINE

The S-shaped spine is unique to humans and allows us to maintain a completely vertical posture and walk more efficiently on two legs.

Gorillas' hind limb muscles prevent them from standing upright with straight legs

POWERFUL LEGS

The heavy musculature of ape hind limbs allows them to maximize power, which helped them to adapt to moving in different habitats.

VARIATION ON A THEME

The evolution of mammals brought diversification in foot shape and number of digits, which helped them to adapt their movement to different habitats.

A horse's foot has evolved over time to have just one digit

A stiff, strong backbone meant Moschops' body was supported above its limbs

EVOLUTIONARY HERITAGE

The evolution of our spine, hips, and legs can be traced back to before the origin of vertebrates. The milestones along the way show the long, complex evolutionary journey of what are now highly specialized adaptations to being habitually upright.

FIGHTING GRAVITY

Sturdier interlocking vertebrae gave the spine more support, allowing heavy land vertebrates such as Moschops to hold their torsos off the ground.

THE LONG WALK TO HUMANKIND

Adaptations to upright climbing behaviours are evident in the fossilized vertebrae of Morotopithecus bishopi

TREE-DWELLING ANCESTORS
Morotopithecus bishopi's vertically adapted spine suggests it may have been one of the earliest apes, and probably a common ancestor for all living apes.

One of the clearest features distinguishing humans from most other primates is uprightness. The evolutionary origins of our posture can be traced back over 20 million years, to the very root of the ape family tree.

Bolt upright

As primates became larger, the dangers of falling from high branches led to the use of more vertical postures

The first apes were those primates whose fossil remains indicate a shift towards moving around the trees in an upright posture (using all four limbs) compared to their monkey ancestors who moved horizontally. The main reason for this postural shift was body size: as apes became larger, balancing on top of branches while on all fours grew harder – it was far safer to hang vertically underneath them or grasp other branches for support. Anatomical adaptations to these upright behaviours, such as long arms and a ribcage that is squashed front-to-back rather than protruding forwards, also helped large apes to climb up vertical tree trunks.

The modern human spine has four curves to ensure it remains directly above the hips and below the skull

Fossilized primate vertebrae found in Moroto, Uganda, estimated to be 20.6 million years old and assigned to *Morotopithecus bishopi*, represent some of the earliest indications of a vertically adapted, apelike spine. While modern human upright posture and movement is highly specialized and arguably unique, upright torsos is a general hallmark of the ape family, and was therefore established long before hominins evolved, and our ancestors left the trees.

BRINGING US UPRIGHT

The curvature of the modern human spine shows the evolutionary journey that our ancestors' bodies took from being smaller, monkeylike animals on all fours, to the tall, habitually upright individuals we are today. Monkey spines curve in only one direction (forwards) and are flexible to allow their bodies to bend while moving around the trees. The hominin spine has gradually incorporated other areas of curvature to bring our bodies upright. The thoracic section of a human spine (the vertebrae with ribs attached) curves forwards in the same direction as monkey spines, while our

Cervical spine consists of seven vertebrae that curve backwards, in what is called "cervical lordosis"

Thoracic spine, made up of 12 vertebrae, attaches to the ribs and curves forwards in "kyphosis"

Lumbar spine consists of five vertebrae that curve backwards, forming the "lumbar lordosis" that is critical for standing upright

Made up of five fused vertebrae, the sacrum attaches to the pelvis and has the coccyx below

CERVICAL

THORACIC

LUMBAR

SACRAL

AHEAD OF THE CURVE

The modern human spine has four regions that curve in different directions, giving it a double S-shape. A backwards curvature is called "lordosis", and a forwards curvature is known as "kyphosis".

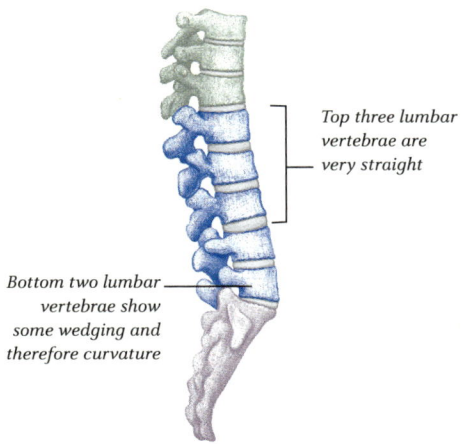

Top three lumbar vertebrae are very straight

Bottom two lumbar vertebrae show some wedging and therefore curvature

Homo neanderthalensis

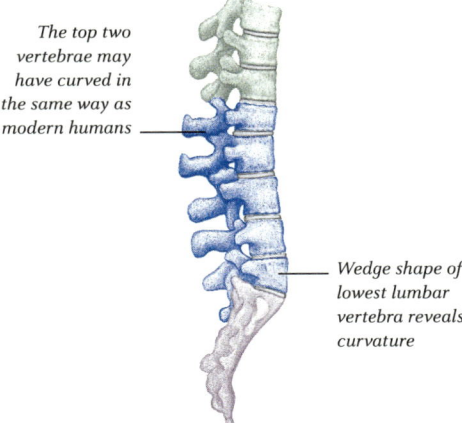

The top two vertebrae may have curved in the same way as modern humans

Wedge shape of lowest lumbar vertebra reveals curvature

Australopithecus sediba

EARLY CURVES

The lumbar vertebrae of *A. sediba* and *H. neanderthalensis* both display some evidence of "lumbar lordosis", although reconstructions vary for the former. The straighter top lumbar spine of Neanderthals may be due to other demands, such as carrying heavy loads.

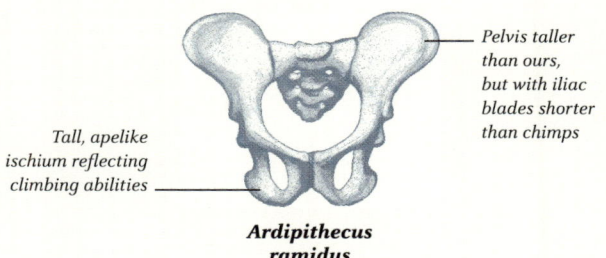

Pelvis taller than ours, but with iliac blades shorter than chimps

Tall, apelike ischium reflecting climbing abilities

Ardipithecus ramidus

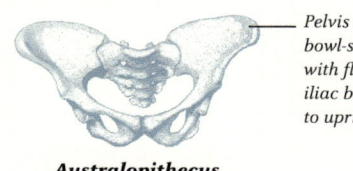

Pelvis short and bowl-shaped, with flaring iliac blades suited to upright walking

Australopithecus afarensis

Pelvis short and bowl-shaped, but iliac blades less broad, ideal for efficient walking

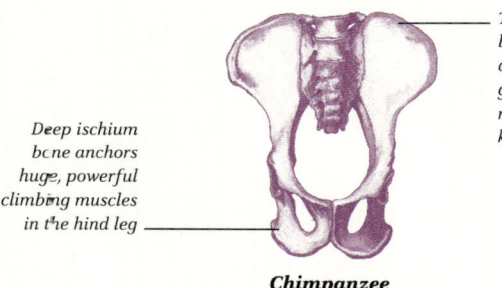

Modern human

Tall iliac blades provide attachments for gluteal muscles needed for knuckle-walking

Deep ischium bone anchors huge, powerful climbing muscles in the hind leg

Chimpanzee

HUMAN PELVIC EVOLUTION

The pelvises of early hominins, modern apes, and modern humans reveal how our pelvis is ideally adapted for the attachment of muscles that allow us to shift our weight from one leg to the other while walking or running.

spines, however, required flexibility to achieve lumbar lordosis. This flexibility came from a lack of the "lumbar entrapment" that is seen in nonhuman apes (see p.69). The human lumbar spine is not trapped by the pelvis, the shape of which is also drastically different from the pelvises of nonhuman apes. This reflects other demands, such as childbirth (see pp.192–97), but is also critical to the way we walk.

Clear evidence of lumbar lordosis in *Australopithecus afarensis*, *A. africanus*, and *A. sediba* (3.7–2 million years ago) demonstrates their reliance on maintaining upright postures, and some (but not all) *Homo erectus* specimens have even more pronounced lumbar lordosis than modern humans. Fossilized spines of Neanderthals, who lived around 250,000–40,000 years ago, imply less pronounced lordosis than either modern humans or the more ancient *Australopithecus* species, resulting in straighter lower backs. Neanderthal spines may reflect demands other than upright movement on their anatomy and, as a result, they might have walked with a more forward-leaning gait and shorter strides than modern humans. However, spinal curvature varies between individuals within each species, because – like so much of our anatomy – it develops in response to the actual demands placed on it as the individual grows. Human infants are born with a C-shaped spine that resembles the single-curved spine seen in monkeys, and only develop lordosis as they begin to lift their heads and move around.

BALANCING THE SKULL

At the top of our spine, the cervical lordosis balances our heavy skull directly on top of the spine. The hole where the spinal cord exits our skull, the foramen magnum (which literally means "large hole"), is right in the centre of the skull's base. In other apes, it is closer to the back of the skull. As hominin species became more adapted to a habitually upright posture and

cervical (neck) and lumbar (lower) spines curve backwards. This backwards curvature, which is called "lordosis", is critical to our ability to stand upright and walk on two legs.

Lumbar lordosis positions our torso directly above our pelvis, and its evolution has been complex. When our early ape ancestors (such as *Morotopithecus*) shifted towards more upright postures, their lumbar spines became short and stiff, which provided support for behaviours such as bridging gaps between trees in a forest that put the spine under substantial pressure. Hominin

Our early ape ancestors had stiff, straight lumbar spines suited to life in the trees

THE HOLE STORY

The position of the hole where the spinal cord exits the skull (the foramen magnum) is one of the clearest skeletal indicators of body posture, as seen here in a modern human and a chimpanzee.

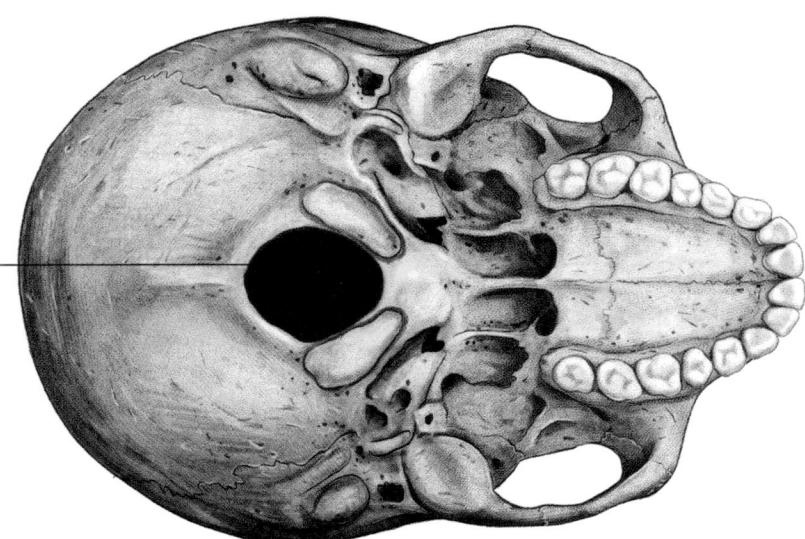

Modern human

Foramen magnum in the centre of the base of the skull, reflecting upright posture

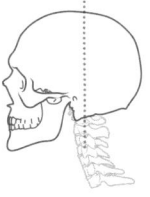

BALANCING ON TOP
The human spinal column needs to balance the skull vertically, so is in the centre of the skull's base.

Foramen magnum near the back of the skull, reflecting frequent horizontal postures

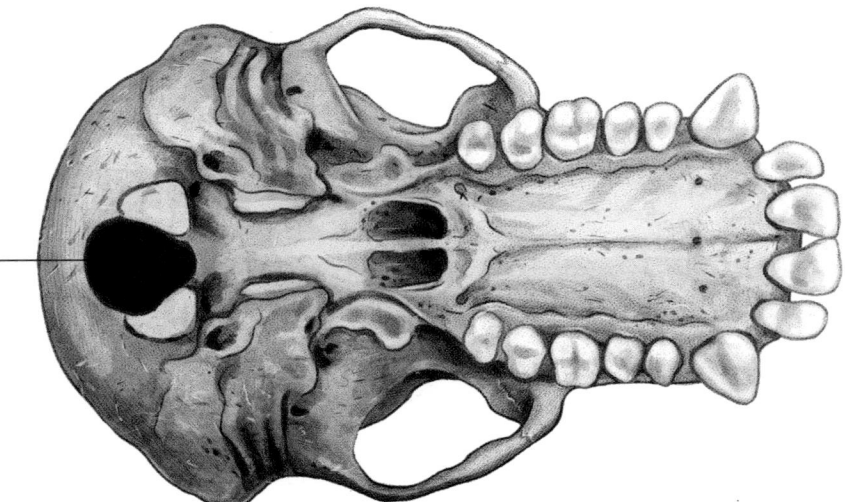

Chimpanzee

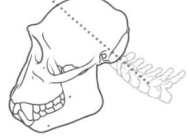

IN AT AN ANGLE
A chimpanzee spinal column enters nearer the back of the skull due to its frequent walking on all four limbs.

Position of the foreman magnum is a key piece of evidence for determining hominin species

movement, their foramen magnum position shifted forwards, reducing the muscular strength needed to balance the skull. For *Ardipithecus ramidus*, its foramen magnum position was a critical piece of evidence used to assign it hominin status.

Of course, other species from more distant parts of the animal kingdom stand on two legs, such as kangaroos, birds and, before their extinction, many dinosaurs. However, their weight-bearing and walking mechanisms use very different models, such as cantilevers (where the torso leans forwards but is counter-balanced by a heavy tail). It is only hominins who have developed a method of balancing each body part vertically on top of the next, by modifying a spine that initially evolved for life in the trees.

BOLT UPRIGHT

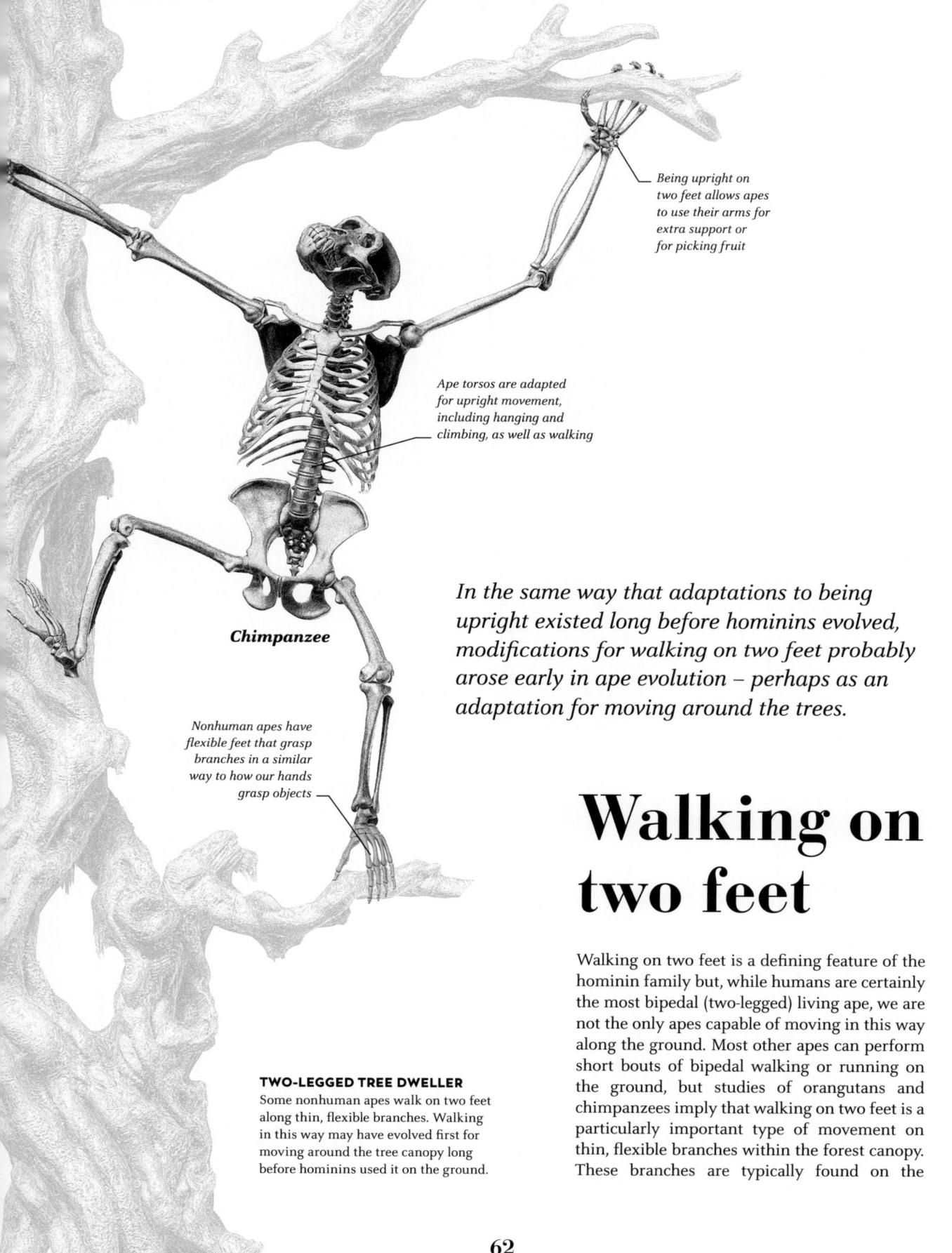

Being upright on two feet allows apes to use their arms for extra support or for picking fruit

Ape torsos are adapted for upright movement, including hanging and climbing, as well as walking

Chimpanzee

Nonhuman apes have flexible feet that grasp branches in a similar way to how our hands grasp objects

TWO-LEGGED TREE DWELLER
Some nonhuman apes walk on two feet along thin, flexible branches. Walking in this way may have evolved first for moving around the tree canopy long before hominins used it on the ground.

In the same way that adaptations to being upright existed long before hominins evolved, modifications for walking on two feet probably arose early in ape evolution – perhaps as an adaptation for moving around the trees.

Walking on two feet

Walking on two feet is a defining feature of the hominin family but, while humans are certainly the most bipedal (two-legged) living ape, we are not the only apes capable of moving in this way along the ground. Most other apes can perform short bouts of bipedal walking or running on the ground, but studies of orangutans and chimpanzees imply that walking on two feet is a particularly important type of movement on thin, flexible branches within the forest canopy. These branches are typically found on the

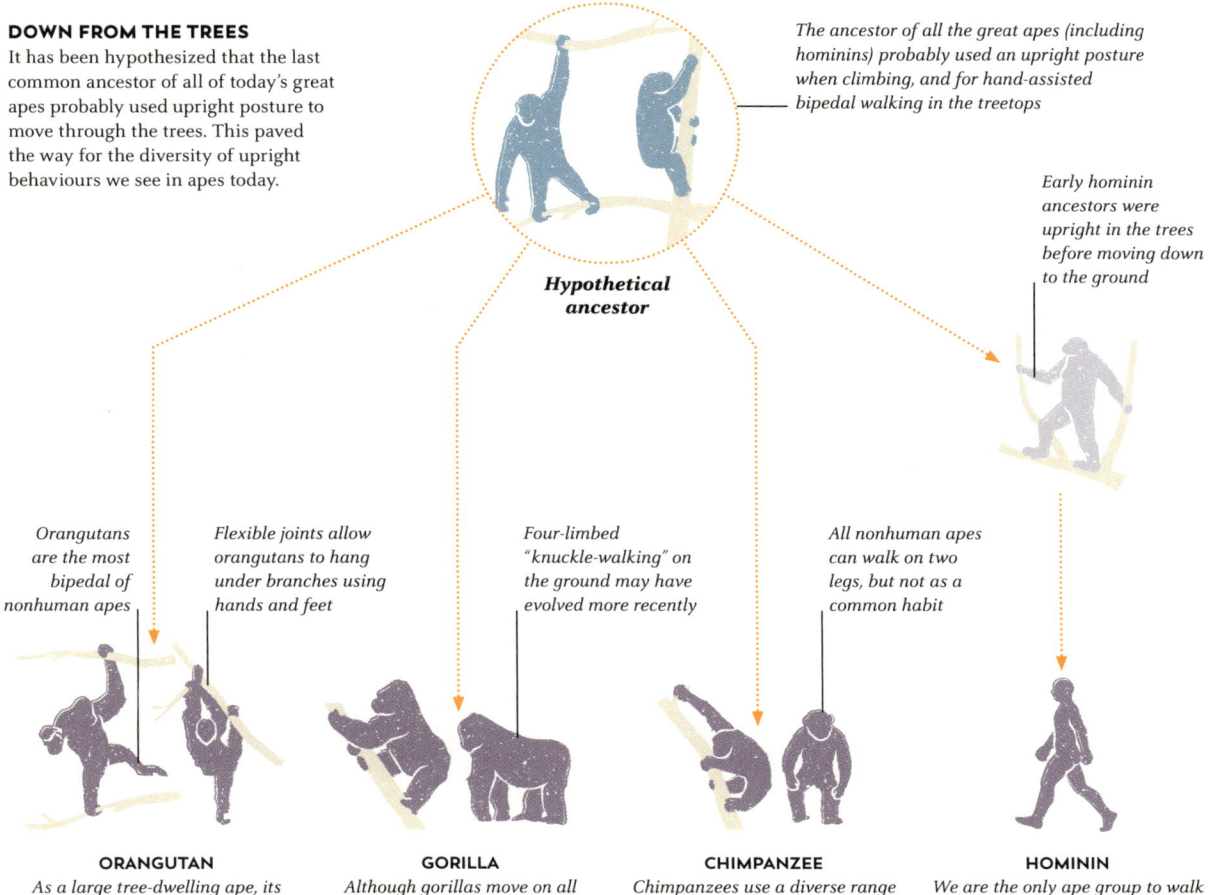

DOWN FROM THE TREES

It has been hypothesized that the last common ancestor of all of today's great apes probably used upright posture to move through the trees. This paved the way for the diversity of upright behaviours we see in apes today.

The ancestor of all the great apes (including hominins) probably used an upright posture when climbing, and for hand-assisted bipedal walking in the treetops

Hypothetical ancestor

Early hominin ancestors were upright in the trees before moving down to the ground

Orangutans are the most bipedal of nonhuman apes

Flexible joints allow orangutans to hang under branches using hands and feet

Four-limbed "knuckle-walking" on the ground may have evolved more recently

All nonhuman apes can walk on two legs, but not as a common habit

ORANGUTAN
As a large tree-dwelling ape, its movement strategies may share similarities with our ape ancestor

GORILLA
Although gorillas move on all fours on the ground, they use upright behaviours in the trees

CHIMPANZEE
Chimpanzees use a diverse range of upright movements, including climbing, hanging, and walking

HOMININ
We are the only ape group to walk upright on the ground habitually, using an energy-efficient gait

peripheries of tree branches and represent an important environment to exploit as they contain precious resources, such as fruit and honey. However, they also present balance challenges for large-bodied apes: thin branches are more likely to behave unpredictably or break under their weight, and the risk of injury after falling from a height increases with body size. This may have driven one of the first shifts towards walking on two feet, as ape ancestors started using their hands to support themselves on higher branches as they collected food in the tree canopies. Orangutans, who live almost exclusively in the forest canopy, walk on two legs more than any other nonhuman ape, and the way they walk along branches reveals interesting similarities to human gait. When branches bend under their weight, orangutans respond by straightening, rather than bending, their legs, in the same way that humans do when running on sprung tracks. This suggests that branch flexibility may have played a particular role in the evolution of bipedal movement in the trees (see p.66).

USEFUL COMBINATIONS

In hominins, it is likely that adaptations to habitually walking on two legs on the ground came later and were combined with tree-dwelling life for a while. The fossilized femur of one of the earliest hominins, six-million-year-old *Orrorin tugenensis*, indicates that it was an upright biped, but its teeth are consistent with a diet of fruit and seeds, and its upper limb bones imply a reliance on climbing trees. Combinations of adaptations for climbing and two-legged

Fossil bones and teeth from our early hominin ancestors suggest a lifestyle that was still at least partially based in the trees

63

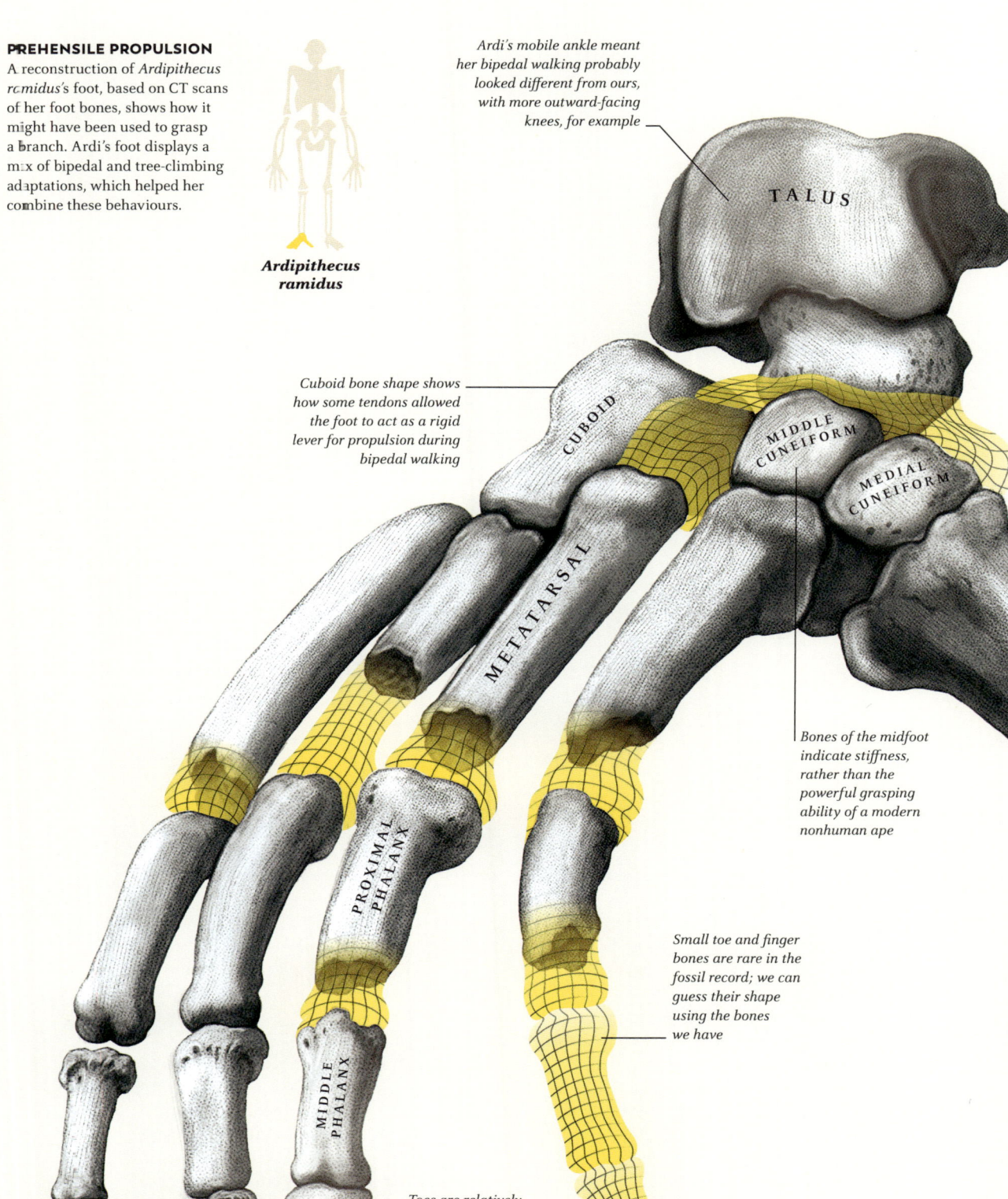

PREHENSILE PROPULSION

A reconstruction of *Ardipithecus ramidus*'s foot, based on CT scans of her foot bones, shows how it might have been used to grasp a branch. Ardi's foot displays a mix of bipedal and tree-climbing adaptations, which helped her combine these behaviours.

Ardipithecus ramidus

Ardi's mobile ankle meant her bipedal walking probably looked different from ours, with more outward-facing knees, for example

TALUS

Cuboid bone shape shows how some tendons allowed the foot to act as a rigid lever for propulsion during bipedal walking

CUBOID

MIDDLE CUNEIFORM

MEDIAL CUNEIFORM

METATARSAL

Bones of the midfoot indicate stiffness, rather than the powerful grasping ability of a modern nonhuman ape

Small toe and finger bones are rare in the fossil record; we can guess their shape using the bones we have

PROXIMAL PHALANX

MIDDLE PHALANX

DISTAL PHALANX

Toes are relatively short compared with other nonhuman apes, but curved allowing Ardi to grasp branches

walking on the ground appear in many other hominins who lived between two and five million years ago, including *Ardipithecus kadabba*, and the *Australopithecus* species *A. anamensis*, *A. afarensis*, and *A. sediba* – although for some of these species, sparse fossil records make reconstructing their behaviour difficult.

PRE-ADAPTED FOR WALKING

We do have a near-complete female skeleton of the 4.4-million-year-old *Ardipithecus ramidus*, however, known as "Ardi", whose foot bones show a remarkably detailed collection of features, some adapted for moving around the trees and some adapted for walking bipedally on the ground. Her big toe (hallux) is turned outwards enough to have enabled her to grasp branches during climbing, but her other toes are shorter. Her midfoot, however, appears rigid when compared to the feet of chimpanzees, which are built for grasping and have substantial flexibility at both the toes and midfoot.

Ardi's foot rigidity probably helped her to walk upright by allowing a more efficient "push-off" from the ground. But, while this foot stiffness certainly represents an important evolutionary milestone in the history of our own walking, it is not necessarily a novel adaptation to bipedal walking on the ground. Some researchers have suggested it is a primitive trait inherited from monkeys, who need a rigid foot to push off when leaping between branches. At some point nonhuman great apes, whose large bodies prohibit much leaping, lost this ability in favour of flexible, grasping feet that were more useful for climbing. Inheriting Ardi's monkeylike foot stiffness, alongside adaptations from older ape ancestors for being upright in the trees, such as a mobile upper body, probably helped hominins to hone their bipedal gaits on the ground later.

Ardi combined ape and hominin features with what may have been older traits inherited from monkeys

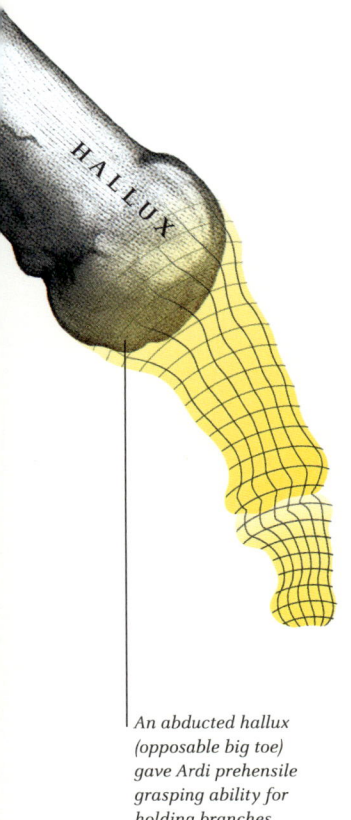

An abducted hallux (opposable big toe) gave Ardi prehensile grasping ability for holding branches

COMPARISON WITH MODERN HUMANS

Below are the estimated positions of *Ardipithecus ramidus*'s foot bones compared with those of a modern human. It shows how we have lost the prehensile big toe and now have rigid feet that maximize propulsion.

● PHALANGES
● METATARSALS
● TARSALS

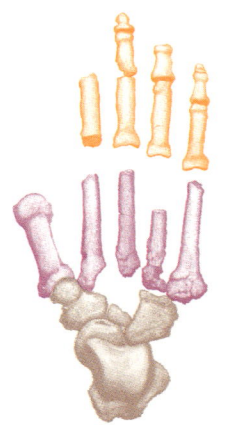

Ardipithecus ramidus

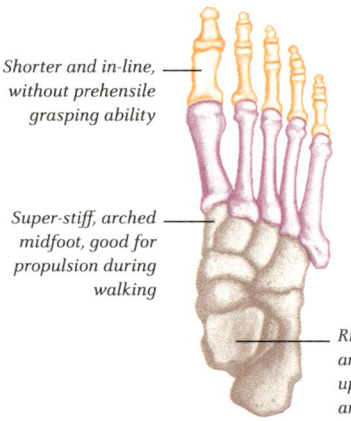

Shorter and in-line, without prehensile grasping ability

Super-stiff, arched midfoot, good for propulsion during walking

Rigid, supportive ankle, good for upright walking and standing

Homo sapiens

65

Our ancestors were already bipedal when they became ground-dwelling. But then a shift away from the trees allowed some to develop a unique bipedal gait, which made walking and running extremely efficient.

Life on the ground

Even though two-legged walking was already established before our ancestors left the trees, moving down to the ground was a key milestone in our evolution. Reconstructing the evolutionary journey of these ground-dwelling hominins relies mostly on a sparse record of fragmented fossil remains, but occasionally, these species left other clues behind: footprints. Some remarkably well-preserved prints discovered in Laetoli, Tanzania, are thought to have been left by two or three hominins walking through freshly fallen volcanic ash 3.7 million years ago. Scientists surmise they were left by *Australopithecus afarensis* and reveal the remarkably modern humanlike gait that this species had, but they also show the separate, grasping big toe that *A. afarensis* still possessed as one of its adaptations to climbing. In fossils, we do start to see evidence of an in-line big toe around the same time

Rare fossilized footprints provide a glimpse of an early stage in the evolution of walking on the ground

MAKING AN IMPRESSION

The Laetoli trails in Tanzania reveal that as early as 3.7 million years ago, some hominins had already begun to walk much like modern humans, with arched feet that were striking the ground with their heels and pushing off with their toes, as we do.

GAIT COMPARISON

The route towards an upright walking stride in hominins involved a diversity in gaits among species, as they adapted to contrasting habitats.

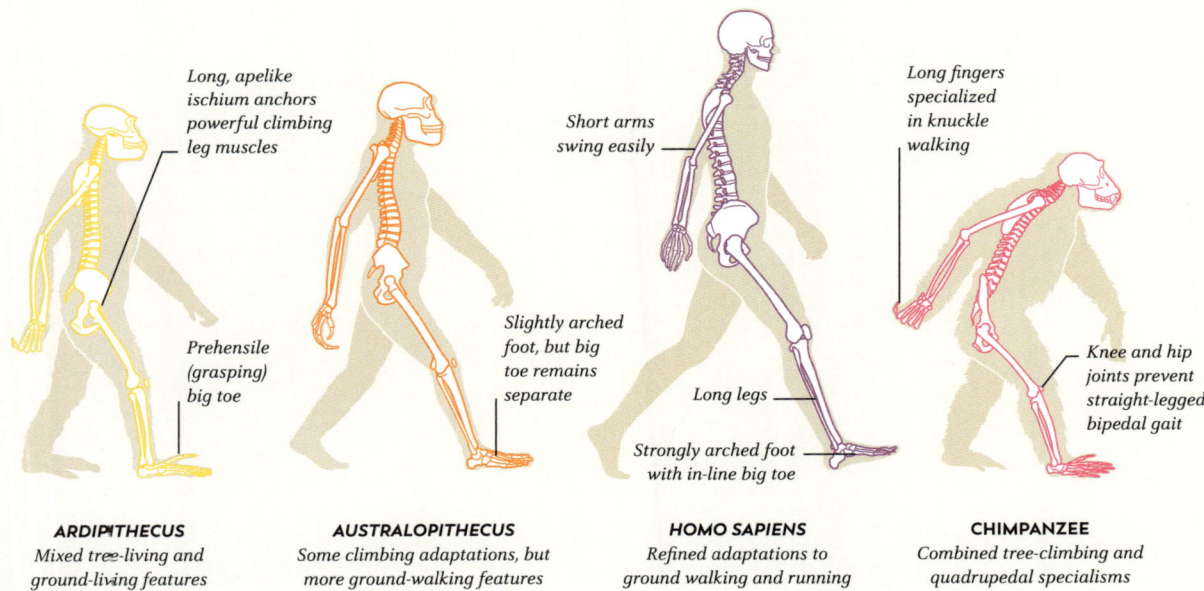

Long, apelike ischium anchors powerful climbing leg muscles

Prehensile (grasping) big toe

Short arms swing easily

Slightly arched foot, but big toe remains separate

Long legs

Strongly arched foot with in-line big toe

Long fingers specialized in knuckle walking

Knee and hip joints prevent straight-legged bipedal gait

ARDIPITHECUS
Mixed tree-living and ground-living features

AUSTRALOPITHECUS
Some climbing adaptations, but more ground-walking features

HOMO SAPIENS
Refined adaptations to ground walking and running

CHIMPANZEE
Combined tree-climbing and quadrupedal specialisms

LIFE ON THE GROUND

in other australopithecine species, as well as other modern humanlike features – but their development among species was rather diverse in nature, indicating a variety of bipedal gaits within the hominin group.

The fossil record of *A. afarensis* includes the famous one-third-complete female skeleton, "Lucy". She was shorter than later hominins, but her pelvis and leg bones tell the same story as the Laetoli footprints – she walked in a comparatively modern way. Her foot bones indicate stiff, arched feet, which acted as shock-absorbers and provided a rigid lever between the ball of the foot and the heel when the heel lifted off the ground. Lucy's thigh and shin bones suggest that, like ours, her thigh bones angled inwards, bringing her knee joints closer together than her hip joints. Her knee joints, again like ours, are wide and flat. These provide support and balance when walking and standing by positioning the knee joints above the ankles.

Fossil skeletons seem to match fossil footprints to confirm that more than 3 MYA, our ancestors walked a little like us

HIPS THAT FREE THE SPINE

Unlike the pelvis of *Ardipithecus ramidus* (see pp.60) from more than a million years earlier, which combines both bipedal and climbing features, Lucy's pelvis is shorter and wider – closer to the flatter, bowl-shaped modern human pelvis that allows rotation between the pelvis and spine during walking. Both species have

FREE-SWINGING HIPS

Unlike other great apes, humans have a flexible lower spine. Our pelvises are free to tilt and rotate in the opposite direction to our torsos and shoulders, keeping us facing forwards. We can do this because our short, bowl-shaped pelvis does not encase the bottom of our spine in the way that a chimpanzee's does.

distinctly humanlike features to their hips in comparison to other apes, but Lucy's are more pronounced. In chimpanzees, the long iliac blades of the pelvis extend upwards, encasing part of the lumbar spine and reducing the ability of the pelvis to tilt and rotate. In humans, the wide-flaring iliac wings leave the lumbar spine free to move – this is why, when we walk, our pelvis rotates and tilts up and down, and our torso and shoulders rotate in the opposite direction to our hips to keep us facing forwards.

RUNWAY WALK

All these features of the leg and pelvis became more exaggerated in later hominins, and today, they are what allow us to put one foot in front of the other while walking, keeping our supporting foot directly underneath our centre of mass

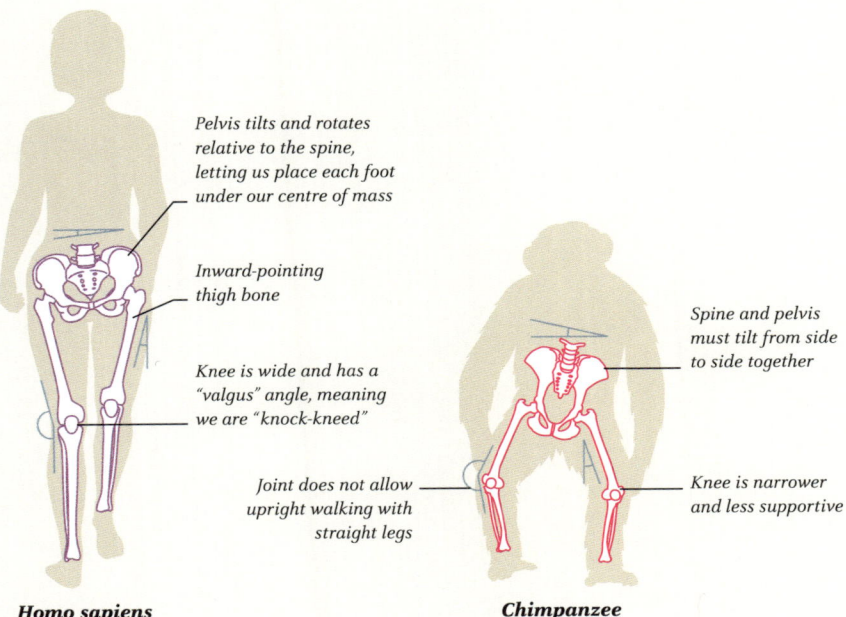

Pelvis tilts and rotates relative to the spine, letting us place each foot under our centre of mass

Inward-pointing thigh bone

Knee is wide and has a "valgus" angle, meaning we are "knock-kneed"

Joint does not allow upright walking with straight legs

Spine and pelvis must tilt from side to side together

Knee is narrower and less supportive

HIP AND KNEE ANGLES

Unlike chimpanzees, we can put one foot in front of the other. Our pelvis tilts and our thigh bones point inwards, making our hip joints further apart than our knee joints and putting our knees directly over our ankles.

Homo sapiens

Chimpanzee

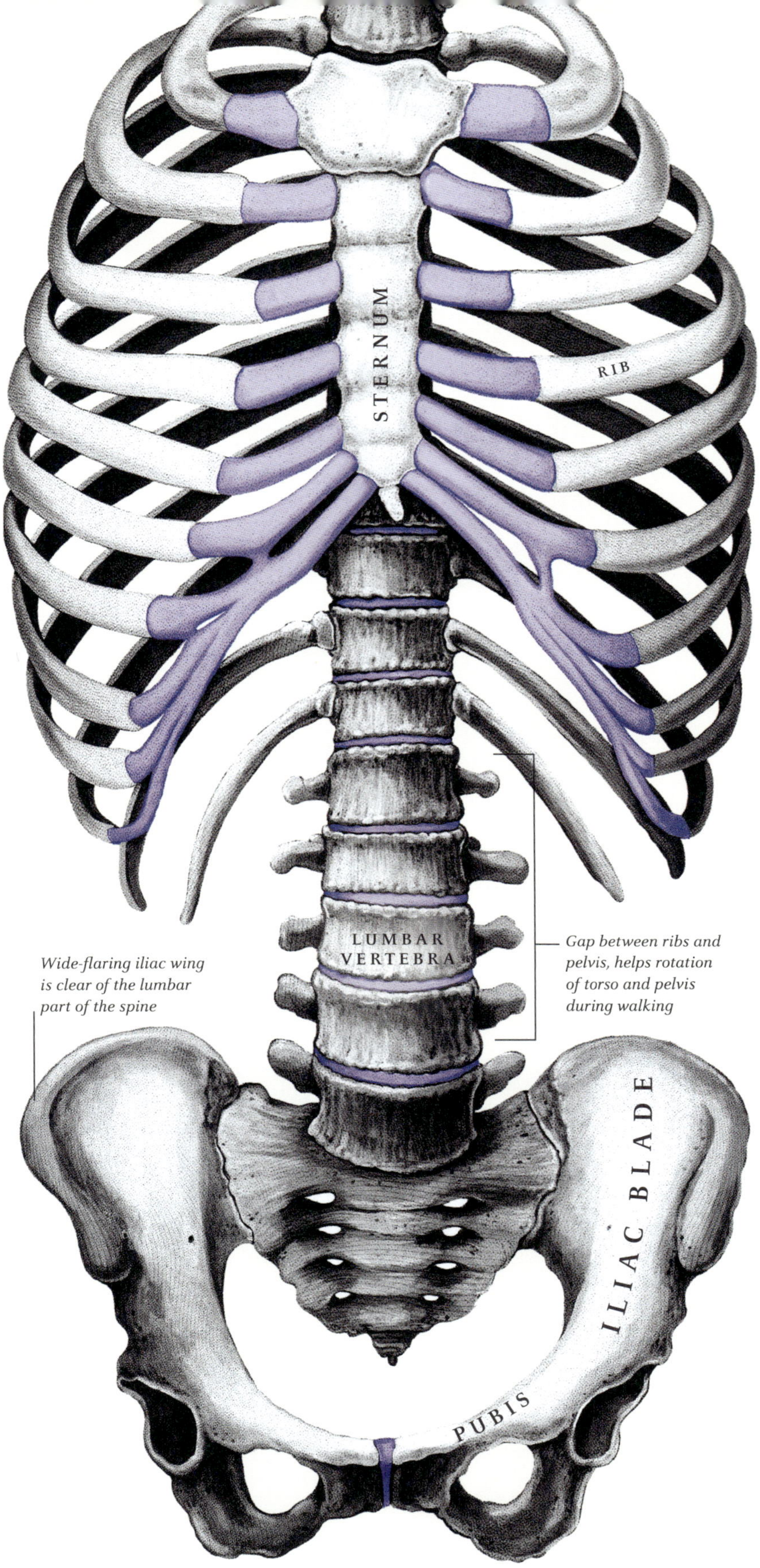

STERNUM

RIB

LUMBAR VERTEBRA

Wide-flaring iliac wing is clear of the lumbar part of the spine

Gap between ribs and pelvis, helps rotation of torso and pelvis during walking

ILIAC BLADE

PUBIS

STIFF AND STRONG

As apes evolved to become heavier, their spines lost the flexibility associated with their smaller monkey relatives. Their spines were further stabilized by wide pelvic iliac blades surrounding the lumbar region.

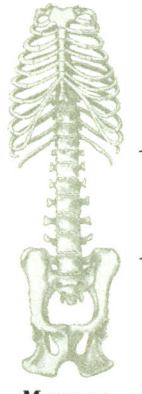

Flexible lumbar spine allows monkey to be super-mobile around the treetops

Macaque, flexible

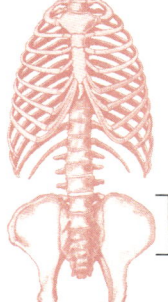

Pelvis encases the lumbar spine, providing support but restricting movement

Chimpanzee, stiff and strong

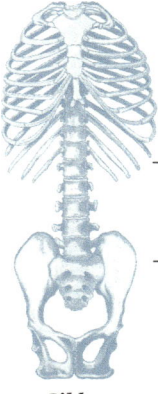

Lightweight, agile gibbon has a flexible lumbar region, like the macaque's

Gibbon, flexible

(imagine a model walking on a catwalk – this is an extreme example of the hip rotation and foot placement that our gait relies on).

The way we walk also exploits the force of gravity. Walking in humans is essentially controlled falling with the heel-strike acting as the brake (notice that as you walk, if you failed to plant your heel down you would continue falling forwards). This mechanism of walking is known as an inverted (upside-down) pendulum: once it hits the ground, the foot is a fixed point over which the hip rotates in an arc. It requires a push-off from the other foot (which is where the rigid lever of the foot becomes important), but once the hip reaches the top of the arc (directly above the foot) it makes use of gravity to complete the stride. This mechanism relies on being upright with a straight, stiff knee joint, which allows the hip to rotate over a straight leg. Our wide iliac blades allow our leg muscles to attach in the right position for this posture; something that other apes are unable to achieve, since their muscles (which are built to maximize power for climbing) stop them from maintaining an upright posture at both the hip and the knee simultaneously. This is why, when chimpanzees

The optimization of other apes' hip and leg muscles for powerful, often rapid, movement results in an ungainly and energy-inefficient upright gait

and gorillas walk on two feet on the ground, they have bent knees and lean forwards at the hip. Their lack of pelvic rotation means they cannot place one foot in front of the other, but must walk by rocking from side to side. You can feel the inefficiency of this gait by trying to walk with bent knees and hips – the extra work done by your leg muscles rapidly becomes apparent.

BENEFITS OF LIFE ON TWO LEGS

Although the shift to efficient two-legged walking on the ground may have caused hominins problems in terms of back pain or other medical issues in the long term (see pp.76–77), it gave them advantages. It freed up their hands to engage in other tasks, such as making tools and carrying infants (which may have been important given that hominin infants, having lost their grasping climbers' feet, would have been less able to cling onto their mothers). A taller, more upright body also allows more efficient temperature regulation in hot climates by increasing heat loss and lowering exposure to the sun. A higher torso position may also have made it easier to defend against other hominins or animals, and to throw projectiles. None of these advantages are likely to have been the main drivers behind the evolution of bipedalism, but they may all have contributed additional benefits that added small evolutionary pressures along the way.

While evolution has fine-tuned our anatomy for bipedalism, it must be noted that we have not lost the capacity to move in other ways. The athletic abilities of gymnasts, rock-climbers, tree-climbers, and Parkour athletes demonstrate the remarkable diversity of movement that a human body can achieve to this day.

INVERTED PENDULUM

Humans use a pendular mechanism of walking, in which the hip rotates in an arc over the foot and exploits the force of gravity. This allows us to walk for a long time using little energy.

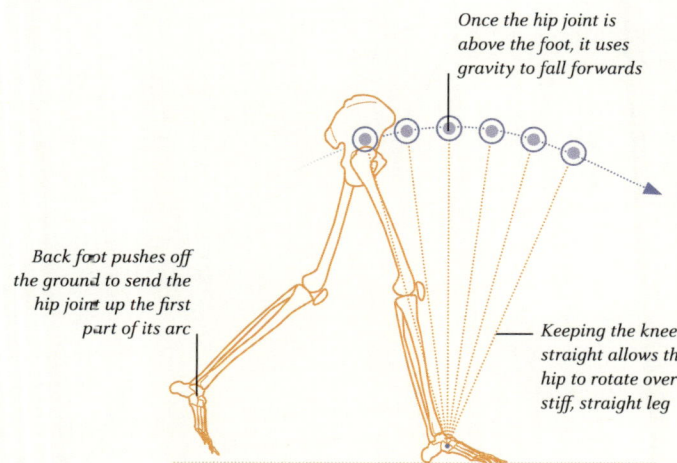

Once the hip joint is above the foot, it uses gravity to fall forwards

Back foot pushes off the ground to send the hip joint up the first part of its arc

Keeping the knee straight allows the hip to rotate over a stiff, straight leg

While evolution has finely tuned our anatomy for two-legged walking on the ground, we have not lost the capacity to swing from the trees or to climb vertical rock faces.

FREE-SWINGING ARMS

Our chest and shoulder anatomy lets our arms swing by our sides. This stabilizes the body during running by helping the torso rotate to counteract the rotations of the legs. This action is much easier with shorter arms.

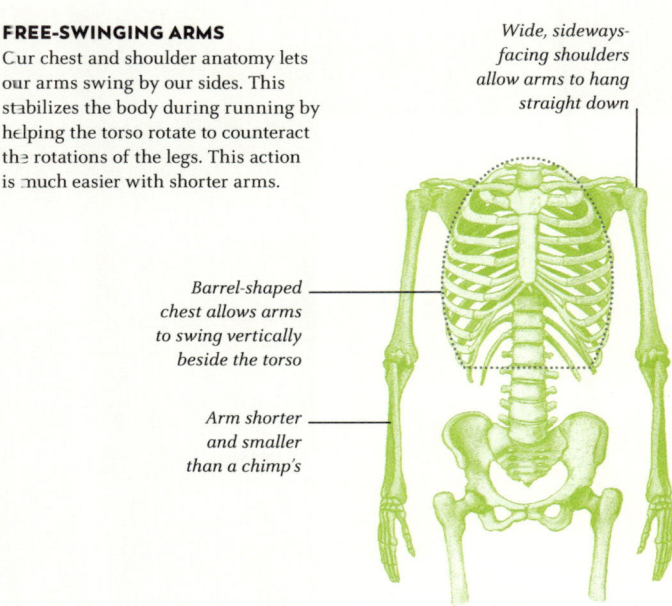

Wide, sideways-facing shoulders allow arms to hang straight down

Barrel-shaped chest allows arms to swing vertically beside the torso

Arm shorter and smaller than a chimp's

Human

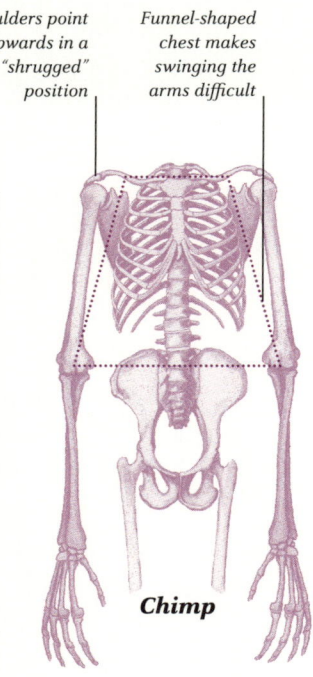

Shoulders point upwards in a "shrugged" position

Funnel-shaped chest makes swinging the arms difficult

Chimp

Endurance runners

Compared with other mammals, humans are natural endurance runners rather than sprinters. As well as our specialized walking gait, our capacity for long-distance running also sets us apart from other apes.

Humans have a slower maximum speed than many other mammals, but we can often outrun them over long distances

Nonhuman apes are extremely poor runners, and while some monkeys can sprint for short distances, no other primate can achieve endurance running in the way that humans can. This ability comes with its own suite of anatomical adaptations that are first visible in the hominin species known as *Homo erectus*. However, exactly when true endurance running emerged in hominins, as well as the reason why it evolved, is debated.

A noticeable feature of the human body is the length of our legs, which can take longer strides to increase running efficiency. Researchers previously thought that long lower limbs arose in early African *Homo erectus* (sometimes called *Homo ergaster*) 1.5 million years ago, after the discovery of an almost-complete specimen in Kenya. Nicknamed "Turkana Boy", he had remarkably modern humanlike anatomy and a tall stature. He was an estimated 1.6 m (5 ft 2 in) tall and, at just 7–12 years old, he probably hadn't reached his full adult height. However, some reconstructions of *Australopithecus* fossils imply that longer legs may have started to evolve earlier than African *Homo erectus*.

WIDE SHOULDERS AND NARROW HIPS

One running adaptation that probably didn't emerge until *Homo erectus* (or later) is swinging arms. Earlier hominins retained shoulders adapted for climbing (see p.30), but later species developed wide, mobile shoulders that were more independent from the head and neck, combined with narrow waists and barrel-shaped chests that let their arms swing freely by their sides. During running, swinging arms help stabilize the body by providing counter-rotations

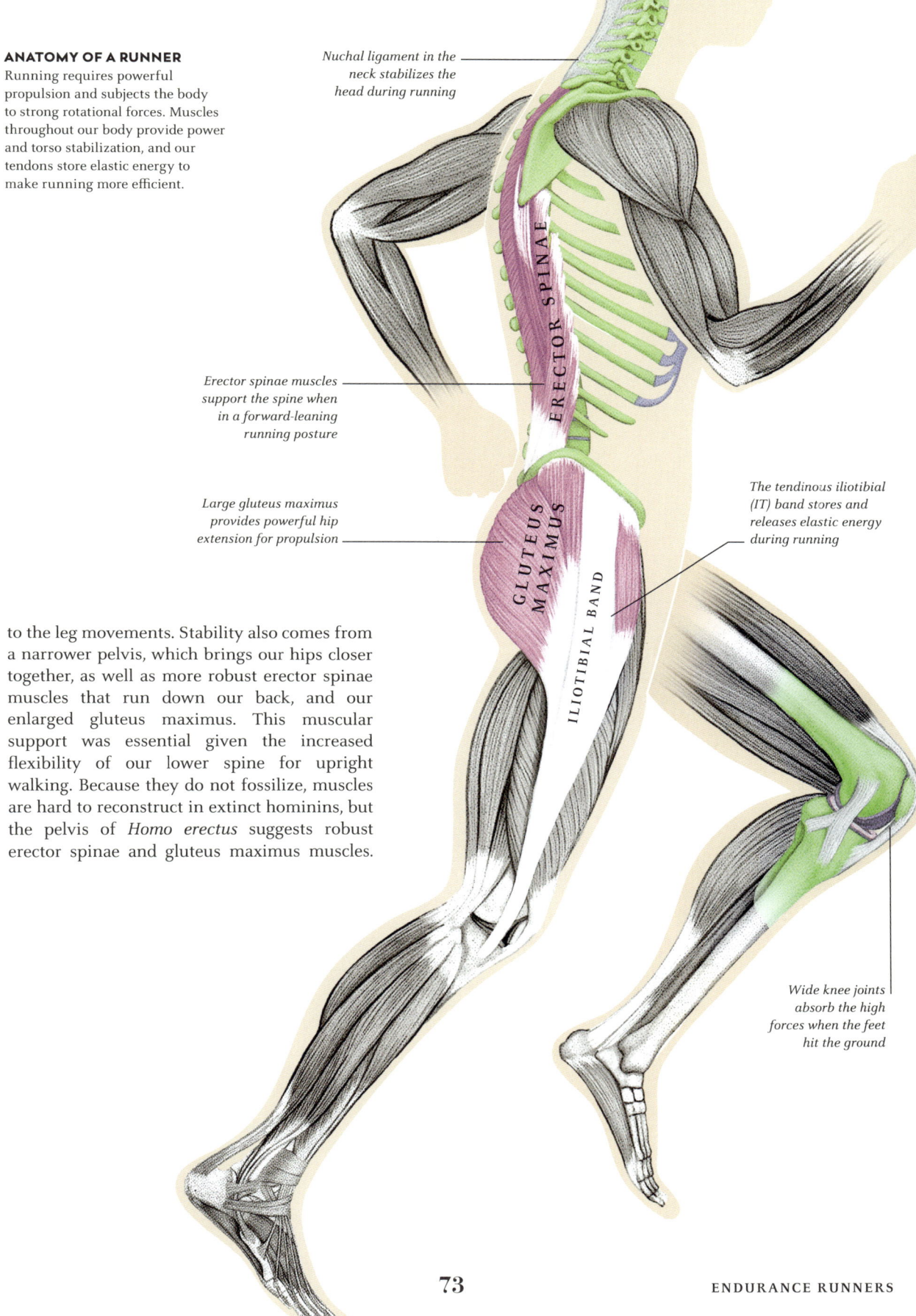

ANATOMY OF A RUNNER

Running requires powerful propulsion and subjects the body to strong rotational forces. Muscles throughout our body provide power and torso stabilization, and our tendons store elastic energy to make running more efficient.

Nuchal ligament in the neck stabilizes the head during running

Erector spinae muscles support the spine when in a forward-leaning running posture

ERECTOR SPINAE

Large gluteus maximus provides powerful hip extension for propulsion

GLUTEUS MAXIMUS

ILIOTIBIAL BAND

The tendinous iliotibial (IT) band stores and releases elastic energy during running

Wide knee joints absorb the high forces when the feet hit the ground

to the leg movements. Stability also comes from a narrower pelvis, which brings our hips closer together, as well as more robust erector spinae muscles that run down our back, and our enlarged gluteus maximus. This muscular support was essential given the increased flexibility of our lower spine for upright walking. Because they do not fossilize, muscles are hard to reconstruct in extinct hominins, but the pelvis of *Homo erectus* suggests robust erector spinae and gluteus maximus muscles.

73

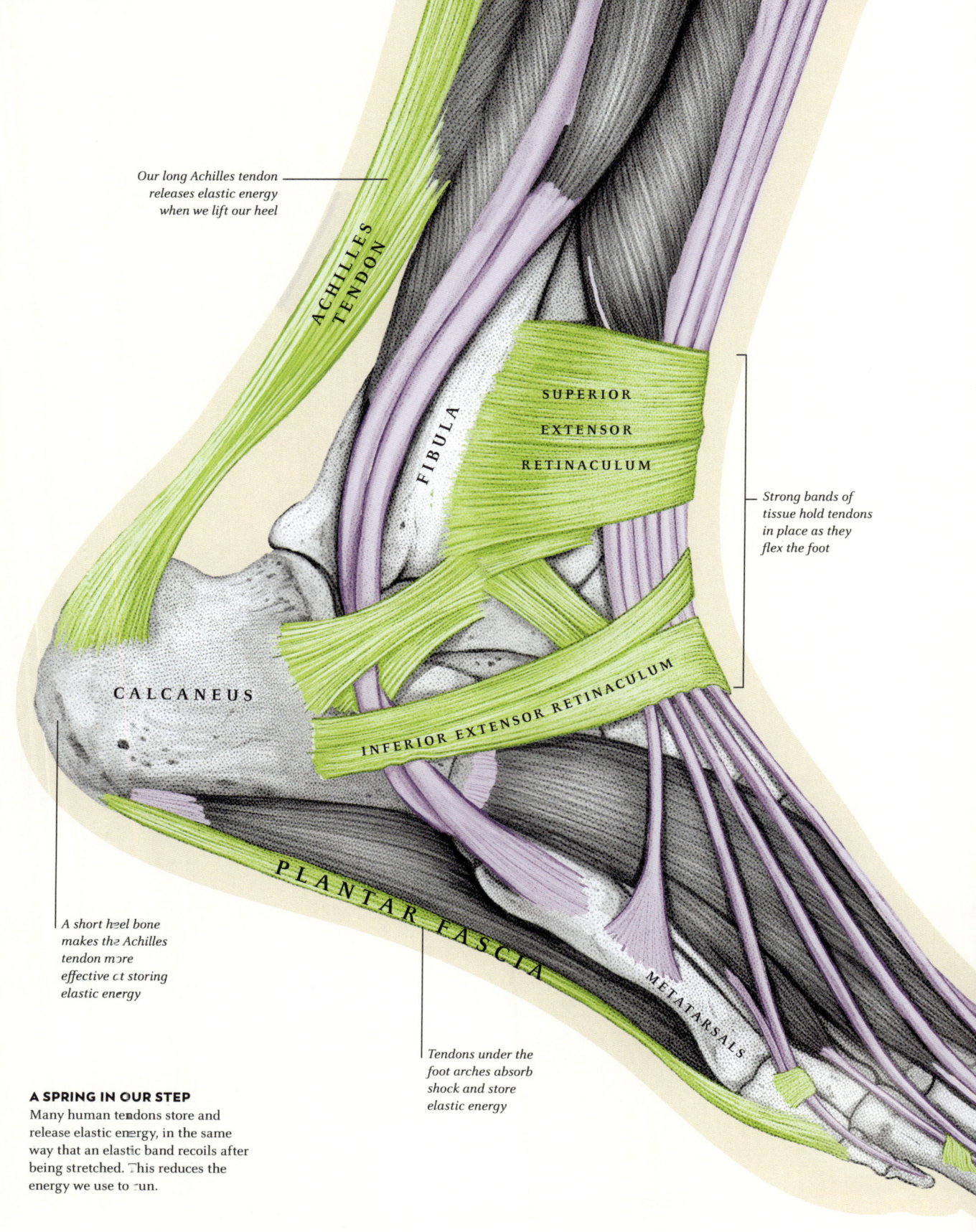

Our long Achilles tendon
releases elastic energy
when we lift our heel

ACHILLES
TENDON

FIBULA

SUPERIOR

EXTENSOR

RETINACULUM

Strong bands of
tissue hold tendons
in place as they
flex the foot

CALCANEUS

INFERIOR EXTENSOR RETINACULUM

PLANTAR FASCIA

METATARSALS

A short heel bone
makes the Achilles
tendon more
effective at storing
elastic energy

Tendons under the
foot arches absorb
shock and store
elastic energy

A SPRING IN OUR STEP

Many human tendons store and
release elastic energy, in the same
way that an elastic band recoils after
being stretched. This reduces the
energy we use to run.

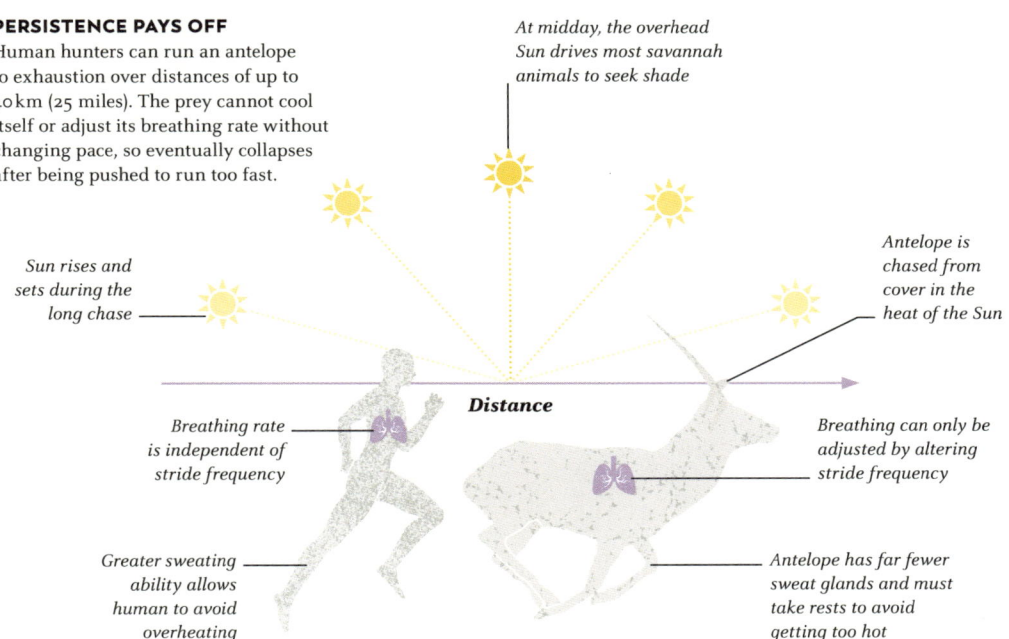

PERSISTENCE PAYS OFF
Human hunters can run an antelope to exhaustion over distances of up to 40 km (25 miles). The prey cannot cool itself or adjust its breathing rate without changing pace, so eventually collapses after being pushed to run too fast.

At midday, the overhead Sun drives most savannah animals to seek shade

Sun rises and sets during the long chase

Antelope is chased from cover in the heat of the Sun

Distance

Breathing rate is independent of stride frequency

Breathing can only be adjusted by altering stride frequency

Greater sweating ability allows human to avoid overheating

Antelope has far fewer sweat glands and must take rests to avoid getting too hot

Some researchers suggest that *Australopithecus afarensis* and *A. africanus* may have had similarly robust gluteal muscles, but it is unclear whether they functioned in the same way.

When we run, our legs and feet store and release elastic energy, which helps to propel us forward. The Achilles tendon is the thickest tendon in the body and attaches two major calf muscles (gastrocnemius and soleus) to the heel. In the same way that a stretched elastic band automatically shortens when released, so the Achilles tendon lengthens at the start of each step and then recoils when your foot pushes off the ground, reducing the work required of your leg muscles. Humans have a very long Achilles tendon, comprising 60–65 per cent of the total "muscle-tendon unit", compared with less than 10 per cent in chimpanzee anatomy. Our short heel bone – where the Achilles attaches – also increases the amount of elastic energy the tendon can store. Analysis of *A. afarensis* heel bones suggest that its Achilles tendon may have

The way in which our leg muscles attach to the bones in the foot and knee helps to propel us forward and lessens the shock of impact with the ground

comprised around 63 per cent of the muscle-tendon unit – so while some adaptations are not as developed as in later hominins, perhaps they were able to run more than we thought.

BEATING THE HEAT
Although some researchers have suggested that running is too energetically costly to have underpinned much of our evolution, the endurance pursuit hypothesis argues that several running adaptations arose for chasing prey to exhaustion. This strategy is employed occasionally by some hunter-gatherers today for chasing antelope or deer for a few hours (or occasionally even a whole day) over distances of up to 30–40 km (19–25 miles). Humans can run for a long time in hot temperatures, partly because we produce much more sweat than other mammals do (see p.30). We also have a higher proportion of lower-power, "slow-twitch" fibres in our pelvic and leg muscles, helping resist fatigue over long distances. We cannot assess muscle fibre types or sweat production in ancient hominins, but some researchers argue that running may have given them advantages, allowing them to catch and eat animals that would otherwise be out of reach.

ENDURANCE RUNNERS

The stresses placed on our curved lower spines can cause uniquely human conditions like spondylolysis, which is a common ailment in athletes such as gymnasts who routinely "hyperextend" their backs.

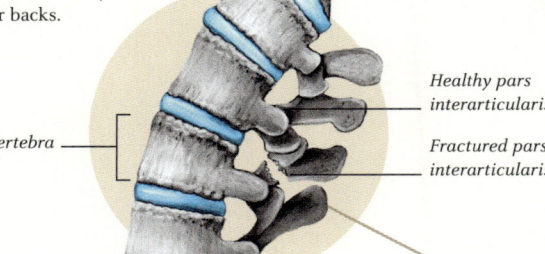

Vertebra

Healthy pars interarticularis

Fractured pars interarticularis

*While our spine represents an impressive combination of **stability and flexibility**, it also reflects the ultimate trade-off between these two things, and therefore comes with compromises.*

Balancing act

All skeletal joints reflect a balance between stability and flexibility: the most stable do not allow any movement, while the most flexible (mobile) joints offer much less stability and are more likely to dislocate. Compared with the short, stiff lumbar spines of other apes, our longer, more flexible lumbar spine is fundamental to the way that we move around. However, because it supports the torso and spinal cord, a lack of stability at the spinal joints can have severe consequences. These demands on our spine mean that we are prone to back pain and our vertebral joints degrade as we age – issues that we think can be traced back to the adoption of habitual bipedal walking in *Australopithecus afarensis*.

A relatively common cause of back pain in adults is spondylolysis, which is a fracture of the pars interarticularis (a bony area at the back of a vertebra). Spondylolysis appears to be unique to humans and is caused by the stress of curving the spine, as happens during extreme arching of

The flexibility of the spine that allows humans to walk upright also makes them susceptible to back problems

the back. It is also related to more pronounced lumbar lordosis (the forwards curvature of our spine in the lumbar region that allows upright standing and walking), suggesting that it arose in hominins as they became adapted to bipedal life on the ground. Indeed, among modern humans, spondylolytic spines tend to be least like the shape of nonhuman ape spines, so could be considered "hyperadapted" to being upright. During pregnancy, the female spine develops an even more pronounced lumbar lordosis. This increased curvature stops the mother's centre of mass shifting forwards due to her pregnancy "bump", and is thought to have been a feature of female hominin spines for at least 2.5 million years, based on *Australopithecus africanus* fossils. However, this adaptability of the female spine does not reduce lower back pain, which usually increases during pregnancy.

LIFESTYLE SHIFTS

Our spines have borne most of the load of various changes in lifestyle throughout our species' history. Starting around 11,000 years ago, some human communities stopped relying as much on hunting and gathering to provide food and began to depend more on farming, which required less varied physical activity. While this lifestyle change probably reduced certain spinal conditions, the strain on the spine would not have necessarily decreased, as people started performing more repetitive activities, often in stooped postures. Much more recently, the industrial revolution led to another shift away from physical labour, yet back pain has endured, partly because we now have weaker back muscles to support our spine, especially the ones that maintain lumbar lordosis.

THE HISTORY OF BACK STRAIN

Each change in human lifestyle and activity throughout history placed different demands on the spine, leading to continued back strain.

CHANGES IN PREGNANCY

The forward curvature of the lower spine (lumbar lordosis) increases in pregnancy to stop the mother's centre of mass (COM) shifting too far forwards.

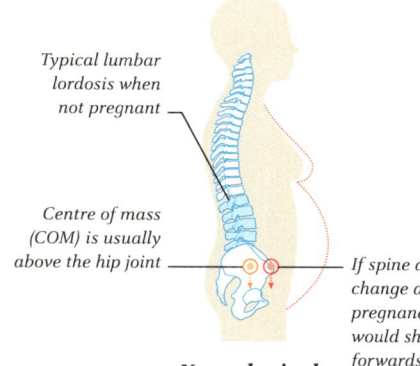

Typical lumbar lordosis when not pregnant

Centre of mass (COM) is usually above the hip joint

If spine did not change during pregnancy, COM would shift forwards

Normal spinal curvature

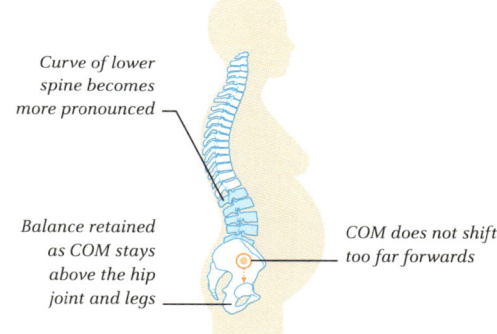

Curve of lower spine becomes more pronounced

Balance retained as COM stays above the hip joint and legs

COM does not shift too far forwards

Spinal curvature during pregnancy

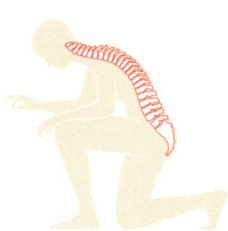

FORAGING

Frequent and varied activity exercised back muscles, but put other strains on the spine

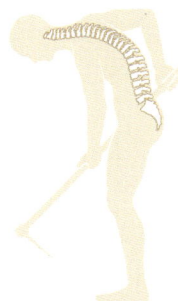

FARMING

Strain from longer hours of activity and more repetitive actions in stooped postures

FACTORY WORK

Reduction in physical labour and muscular spinal support leads to lower back problems

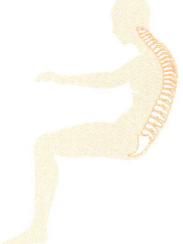

OFFICE WORK

Often sitting for long periods with little physical activity to strengthen spinal muscles

BALANCING ACT

A COMPLEX JOURNEY
This network shows how the evolution of hominin legs and spines, and the development of upright walking, is linked to other parts of hominin anatomy and behaviour, including tool-use, hunting, and even the size of their brains.

Short pelvis and inward-angled femur
S-shaped spine
As hominins spent more time on the ground, they developed adaptations to efficient walking
Large-bodied apes used upright behaviours for stability and foraging in the treetops
Some upright walking
Grasping feet
Short, encased lumbar spine
Bipedal ground dwellers

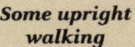

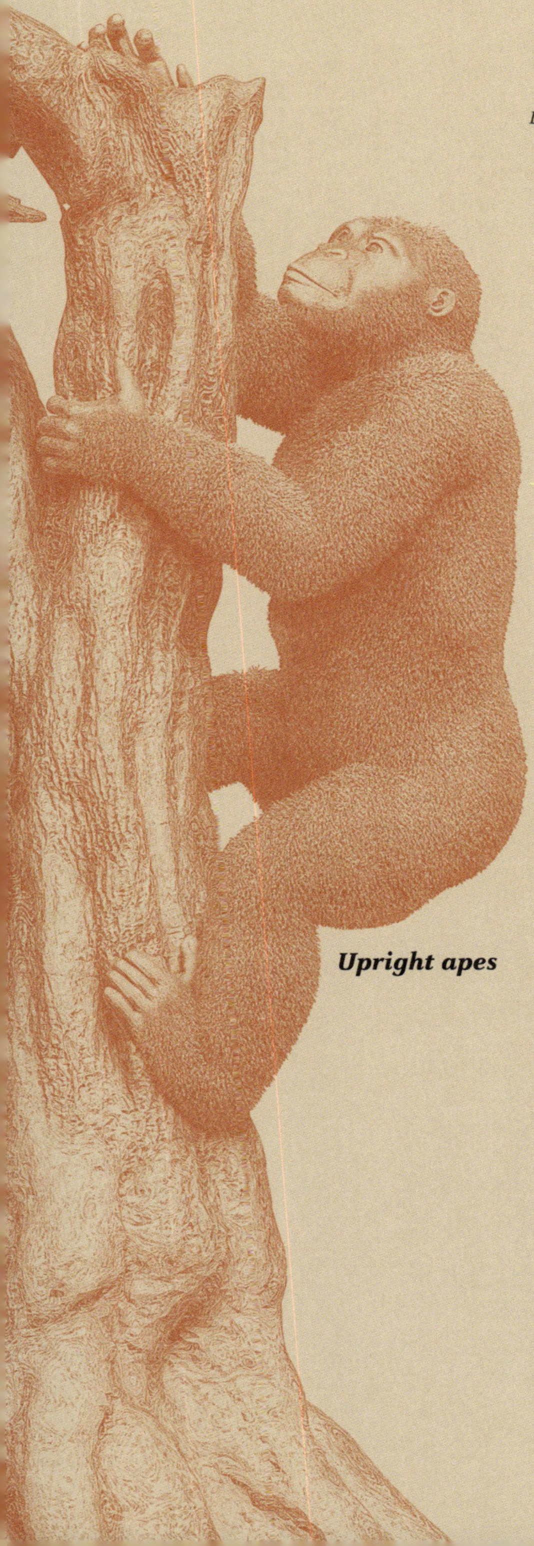

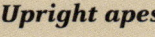

Upright apes

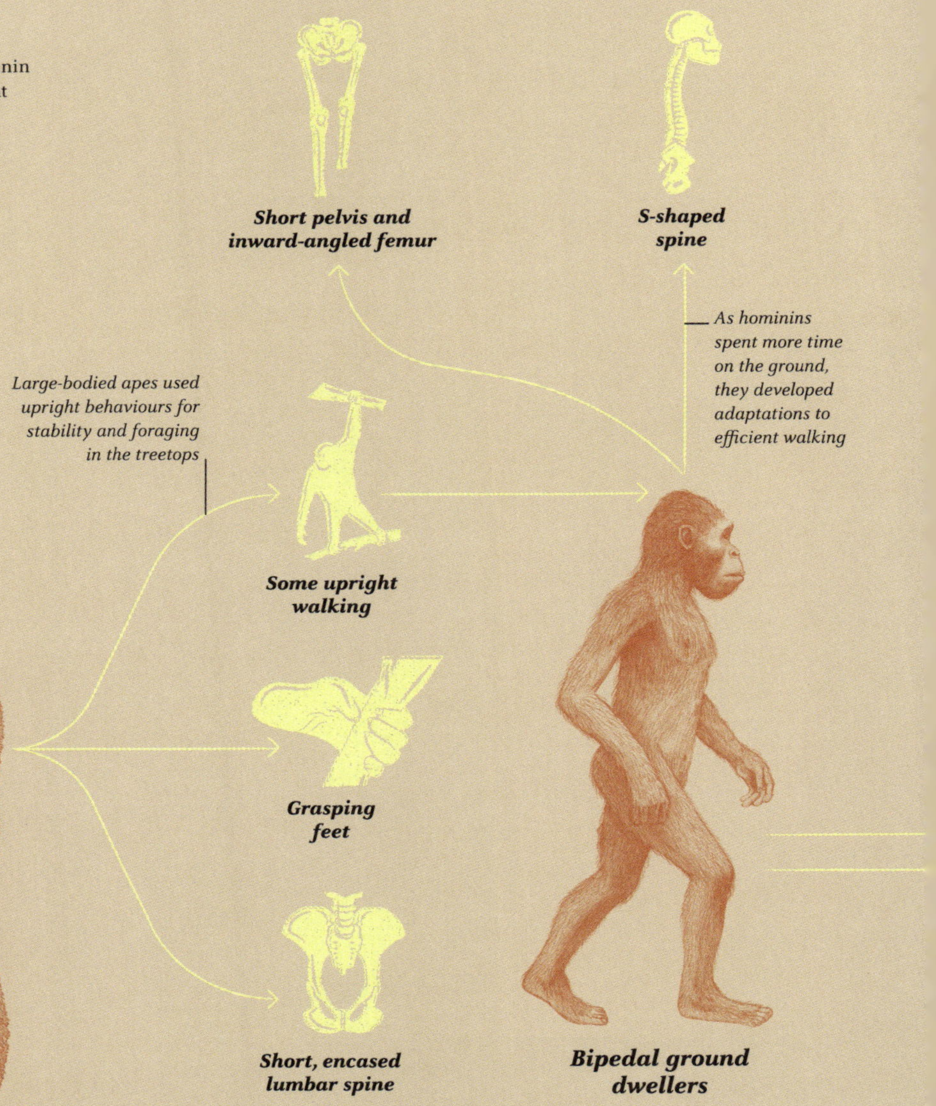

A walk through time

From apes that began moving upright in the trees to hominins capable of running marathons, the evolution of human movement was closely linked to changes in foraging and social behaviour as well as changes in brains, hearts, and lungs.

Greater manual dexterity

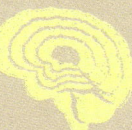

Bigger brains

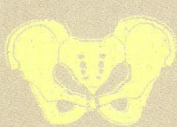

Female pelvis shape

CH2 *Hands were used less for movement and more for foraging, communication, and tool-making.*

CH8 *Feedback loops between tool-making and use, and brain size, have driven the evolution of both.*

CH7 *The female pelvis must meet the needs of birthing large-brained offspring.*

Upright walking freed the hands to hone more complex manual behaviours

Making tools and throwing weapons

Hunting

Development of efficient walking and running has helped to shape the pelvis

CH6 *Running adaptations meant hominins could outrun prey and diversify their diets.*

Hominins developed lower-limb adaptations that aided efficient running, allowing them to cover long distances

CH5 *The human heart and lungs have evolved to provide oxygen efficiently during running.*

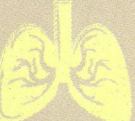

More efficient lungs

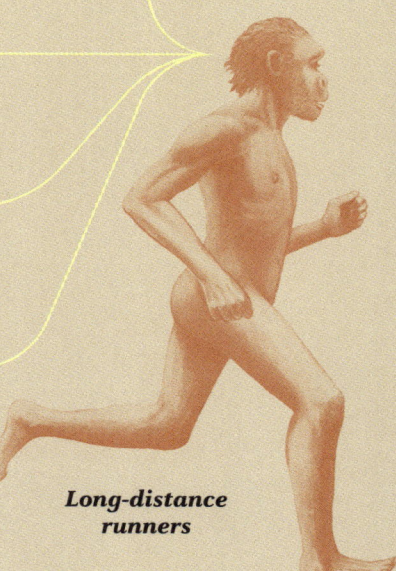

More powerful heart

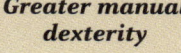

Arched feet with springy tendons

CH4 *Later hominins were able to avoid overheating during running by sweating.*

Sweating, loss of body hair

Long-distance runners

A WALK THROUGH TIME

Skin
and hair

Human skin is uniquely sweaty and relatively hairless compared to our ape ancestors. Changes in the form and function of our outer layer may have been influenced partly by mobility and migration in an ever-expanding range of habitats.

SLIMY BEGINNINGS

In contrast to the smooth, dry, water-resistant skin humans have today, our amphibian ancestors would have had a moist skin, designed to allow certain substances in and out of their bodies. As our distant ancestors migrated onto land, skin developed protective and sensory abilities, ever-changing and adapting to our shifting environments.

WHISKERS

It is likely that hair first evolved because it was useful in sensing touch (like modern whiskers), rather than keeping warm, and also may have protected or cushioned the animal, or camouflaged it.

Scaloposaurus was a distant ancestor of mammals that likely had sensory whiskers

Spinolestes was an early mammal that had fur like modern mammals

FUR

Mammals maintain a high body temperature with their high metabolic rate. A continuous insulating coat of fur would have made it easier to retain their body heat.

MAMMALS

220 MYA–55 MYA

Nails are made of keratin, the same protein found in hair and skin

NAILS

Primates have flat nails instead of sharp claws, which allow for better grip and precision. Nails protect the sensitive fingertips while keeping them free for touching, grooming, and manipulating objects.

PRIMATES

55–15 MYA

HOMINIDS

15–7 MYA

HOMININS

Skin apocrine glands secrete oily sweat through the hair follicle

GLANDS

Most animals have glands that release substances, but mammals are unique in possessing sebaceous, sweat, and, of course, mammary glands.

Homo erectus probably had numerous eccrine sweat glands, like modern humans

The ancestors of great apes probably had thinner body hair than earlier apes

NAKED SKIN

Some human ancestors underwent an even more dramatic change – their body hair reduced to a thin, fuzzy vellus layer and their skin acquired a huge number of eccrine sweat glands, giving them great cooling potential.

THINNING HAIR

As apes and humans evolved, body hair thinned – a cooling adaptation that allowed sweat to evaporate quickly and helped regulate body temperature in hot, open African environments.

Human skin can be traced back to reptile ancestors, as they adapted to life on dry land

800–375 MYA

375–220 MYA

Early vertebrates had skin coated in a protective, aqueous covering of mucus, secreted by specialized cells

OUTER LIVING LAYER
All marine animals had a protective outer layer that let certain substances in and out.

Tiktaalik was a lobe-finned fish that may have been an ancestor of land vertebrates

MOVING ONTO LAND
As marine vertebrates migrated onto dry land, their previously porous, mucus-covered skin evolved into a protective, waterproof layer that prevented water loss.

Reptile scales are folds in the epidermis covered with a layer of keratin

DEAD, WATERPROOF LAYER
On land, vertebrates developed new epidermal structures better suited to terrestrial living. Waterproof skin became keratinized – it developed a protective outer barrier made of the tough protein, keratin.

Skin is one of our oldest organs. From the earliest marine invertebrates to mammals and modern humans, skin has always stood between the body and the outside world. Even simple animals like sea sponges have protective outer layers, and as life evolved so did more complex skins that could shield, sense, and secrete protective substances.

In land animals, skin that prevented water loss became vital. As vertebrates moved onto dry land, they developed thickened skin with a surface layer of dead cells filled with keratin – a tough, water-resistant protein that forms scales, feathers, hair, and claws. This layer helped prevent drying out while still letting in essential substances.

Mammals evolved insulating fur, glands that produced sweat and scent, and sensitive whiskers. Over time, some human ancestors began to lose their body hair, which continued to evolve into the thin, flexible skin we have today, with sweat glands that keep us cool, a bacterial and fungal microbiome that protects us against infection, and melanin that shields us from the sun.

NEWLY NAKED SKIN

Humans have more sweat glands than any other animal – millions of them – allowing us to release heat efficiently through evaporative cooling. With less fur to trap heat, breezes and sweat could cool us quickly. Naked skin may have looked vulnerable, but it helped us turn into long-distance walkers and runners.

The story of our outer layer

Our skin tells the story of our past: from ancient fish slime and waterproof scales to the bare, active bodies of humans. It is not just a protective covering – it has helped shape how we live, move, and survive.

Naked apes

Humans are often called the "naked ape", but we are not truly hairless. The loss of our fur evolved over millions of years as we adapted to life on hot, open grasslands in Africa.

Starting from around 8–5 million years ago (MYA), Africa's climate became cooler and drier, with dense forests giving way to more open tropical savannahs and wooded grasslands. This shift in climate changed how our ancestors lived and moved. Instead of climbing through shady trees, they began walking and running upright over long distances, often in full sunlight. In this new landscape, thick fur was a liability – it trapped heat and slowed

Thinner body hair became an evolutionary advantage in hotter, drier climates

evaporation. Natural selection favoured individuals who could keep cool. A thinner coat of body hair combined with a greater density of eccrine sweat glands (see pp.92–93) offered an advantage in endurance, mobility, and survival. Over time, these traits became hallmarks of the evolving human body.

SHORTER AND THINNER

Each strand of human hair is made from keratin, a tough, fibrous protein also found in the skin and nails. This strand grows from a tiny pocket in the skin that anchors each strand – the hair follicle. While humans have around the same number of follicles as chimpanzees, each human hair is shorter and thinner, resulting in the vellus, or "peach fuzz", layer of fine hair that covers our bodies today. This is dictated by our

ENVIRONMENTAL SHIFTS

As hominins expanded their habitats from shady forests to exposed grasslands, their bodies adapted to new environmental challenges by shedding their thick, insulating body fur and possibly keeping hair on the head as protection from the sun.

Our ancestors probably had a thick layer of body hair similar to modern chimpanzees

A woody habitat provided some shelter from the heat of the sun

Hair on the head retained as vital protection – the sun's rays were now concentrated on the top of upright hominins

Open grasslands became more widespread

Hominins walked across the expanding grasslands in search of food and safety

8–4 MYA

4–3 MYA

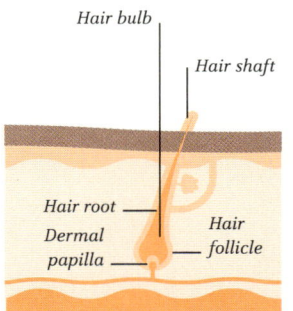

Hair bulb

Hair shaft

Hair root

Dermal papilla

Hair follicle

EARLY ANAGEN
Follicle produces new cells in hair root, which die and push upwards to form the shaft.

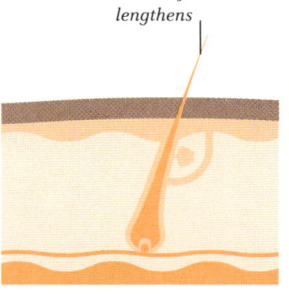

Shaft lengthens

LATE ANAGEN
The hair shaft elongates over a period spanning a few weeks to a few months.

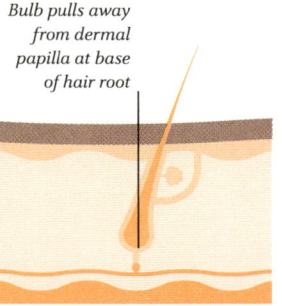

Bulb pulls away from dermal papilla at base of hair root

CATAGEN
Follicle shrinks and hair stops growing. The bulb pulls away from the dermal papilla.

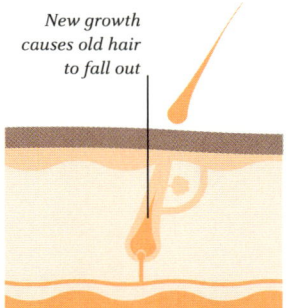

New growth causes old hair to fall out

TELOGEN
Loose hair is shed, dislodged, or pushed out by new growth, and the follicle starts a new cycle.

hair's growth cycle, which is significantly shorter than that of our hairier relatives.

Human hair follows a growth cycle with three key phases: anagen (growth); catagen (transition), where growth slows and the follicle shrinks; and telogen (resting), when the hair falls out and the cycle restarts. All body hair follows this cycle, but its length and thickness depend on how long it stays in the anagen phase. For example, scalp hair grows longer because its growth phase lasts longer than that of eyebrows or body hair. In humans, thinner body hair developed because it improved

The study of human lice can help date when hominins began to lose their thick body hair

SHORT GROWTH CYCLE

Human body hair is shorter and finer than that of other apes because our follicles have a shorter anagen phase during the growth cycle. A full cycle for vellus hair, from early growth to detachment, lasts for a matter of weeks. In contrast, human head hair has a cycle of 2–7 years.

INFECTED TWICE WITH LICE

Human lice hold clues to the loss of body hair. DNA studies have revealed that our head lice and pubic lice are genetically different, showing that they were distinct species in isolated regions of our body by 3 MYA.

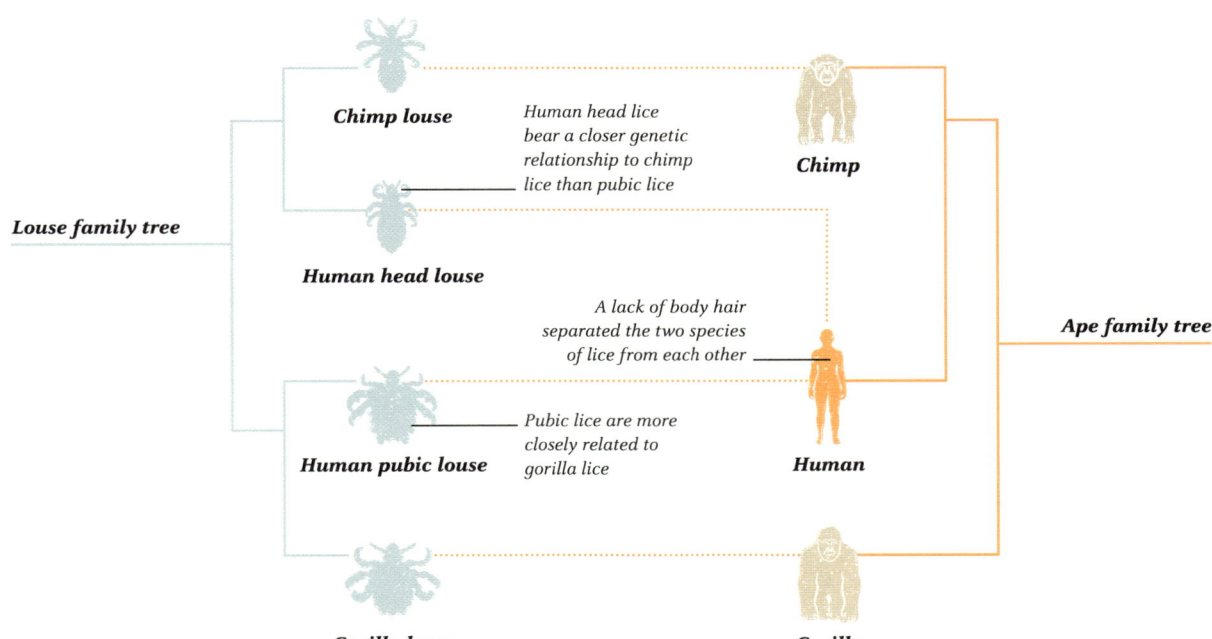

Chimp louse

Human head lice bear a closer genetic relationship to chimp lice than pubic lice

Louse family tree

Human head louse

A lack of body hair separated the two species of lice from each other

Human pubic louse

Pubic lice are more closely related to gorilla lice

Gorilla louse

Chimp

Human

Ape family tree

Gorilla

Thinner body hair offered an advantage in endurance, mobility, and survival.

HEAT REGULATION

Studies comparing different hair shapes found that curly hair helped insulate the head from solar heat while limiting temperature swings. Though less efficient at cooling when wet, it reduced overall heat gain in hot, sunny climates.

● HEAT GAINED
● HEAT LOST THROUGH EVAPORATION

The insulating effect of tight curls retains a steady temperature close to the scalp

Hairlessness allows for maximum evaporative cooling, however offers no protection from the sun

Tightly curled hair

Wavy hair

Straight hair

Bare head

evaporative cooling – sweat evaporates from the skin much more quickly if air currents are not hindered by thick hair. This conferred an evolutionary advantage onto those running around in hot environments.

HAIRY HIGHWAY

We may think lice tell a story of infestation, but they also tell a story of our evolution. Humans host two types of lice that live in different hair zones – head lice and pubic lice. Surprisingly, these two parasites are not closely related. Genetic studies show that head lice (genus *Pediculus*) are closely related to lice found on chimpanzees, while pubic lice (*Pthirus pubis*) are genetically similar to gorilla lice (*Pthirus gorillae*). This divergence points to a fascinating episode in early human evolution, beginning around 6 MYA when the human and chimpanzee lice lineages split. Then, around 3 MYA, early hominins picked up pubic lice – most likely from sharing sleeping nests recently abandoned by gorillas. The fact that these lice remained separate populations suggest that by 3 MYA, human ancestors already had significant areas of bare skin between the head and groin. Body hair had become so sparse that lice could not travel from one region to the other. For lice, the head and pubic regions had become isolated habitats – a clue that our fur was already thinning fast.

CURLY CANOPY

The shape of our hair is determined by the follicle. Round follicles produce straight hair while oval or flattened follicles create waves and curls. Evolution seems to have favoured curls. An explanation for this may lie in our adaptation to intense equatorial sunlight. Tightly curled hair forms a protective canopy over the scalp, keeping it cool while shielding it from UV rays. Unlike straight hair, which lies flat on the head and conducts heat, curly hair traps a layer of air close to the skin. This insulating effect reduces direct sun exposure and overall heat absorption.

It is possible that curly hair evolved in early African human populations in response to hot, open environments as it offered effective thermal protection. As humans migrated to cooler regions, different hair shapes evolved.

LOSING HAIR

As we age, hair changes. Melanocytes, the cells that give hair its colour, gradually stop producing melanin, leading to greying. Hair follicles also shrink or become less active, causing hair to thin or fall out in a process we know as balding. The evolutionary basis of balding has been debated – it may have no real purpose – however, some researchers suggest that it lent a male elder a sense of seniority, which elicited respect and high status that was then transferred onto his offspring, bestowing upon them privileges that increased their reproductive (and hence, evolutionary) success. While this helped preserve his genes, it also promoted the survival of his kin, who would have also carried the balding gene.

Curly hair maintains a consistent temperature close to the scalp, and provides protection from intense heat

The gradual thinning of our ancestors' thick body hair to a fine, vellus fuzz was influenced by greater UV exposure and heat stress, as hominins (our immediate ancestors and close relatives) left the shade of the forests around 4–2 million years ago (MYA). In this new environment they became active walkers and runners in the heat of the day (see p.84). As a result, our skin, alongside our hair, evolved into a multi-functional structure that met new challenges, such as exposure to intense sunlight and, therefore, the need to cool the body through sweating.

As hominins began running in hotter climates, their skin developed new cooling mechanisms

Skin is a complex organ composed of three main layers. The epidermis is the topmost layer and is made mostly of flat keratinocytes – specialized skin cells that produce keratin, a tough protein that forms the skin's protective barrier. The epidermis is the first line of

defence against pathogens, toxins, and harmful UV radiation. It is waterproof and constantly renewed. The middle layer, the dermis, is made of collagen and elastin fibres. It also houses nerve endings, glands, hair follicles (see p.85), and blood vessels, giving skin its strength, flexibility, and sensory capabilities. Skin sensors detect touch, pressure, temperature, and pain – all crucial for survival. The final, deepest layer, the hypodermis, is made of adipose (fat) and connective tissue, providing insulation, energy storage, and cushioning.

This three-layered structure evolved early in amniotes – our common ancestor with reptiles – as a vital barrier to reduce water loss while still allowing limited absorption and gas exchange. Although humans initially inherited this structure, evolution refined it, modifying our skin – and simultaneously our hair – to respond effectively to the demands of our shifting environment. Together, these

Human skin was refined from an earlier prototype that is still worn by some reptiles

On the surface

Human skin is distinctive in being almost naked, with only a fine covering of thin "vellus" body hair. This state may have developed as a cooling adaptation, connected with an important period of habitat change that eventually led to humans becoming the "sweatiest ape".

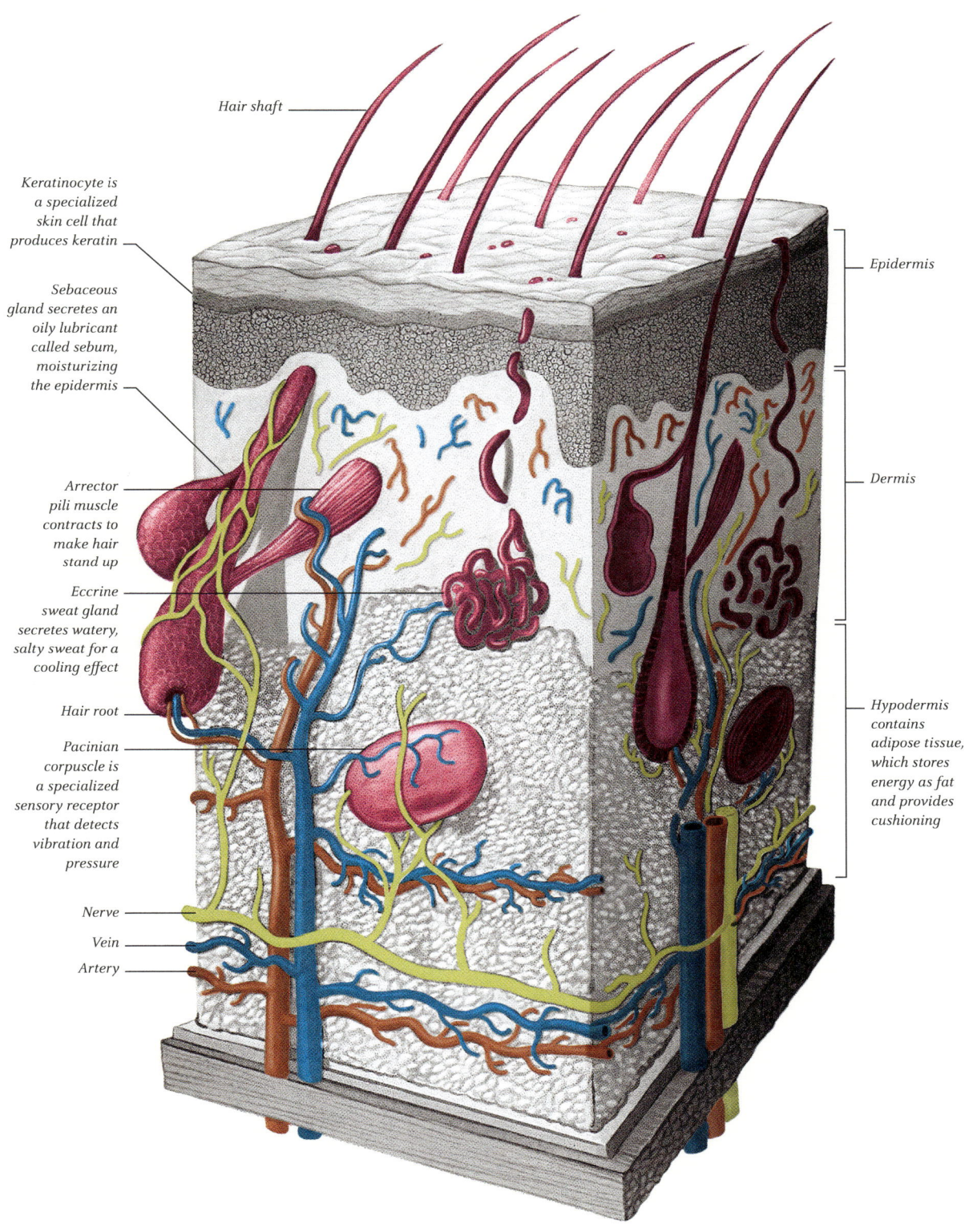

Hair shaft

Keratinocyte is a specialized skin cell that produces keratin

Sebaceous gland secretes an oily lubricant called sebum, moisturizing the epidermis

Arrector pili muscle contracts to make hair stand up

Eccrine sweat gland secretes watery, salty sweat for a cooling effect

Hair root

Pacinian corpuscle is a specialized sensory receptor that detects vibration and pressure

Nerve

Vein

Artery

Epidermis

Dermis

Hypodermis contains adipose tissue, which stores energy as fat and provides cushioning

89

ESSENTIAL PIGMENT

Many researchers believe skin tone evolved according to the intensity of UV exposure. This map shows the skin tone calculated to balance protection against harmful UV rays and the manufacture of sufficient vitamin D in the skin.

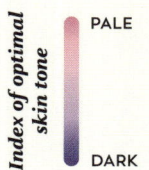

Index of optimal skin tone

PALE

DARK

NORTH AMERICA

Indigenous Americans show a gradient in skin tone due to climate variation, from dim northern areas to sunny tropics

ATLANTIC OCEAN

PACIFIC OCEAN

EQUATOR

SOUTH AMERICA

High-altitude areas such as the Andes mountains are exposed to intense UV light

responses have shaped the skin we have today, influencing factors such as its colour, texture, and living microbiome.

DARKER SKIN

Early members of our genus, *Homo*, would have been under evolutionary pressure to have dark skin. Experts suspect this because evidence suggests that hominin body hair had already thinned by 3 MYA (see p.84). Now exposed to the sun's glare, if hominin skin was once pale under their hair (as chimp skin is today) it became crucial that it was dark to protect the skin from ultraviolet (UV) light. Too much UV radiation causes skin cells to be damaged or killed, resulting in sunburn, and the ionizing effects of the UV harms the cells' DNA, eventually risking skin cancer.

Human skin protects itself with the dark pigment melanin. Melanin is used throughout the animal kingdom when darkness or opacity

Darker skin tone provided essential sun protection for newly exposed skin

COMPARING SHADES

Darker skin contains more active melanocytes, which distribute melanosomes that burst to disperse granules of melanin in the skin. Pale skin, in contrast, produces fewer melanosomes that do not burst, and therefore provides less melanin to act as a protective cover.

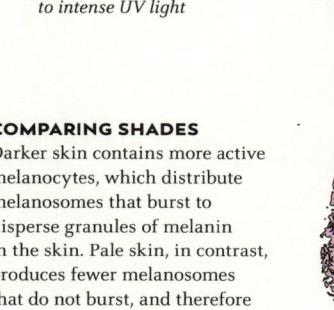

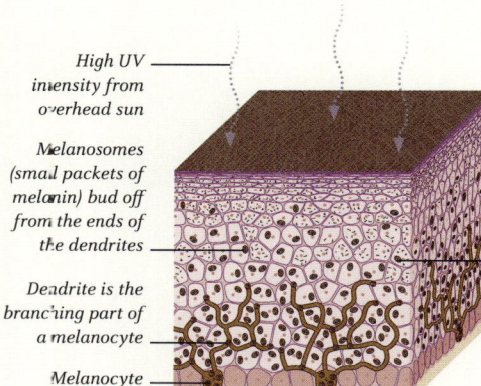

High UV intensity from overhead sun

Melanosomes (small packets of melanin) bud off from the ends of the dendrites

Dendrite is the branching part of a melanocyte

Melanocyte

Melanosomes burst, spreading melanin granules

Dark skin

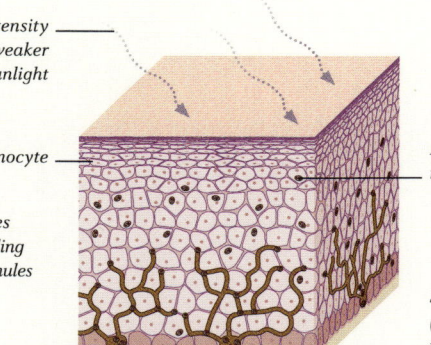

Low UV intensity from weaker sunlight

Keratinocyte

Melanosomes remain in intact

Stratum basale (base layer of the epidermis)

Pale skin

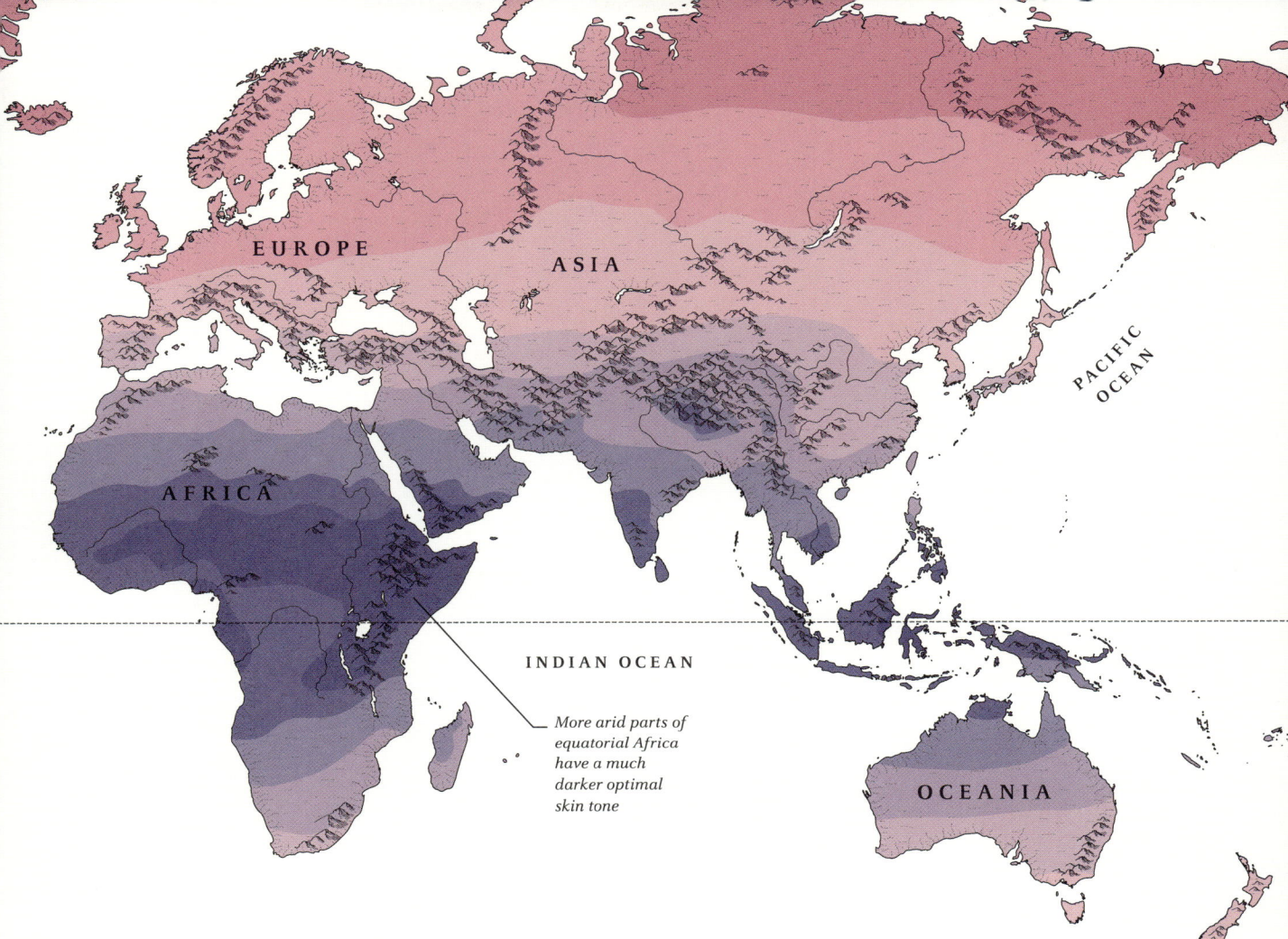

EUROPE

ASIA

PACIFIC OCEAN

AFRICA

INDIAN OCEAN

More arid parts of
equatorial Africa
have a much
darker optimal
skin tone

OCEANIA

is needed. The dark feathers of birds, for instance, are full of melanin, as is squid ink, where it is used to create a protective "smoke screen" effect to confuse potential predators. In mammals, melanin is made by melanocytes – specialized skin cells at the base of the epidermis. They transfer melanin to nearby keratinocyte cells, which carry it towards the surface as the skin renews. There is more than one variant of melanin. Pheomelanin is reddish-yellow and creates the shades in pale skin and red hair. Eumelanin is dark brown, and it is this type that protects the skin. The melanocytes of darker-skinned people are more active, making more melanin that acts as a microscopic umbrella, shielding the DNA inside skin cells.

Melanocytes can actively respond to UV danger. They detect UV damage in the skin and trigger an increased production of melanin, resulting in what we see as a tan. Tanning ability is also seen as an ancestral human trait.

LOSING PIGMENT

When *Homo sapiens* spread from Africa through Eurasia around 70,000–60,000 YA (see pp.270–71), they would have had dark skin inherited from their immediate tropical ancestors. As humans migrated into higher latitudes – where sunlight is weaker and more seasonally variable – dark skin became less of an advantage. Some suggest that a new balance had to be struck between protecting the skin from excessive sunlight and allowing a minimum amount of UV to reach its lower layers to trigger a vital metabolic process – the manufacture of vitamin D.

Skin contains a type of cholesterol called 7-dehydrocholesterol, which is transformed by UV into previtamin D_3 and finally into a form of vitamin D the body can use. The

Varied skin tones arise from the need to balance sun protection and vitamin D production

vitamin helps the body absorb calcium and phosphorus and is essential for strong bones and teeth. Without enough UV exposure, humans with dark skin would have risked vitamin D deficiency.

Homo sapiens reached the high-latitude regions of Europe and Asia around 45,000 years ago, and over the ensuing millennia their skin underwent depigmentation – it lost melanin. This change can be traced in the genetics of modern populations. These studies suggest that pale skin evolved several times in northern Europe and northern Asia and by different mechanisms. Some populations developed melanocytes that tend to produce pheomelanin rather than eumelanin, resulting in very pale skin and a much reduced tanning ability. On one occasion around 18,000 years ago, a mutation in the KITLG gene (which regulates pigmentation) in northern Eurasia led not only to pale skin, but also blond hair and blue eyes, and this trait spread rapidly into Europe via the Baltic Sea region. Other depigmentation genes became common in Europe only in the last 8,000–6,000 years and seem to have been introduced by Neolithic farmers and Steppe herdspeople.

THE SWEATIEST APE

Humans are uniquely sweaty animals. This is because our skin evolved in a way to help us regulate our body temperature, using sweat glands and blood flow to maintain a stable internal environment. We have 2–4 million eccrine sweat glands embedded throughout the dermis. This is a far greater number than can be found in other primates or, in fact, any other mammal. Eccrine glands are not the only type of sweat gland. The other – apocrine glands – occur at the base of hairs as part of the hair root. Their function is partly to create scent and they're mainly found in the armpit and groin. Eccrine glands, in contrast, are especially dense on the forehead, palms, feet, and torso, and they release watery, salty sweat through pores onto the skin's surface. As the sweat evaporates, it cools the skin, assisted by our lack of body hair, which allows breezes to increase the evaporative cooling effect. This adaptation helped early humans thrive in hot,

Sweat glands take advantage of naked skin to achieve a maximum cooling effect

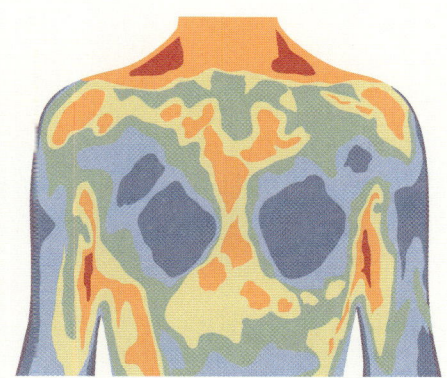

Before exercise

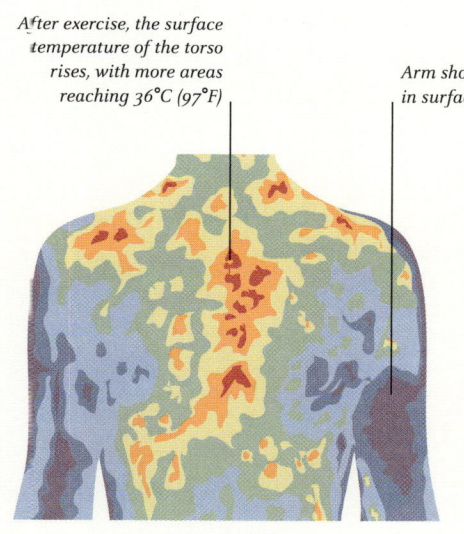

After exercise, the surface temperature of the torso rises, with more areas reaching 36°C (97°F)

Arm shows a reduction in surface temperature

After exercise

NAKED ADVANTAGE

These heat scans show the temperature fluctuations of a naked torso before and after exercise. Core body temperature increases after exercise, due to greater metabolic activity. Surface temperature at the arms and parts of the torso decreases due to evaporative cooling through sweating.

- 🔴 36°C (97°F)
- 🟠 35°C (95°F)
- 🟡 34°C (93°F)
- 🟢 33°C (91°F)
- 🔵 32°C (90°F)
- 🔵 31°C (88°F)
- 🟣 29°C (84°F)

SWEAT GLAND DISTRIBUTION

A study comparing the density of eccrine sweat glands on the forehead, chest, and forearms of macaque monkeys, chimpanzees, and humans revealed that humans have a far greater number of sweat glands than other primates. Animals such as macaques and chimps rely more on panting to cool down.

- 🟢 FOREHEAD
- 🟠 CHEST
- 🟣 FOREARM

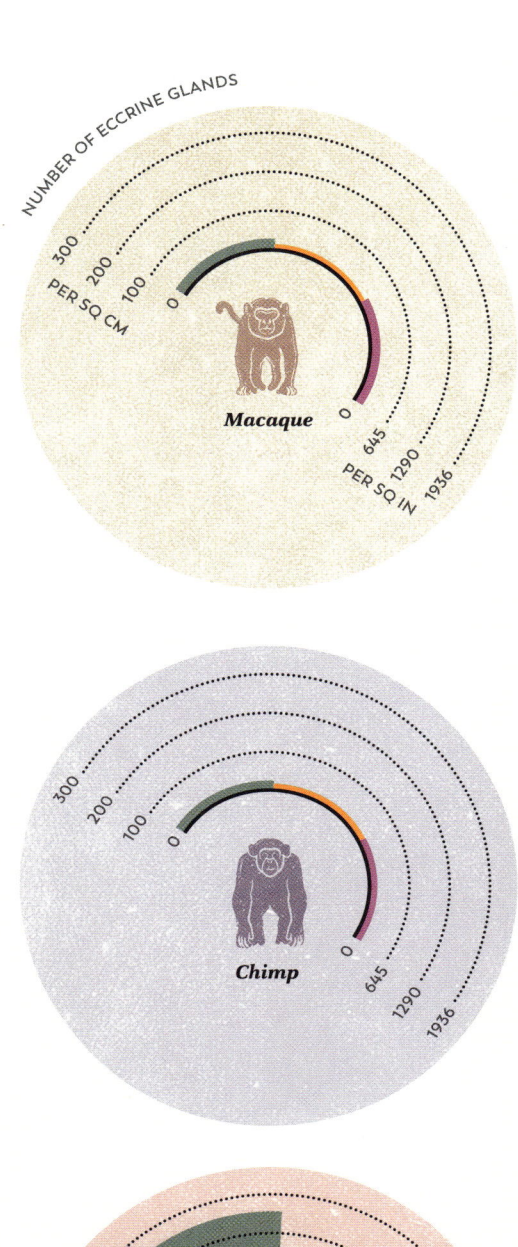

exposed environments and may have even enabled the development of endurance hunting (see pp.72–75), something that early *Homo* was well adapted to do.

Our uniquely sweaty skin reflects a powerful evolutionary trade-off – we had greater vulnerability to dehydration, but our ability to offload heat helped us to adopt an active lifestyle, including walking and running for hours in an open environment. This specifically human response to heat may explain why we adapted so well to open savannah habitats, while apes have remained in shaded forests.

FIRST LINE OF DEFENCE

Skin is more than just a covering – it is also our first line of defence against the outside world. As a physical barrier, it prevents harmful substances, microbes, and irritants from entering. The epidermis is waterproof and tightly packed with keratin-filled cells that block out most bacteria, viruses, and fungi. Meanwhile, the skin's selective permeability allows in useful substances, such as oxygen and topical medicines, while keeping out toxins and allergens. Beneath this barrier lies a complex defence system. Skin houses its own immune cells that can detect invading microbes and trigger wider immune responses. These specialized cells coordinate with the body's T-cells (another type of white blood cell) to remove threats and repair damaged tissue.

Skin is selectively permeable. It blocks harmful bacteria while admitting beneficial substances

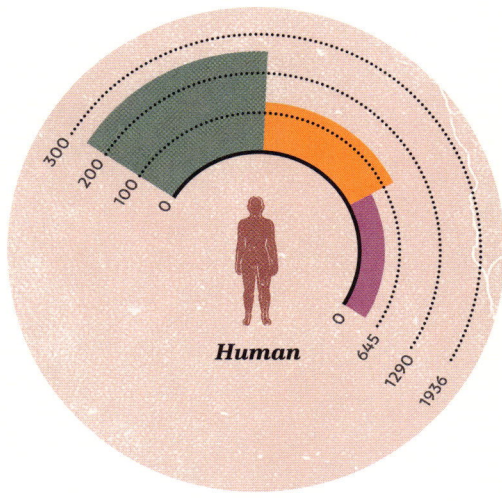

SKIN AS A HABITAT

Our skin is a thriving ecosystem that supports a diverse microbiome – a community of beneficial bacteria, fungi, and viruses that compete with harmful microbes for space and resources. These tiny organisms are not invaders; most are beneficial partners that protect and balance the skin. They help regulate inflammation, train the immune system, and produce antimicrobial substances that keep pathogens in check. Different microbes prefer different bodily environments. Oily areas, such as the face and chest, host bacteria such as *Cutibacterium acnes*. Moist regions, including armpits and feet, support bacteria such as *Staphylococcus* and *Corynebacterium*. Even dry zones such as the forearms may host unique species of microbes. Fungi, such as *Malassezia*, also play a role – especially in oily areas – while viruses, which make up the skin's "virome", may regulate bacterial populations by acting like natural antibiotics. The skin virome constitutes a new frontier in research that may yield breakthroughs in fighting infection.

UNWELCOME GUESTS

Skin health hinges on a harmonious balance between the microorganisms that live on its surface. The skin microbiome composition changes with age, health, and hygiene habits. It has developed historically with dramatic changes in the environment, including the emergence of disinfectants and antibiotics. Many of these changes may affect the balance between microbes and skin. Some microbiome members may be weakened so that they admit invasive pathogens, or they may "overgrow", causing skin conditions. *Staphylococcus* bacteria can cause boils or abscesses in the skin, and some *Corynebacterium* infected with viruses can cause diphtheria, which may be fatal. For reasons still unknown, *Cutibacterium* begins to colonize the skin 1–3 years before puberty and can grow exponentially during this time, which is why so many teenagers and young adults struggle with acne. However, these bacteria remain as permanent lodgers on adult humans, normally ceasing to cause symptoms.

Scientists now believe that the dynamic world of the skin microbiome is crucial not just for skin health, but also for inflammation

A HOME TO MANY
The skin microbiome is composed of various microorganisms, including fungi, viruses, and even small arthropods that all live in and on the skin. This micrograph of a human hand shows bacteria living on its surface.

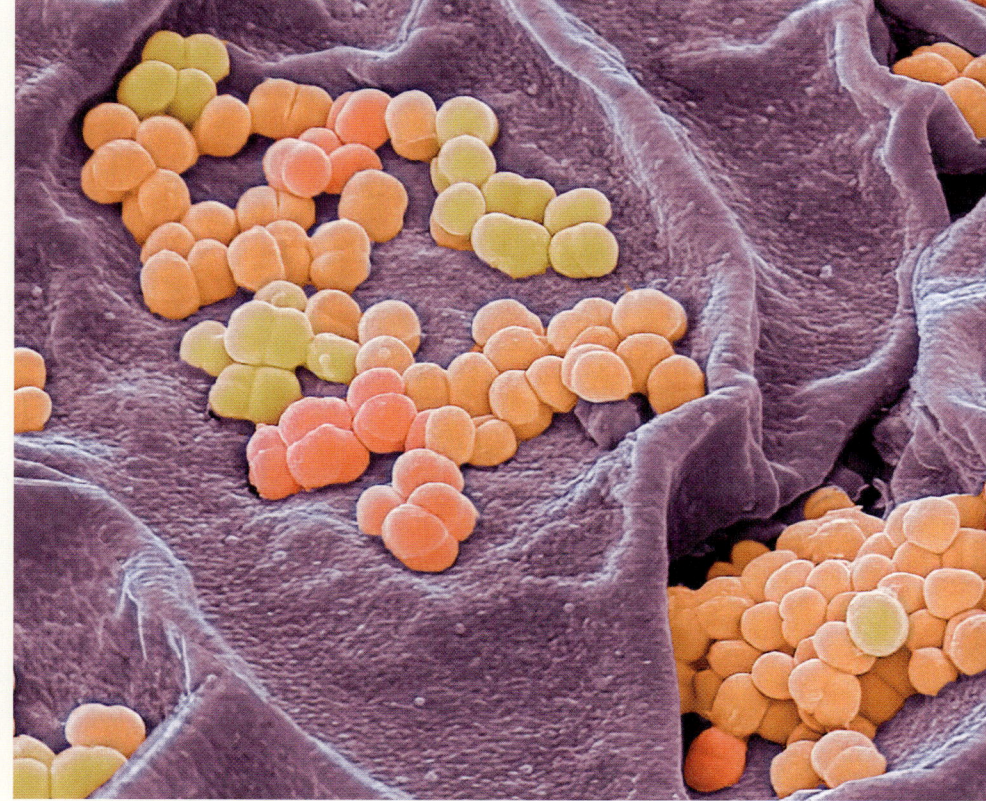

THE SKIN BARRIER

A variety of health and environmental factors can disturb the skin barrier, allowing usually harmless microbes that live on the skin's surface into the body. This can cause infections or issues such as eczema or dermatitis.

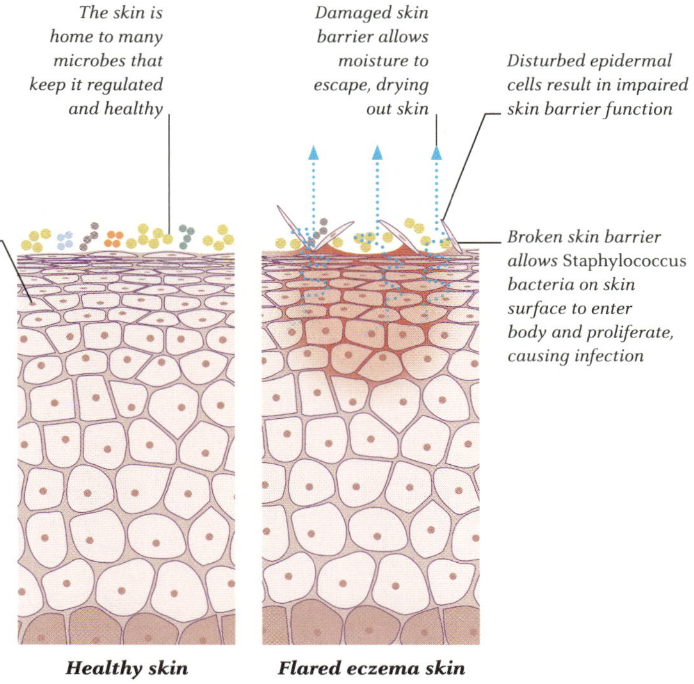

The skin is home to many microbes that keep it regulated and healthy

Damaged skin barrier allows moisture to escape, drying out skin

Disturbed epidermal cells result in impaired skin barrier function

Normal epidermal cell

Broken skin barrier allows Staphylococcus bacteria on skin surface to enter body and proliferate, causing infection

control, immune balance, disease resistance, and perhaps even mood. In fact, it may also have influenced the evolution of human skin itself. As the microbiome evolves, the skin needs to evolve in response, to maintain the productive collaboration. Skin and microbes have been evolving together for millions of years – and there is no reason why it would stop now.

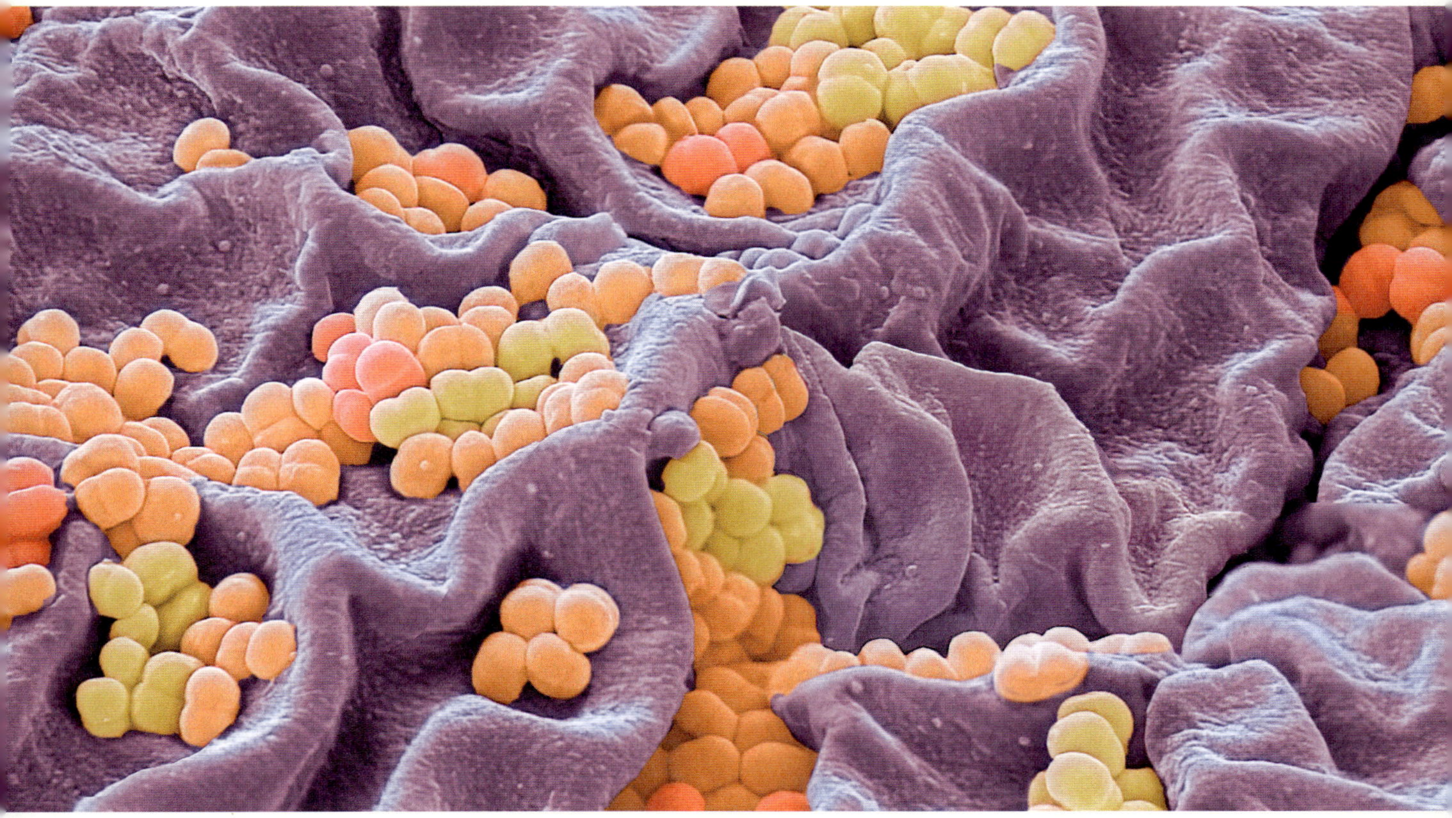

Healthy skin   **Flared eczema skin**

ON THE SURFACE

Getting dressed

The loss of their thick coat of body hair left human ancestors vulnerable to the elements. Clothes became an innovative solution to this new evolutionary challenge, helping them survive changing weather – from strong sun to freezing cold.

By around 2 million years ago, most hominins had replaced what was previously a thick coat of insulating body fur with a thin, fuzzy layer of "vellus" body hair. This was advantageous in hot African climates as it allowed eccrine sweat glands to efficiently cool the body (see p.92), thereby enabling hominins to engage in long-distance endurance activity in warm, sunny, exposed environments. However, as our ancestors spread into colder and harsher places, hairlessness became a vulnerability, and new methods of keeping warm, dry, and protected became essential. Using animal hides and plant fibres, early humans began to wrap and stitch coverings for their bodies. While the oldest garments rarely survive, clues remain in the form of ancient tools, ornaments, and even parasites. These clues indicate that

Although advantageous in hot environments, bare skin was vulnerable in colder climates

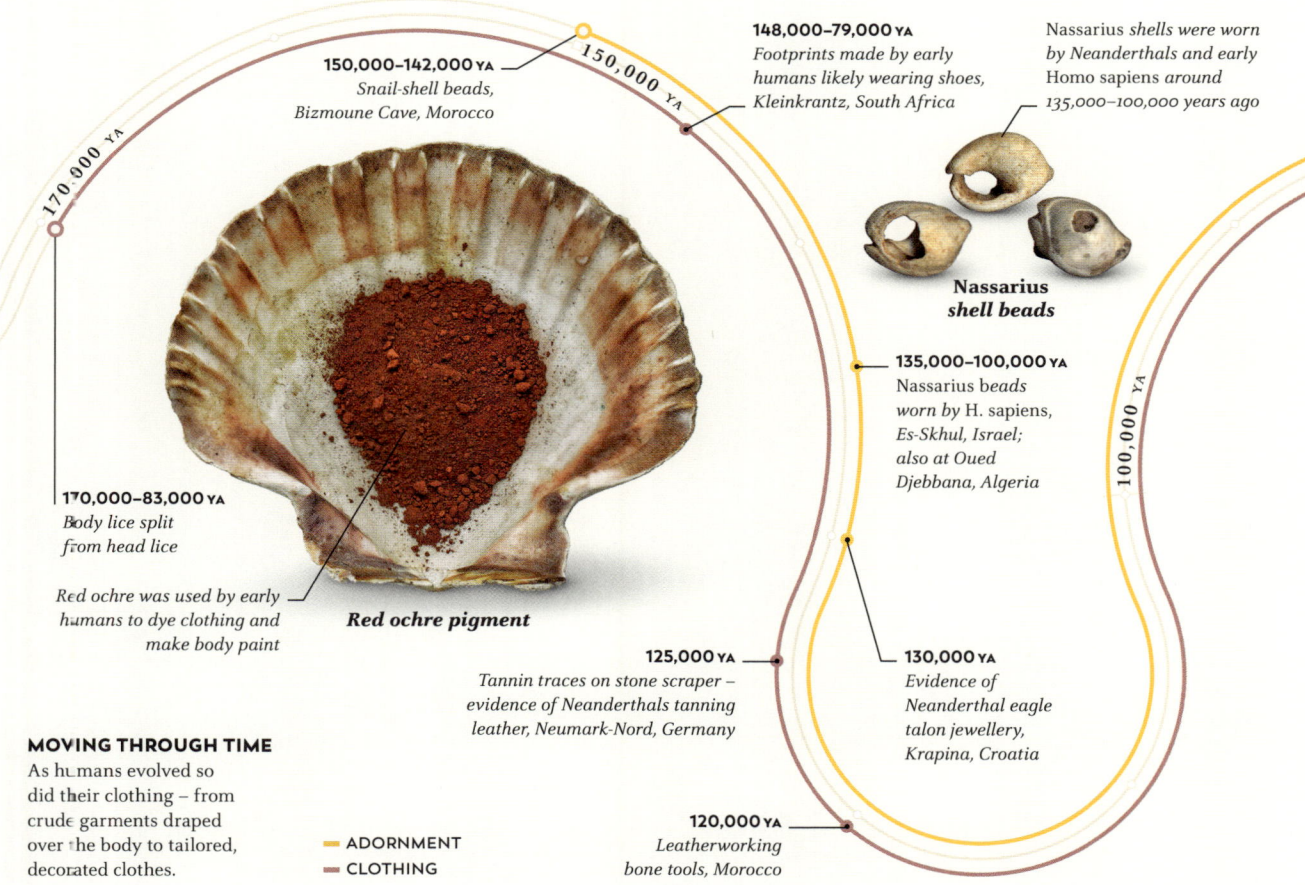

170,000 YA

150,000 YA

150,000–142,000 YA
Snail-shell beads, Bizmoune Cave, Morocco

148,000–79,000 YA
Footprints made by early humans likely wearing shoes, Kleinkrantz, South Africa

Nassarius *shells were worn by Neanderthals and early* Homo sapiens *around 135,000–100,000 years ago*

Nassarius shell beads

135,000–100,000 YA
Nassarius beads *worn by* H. sapiens, *Es-Skhul, Israel; also at Oued Djebbana, Algeria*

100,000 YA

170,000–83,000 YA
Body lice split from head lice

Red ochre was used by early humans to dye clothing and make body paint

Red ochre pigment

125,000 YA
Tannin traces on stone scraper – evidence of Neanderthals tanning leather, Neumark-Nord, Germany

130,000 YA
Evidence of Neanderthal eagle talon jewellery, Krapina, Croatia

120,000 YA
Leatherworking bone tools, Morocco

MOVING THROUGH TIME
As humans evolved so did their clothing – from crude garments draped over the body to tailored, decorated clothes.

▬ ADORNMENT
▬ CLOTHING

clothing was part of human life far earlier than any surviving remains of the clothes themselves suggest.

THE THIRD LOUSE

One of the best indirect clues for the invention of clothes comes from lice. Human head lice live on the scalp, however clothing (or body) lice live only on clothes and bedding. Genetic studies indicate that clothing lice split from head lice around 170,000 years ago, which, experts conclude, would have roughly coincided with the time that hominins began wearing clothing – head lice would have migrated to clothing and, in this new habitat, evolved separately from their scalp-based relatives.

CRUDE CLOTHING

The very first clothes were probably little more than animal hides draped around the body. Early humans were already hunting large animals for food and hides were a ready by-product that could provide warmth and protection against cold winds and harsh sun. Even these crude coverings could confer a survival advantage on those who had them

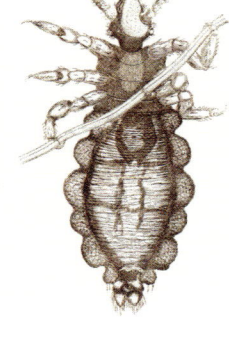

BODY LOUSE

Body lice split from head lice and adapted to live in clothing seams and bedding. They constitute one of three species of lice found on humans, along with head lice and pubic lice.

80,000 YA
Nassarius *shell beads at four sites, Morocco*

75,000 YA
Beads, Blombos Cave, South Africa

77,000 YA
Bone awls, Blombos Cave, South Africa

70,000–50,000 YA
Bone, tusk, and teeth jewellery found alongside intricate chlorite bangle, Denisova Cave, Siberia

63,000–59,000 YA
Bone needles, Sibudu Cave, South Africa

Needles would have been used to stitch rough animal hides into well-fitting clothes

Bone needle

51,000 YA
Lissoirs (bone hide polishers), Pech-de-l'Azé I, France

50,000 YA
Bone needles, Denisova Cave, Siberia

50,000 YA

This polished bangle features a precisely drilled hole, demonstrating that by around 50,000 years ago our close relatives, the Denisovans (H. longi), had developed sophisticated craftsmanship skills

Denisovan chlorite bangle

20,000 YA

26,000–24,000 YA
Venus of Lespugue shows early representation of spun thread, France

27,000–25,000 YA
Fibre impressions on clay, Dolní Věstonice, Czechia

28,000 YA
Around 3,000 mammoth ivory beads, fox canine teeth, and mammoth ivory spears interred in elaborate burial, Sungir, Russia

30,000–23,000 YA
Bone needles, Xiaogushan, China

34,000–28,000 YA
Dyed flax fibres, Dzudzuana Cave, Georgia

GETTING DRESSED

and enabled further exploration into colder climates that bare skin alone could not endure. Simple ties or belts of fibre or sinew may have been used to keep these hides in place.

ANCIENT TOOLS

Fresh animal hides are heavy, stiff, and quick to rot. To make effective wraps, they first had to be preserved through a process called tanning, by which animal skins are stripped of flesh using stone scrapers and immersed in a solution containing tannins. One of the earlier stone scrapers comes from Germany and is around 125,000 years old. Traces of oak acid on its edge reveal that early humans, probably Neanderthals, boiled plant material to create a preservative. After tanning, hides would have been manipulated by tools called lissoirs, with some of the earliest examples dating back to 400,000 years ago. Rubbing or polishing stiff hides with these tools made the leather more supple and waterproof.

The discovery of 77,000-year-old bone awls in South Africa revealed a transition in how humans wore these softened skins. Awls are pointed tools that could be used to punch holes through tough materials. Although the earliest needles appeared much later, around 63,000 years ago, it can be assumed that these holes were used for a primitive form of sewing, perhaps using animal sinew or plant fibres to

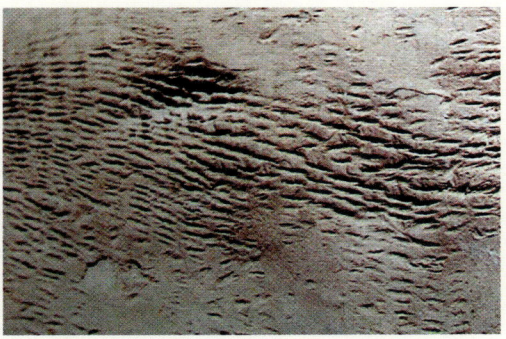

FIRST IMPRESSIONS
Fibre impressions pressed into clay at Dolní Věstonice, Czechia, show that humans were weaving plant fibres into cloth as early as 27,000–25,000 years ago.

stitch the skins together. This method would have achieved tighter, better-fitting, and more protective garments.

Humans used awls and lissoirs to process animal hides

THE FIRST FIBRES

The history of textiles is difficult to reconstruct due to the poor preservation of any evidence (early textiles generally being made from perishable materials, such as plant stems or animal hair). However, in 2009 a team of archaeologists identified microscopic flax fibres within soil samples excavated from Dzudzuana

NEANDERTHAL STONE SCRAPER
This stone scraper was excavated in the Netherlands and possibly belonged to Neanderthals. Scrapers were used to remove hair, fat, and flesh from animal skins before they were tanned (preserved).

FRENCH LISSOIR
Bone tools called lissoirs were used to smooth animal hides so they could be manipulated into soft garments that would drape over the body. This lissoir is around 51,000 years old and was probably used by Neanderthals.

Rounded edge was ideal for pressing and softening animal hides during leatherworking

Eroded surface would have been smooth when new

Front view

Sharp edges would have been used to scrape flesh off animal hides

Neanderthal stone scraper

Side view

The Venus of Lespugue is a carved ivory statue excavated from the Cave of Rideaux in Lespugue, France. Thought to represent fertility and the power of female reproductivity, it is considered to be the earliest known artistic representation of woven clothing.

Carved lines are thought to show a skirt of twisted plant fibres – perhaps flax or grasses

Venus of Lespugue

Cave, Georgia. Further study revealed that many of these fibres were at least 34,000 years old and some had been twisted or dyed. Previously, the only surviving evidence of early textiles were fibre impressions left on clay at Dolní Věstonice, Czechia, dated to around 27,000–25,000 years ago. The discovery in Georgia revealed that some humans had conceived this critical invention at least 7,000 years earlier, and were possibly using these fibres to make ropes or threads (evidenced by their twisted shape), or even coloured garments. By studying the flora around Dzudzuana Cave, researchers were able to predict that the flax would have been collected from the wild before being processed into threads, cloth, clothing, mats, or bags.

The earliest, and only, depiction of cloth garments from this period is the Venus of Lespugue. Carved from ivory, this symbolic figurine is around 26,000–24,000 years old and appears to be wearing a skirt of spun thread. While an extraordinary record of early clothing, the Venus also suggests that by that time clothes were not just a means of survival, but also a way to signal identity, status, and creativity.

GROWING SOPHISTICATION

The invention of needles marked a turning point in the evolution of clothes. The earliest examples are 63,000 years old and come from Sibudu Cave in South Africa. Subsequent needles were also unearthed in Denisova Cave, Siberia (from 50,000 years ago), and in China (from 30,000–23,000 years ago). These tools, often

ÖTZI THE ICEMAN
Mummified in the ice, Ötzi demonstrates how some early Alpine humans dressed to combat the cold and wet.

Pieces of bearskin were stitched together to make a cap

Made from woven alpine grass, this cape or mat would have been used to shelter Ötzi from rain

Hide coat was made from strips of goat and sheep skins stitched together with sinew, over loincloth of sheepskin

Leather pouches held tools, such as a flint scraper, drill, and fire-lighting tinder – all essential for survival

Belt was made of leather and used to fasten Ötzi's coat; it also had pouches attached to it

Leggings were made from strips of goat leather tied together with hide thongs, providing warmth and flexibility for climbing; loops at the bottom fastened to shoes

Shoes were made from deerskin uppers, bearskin soles, and grass padding inside – warm and waterproof

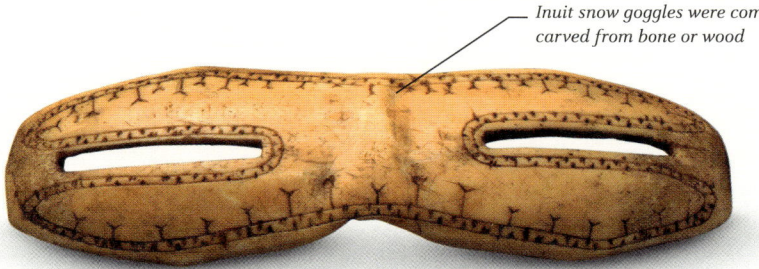

Inuit snow goggles were commonly carved from bone or wood

EYE PROTECTION

Far from being purely practical, snow goggles were decorated with colour and symbols to represent protection or identity. These goggles from around 1,200–800 years ago were tied to the face using leather or sinew, which held the goggles close to the face.

carved from bone or ivory using stone flakes, allowed hides and furs to be stitched together using sinew or threads made from hair or plant fibres. This rudimentary form of tailoring produced closer fits and, therefore, better insulation and warmth.

PROTECTIVE COVERINGS

As humans spread into new and often harsh environments, they created specialized garments to protect their most vulnerable areas – the head, eyes, hands, and feet. Whether against sun, snow, wind, or dust, each innovation reflected the local challenges encountered by humans, and their creative problem-solving.

Evidence of early clothing adapted to hotter weather is scant, although historians theorize that humans may have worn cloth wraps or veils around their heads to shield their skin and eyes from heat and sandstorms. In colder regions, however, we have a remarkable record showing how some humans may have adapted to icy climates, most notably in the form of Ötzi the Iceman, a mummy excavated from the Ötzal Alps in Italy. Approximately 5,300 years old and preserved fully dressed, Ötzi was found wearing fragments of a coat and leggings made from goat and sheep skins, a grass cape, and shoes made from animal hides and grass. Despite this fragmentary evidence, Ötzi provides

Extreme heat or cold challenged humans to develop new styles of protective clothing

a snapshot of how some communities may have dressed to combat the low temperatures found in high-altitude areas.

Protecting the eyes and hands was vital in extreme conditions. In the Arctic, humans used stone tools to craft snow goggles from bone or ivory – using sandstones to grind and shape the goggles to the wearer's face and stone drills to create narrow eye slits. These were worn to block harsh light reflected off snow and ice. Hands may have been shielded from frostbite or injury by woollen or hide gloves, materials robust enough to be protective but also soft enough to allow humans to hunt or manipulate tools.

If the evolution of early humans was guided by the environment, their clothing, too, was adapted and refined to suit the demands of whatever new climates they encountered – as humans evolved, so did their clothes.

Clothing was adapted to suit new climates – as humans became more sophisticated, so did their clothes.

101

SAQQAQ SOCK

This 4,000-year-old leather sock from Greenland demonstrates how early Arctic humans layered hides, and sometimes furs, around their feet to insulate them against the cold.

Stitching together layers of sealskin prevented wear and made the sock fit snugly

SHOD HUMANS

In 2023, a group of experts discovered a set of footprints in Kleinkrantz, South Africa. In contrast to barefoot markings, these impressions lacked toe imprints and displayed rounded backs, a clear outline, and possible evidence of strap attachment points. Dating from around 148,000–79,000 years ago, researchers speculate that these footprints may be the earliest evidence of humans wearing shoes. Similar markings were also found in Goukamma, South Africa, from around 136,000–73,000 years ago.

Shoes would have become essential as humans explored rougher terrains. To protect their soles, they may have fashioned shoes from simple wrappings made of hide, bark, or plant fibres, natural materials that would have decomposed after excessive and prolonged use. For this reason, experts rely on indirect evidence to determine when early humans began wearing shoes. Alongside the Kleinkrantz and Goukamma footprints, rock paintings discovered in Baviaanskloof, South Africa, depict hunters wearing shoes – simple sandals with clear ties and soles – suggesting that footwear was favoured by those engaged in long-distance walking and running. The provenance of these paintings is disputed, but some attribute them to the San people who occupied Baviaanskloof at least 2,000 (but possibly as early as 100,000) years ago.

As humans migrated into deserts, forests, and icy tundra, they probably adapted their footwear to suit the local environment. Ötzi, dressed for Alpine weather, was found wearing deerskin and bearskin shoes insulated by grass padding. In Greenland, archaeologists uncovered a 4,000-year-old sock belonging to the Inuit Saqqaq people. Constructed out of layered seal skins, the Saqqaq sock would have provided effective insulation against freezing Arctic temperatures.

In colder weather, humans created shoes from layered animal skins and grasses

ANCIENT ADORNMENT

For some, the emergence of ancient adornment signals a time when early humans began thinking symbolically. There is a wealth of archaeological evidence demonstrating that different human species were experimenting

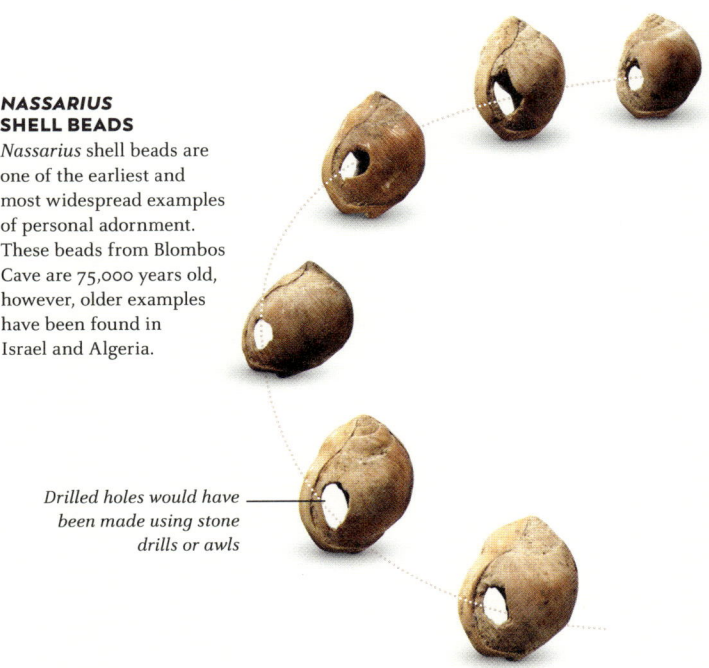

NASSARIUS SHELL BEADS

Nassarius shell beads are one of the earliest and most widespread examples of personal adornment. These beads from Blombos Cave are 75,000 years old, however, older examples have been found in Israel and Algeria.

Drilled holes would have been made using stone drills or awls

HUNTERS ON THE MOVE

These rock paintings from Baviaanskloof, South Africa, depict hunters wearing shoes with soles and tied straps. These paintings are estimated to be at least 2,000 years old, possibly even older.

GETTING DRESSED

A NEW AGE

This reconstruction, made with Bronze Age technology, is based on beads found grouped together at a 3,000-year-old site in England. The originals are assumed to belong to a single necklace, suggested by fragments of thread preserved within the central amber bead. The holes would have been drilled using a bronze-tipped drill.

Uneven arrangement suggests necklace was a collection of an individual's treasures, rather than a single piece of jewellery design

These green glass beads have been traced back to Iran

Black shale beads from Dorset, England

Bronze Age necklace

Early humans used animal parts, including shells and talons, to create the first jewellery

with personal decoration by 142,000 years ago. The earliest examples are a collection of drilled *Tritia* (a genus of mollusc) shells found in Morocco, which experts hypothesize were strung and worn by *Homo sapiens* as necklaces or sewn onto clothing. Later, around 135,000 years ago, *Nassarius* shells appeared in Israel and Algeria. Moving eastwards to Croatia, excavators unearthed a series of 130,000-year-old eagle talons with grooves around their base, suggestive of where string would have tied them into what was probably a necklace worn by Neanderthals. In Denisova Cave, Siberia, archaeologists discovered a fragment of a 50,000-year-old green chlorite bangle that had been polished and perforated with a precisely drilled hole on one side – details that required tools and techniques that experts had previously believed emerged much later.

THE RISE OF SYMBOLIC CULTURE

Some argue that the evolution of adornment demonstrated early humans' ability to give objects symbolic meaning beyond their

practical function. Researchers debate the significance of wearing shell beads, however theories range from a means of personal or community identification to examples of burgeoning self-concept and self-expression.

LIVING CANVAS

The human body itself became a canvas for art. The very first tattoos were probably made by rubbing pigments made from red ochre or charcoal into incisions in the skin made by bone or stone tools. Later, humans adopted a "stippling" method, not too dissimilar to modern tattooing, where the skin was repeatedly poked with a sharp instrument coated in a pigment, permanently marking the skin.

This stippling method was used to create the oldest known tattoos, which belong to Ötzi the Iceman (see p.101). Ötzi's tattoos are a series of lines and crosses clustered around his joints and along his spine, placements that suggest they may have served a healing purpose. However, later examples of tattoos may have been decorative and symbolic, not just therapeutic. In

Grooves worn into spacer plates suggest necklace was frequently worn over many generations

Multiple strands of beads are held together by multi-perforated plates

Poltalloch necklace

Jet beads were a prized material, used to convey high status

Jet bead bracelet

MARK OF HIGH STATUS

This reconstruction, made the Bronze Age way, is based on jet ornaments from around 4,450–3,500 years ago, found in a cist burial in Scotland alongside a flint blade and ochre fragments. Jet was considered symbolic in the Bronze Age.

Needle tip would have been dipped in pigment or soot, a primitive "ink"

TURKEY-BONE TATTOO NEEDLES

These bone tattoo needles are at least 3,600 years old and are some of the earliest evidence of tattooing in the Americas, and also the world.

Siberia, archaeologists unearthed the Princess of Ukok, a 2,500-year-old mummy that had a series of animal tattoos on her shoulder, wrist, and thumb. Unusually intricate for the time, some speculate these tattoos indicated high social status and were a means of family identification. Like jewellery, tattooing adopted its own form of symbolic significance, contributing to the continued development of artistic expression.

ÖTZI'S TATTOOS

61 tattoos were found mostly around Ötzi's spine, legs, and ribs, made by stippling his skin with soot. Their location suggests that they were made for therapeutic rather than decorative reasons.

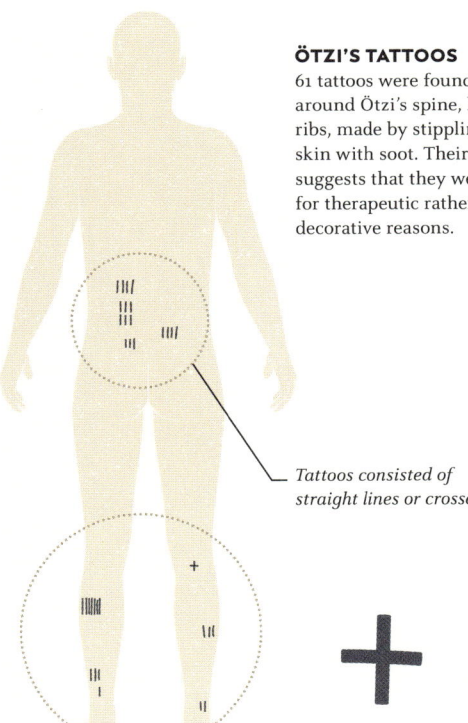

Tattoos consisted of straight lines or crosses

Healing symbols

Dressed to survive

In the heat of the exposed grass savannah, our ancestors shed their thick fur in favour of fine hair, enabling sweat to cool the skin efficiently. In colder, harsher environments, our ancestors would have needed clothes to survive. Later, clothing became more than just protection – together with art and adornment it marked the beginnings of a new symbolic awareness.

Walking on two legs

CH3 *Walking upright encouraged hominins to leave the forests and migrate into grasslands.*

With the onset of the last Ice Age, drier and cooler conditions led to grasslands spreading across Africa

Spread of grasslands

Climate change

The dark pigment melanin protected naked skin from the sun's harmful UV rays

CH3 *Walking and running in hotter environments required more efficient cooling abilities, resulting in a higher density of sweat glands.*

Darkening skin

Thinning body hair

Running in hot conditions

Sweat glands

CH4 *Although beneficial in hotter weather, naked skin left early humans vulnerable to the elements. Primitive clothing made from animal skins served a protective purpose.*

In low UV areas, it was easier to synthesize vitamin D with paler skin

First clothing

Migration to colder climates in Eurasia

Higher latitudes and lower UV levels

Lighter skin and eyes

GETTING NAKED, GETTING DRESSED
The story of our skin and hair can be traced
back to changes in habitat, when hominins
left the shade of the forests for sunny,
exposed grasslands. As they migrated
northwards into colder climates, protecting
naked skin became essential for survival. The
solution to this new challenge was clothing.

Rise of symbolic culture

CH8 *The first forms
of art included
body painting
and jewellery.*

*Birth of
art*

*Clothes became
a way to express
identity and
symbolism*

*Evolution of
motor skills*

Brain evolution

*It is likely that the
development of tools
and human motor
skills influenced
each other*

CH8 *Improved
brain function
enabled the
invention and
refinement
of tools.*

*Development
of tools*

*Refinement
of tools*

*Refinement
of clothing*

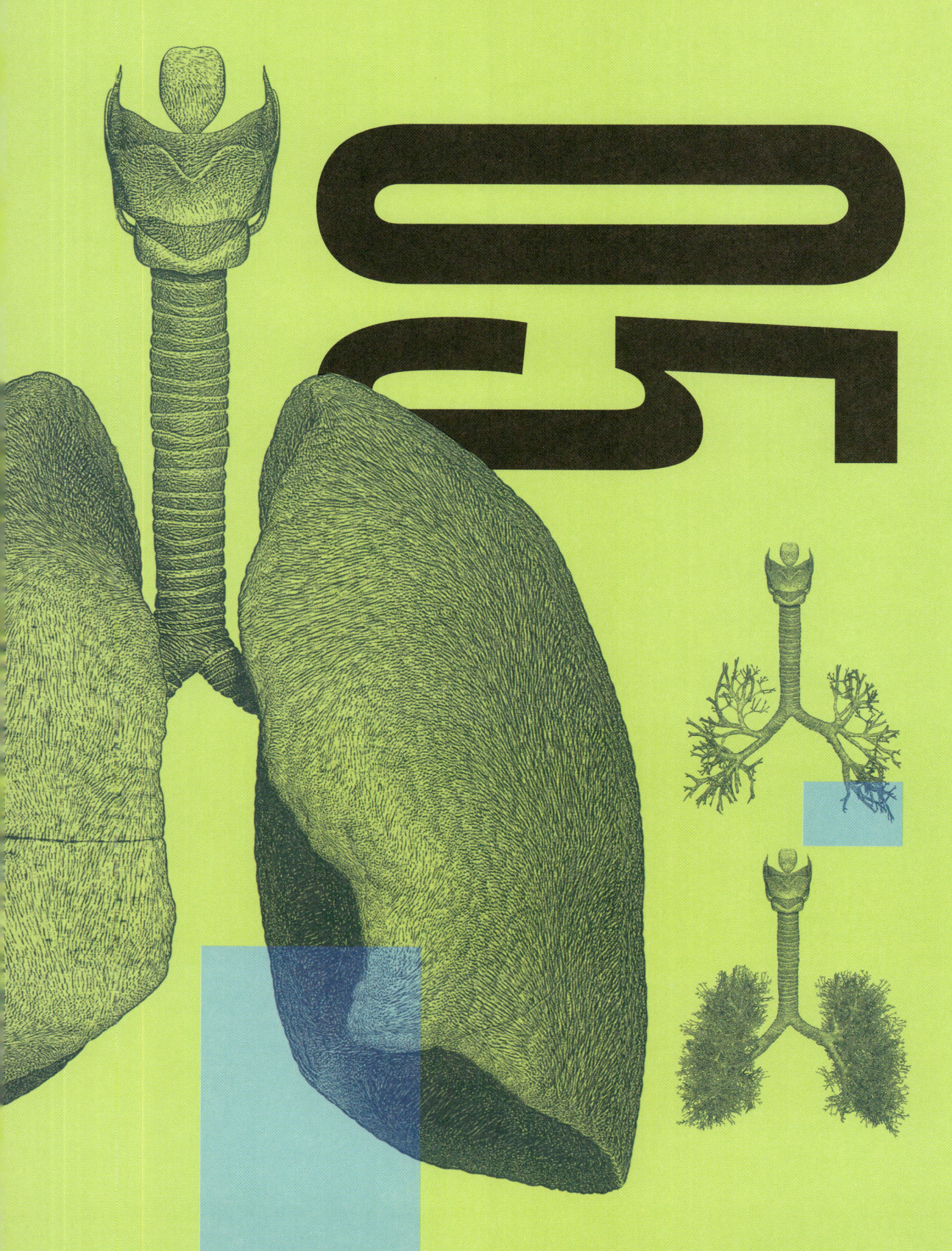

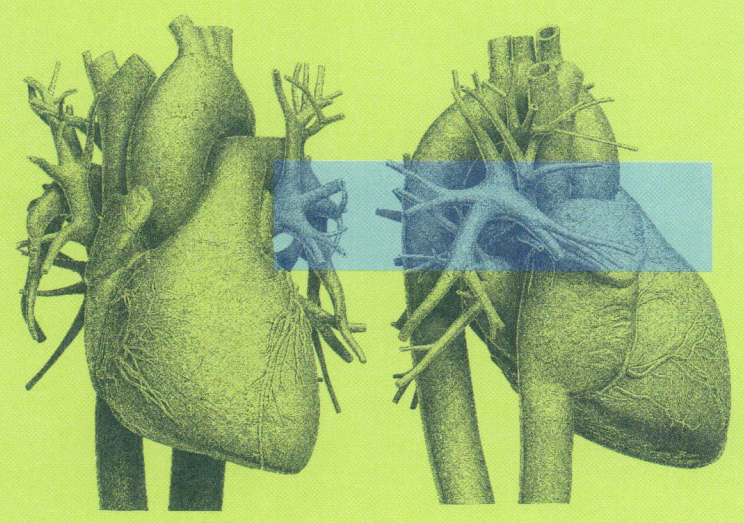

Lungs
and heart

Humans have inherited the efficient mammalian hearts and lungs that are needed to power high-energy warm-blooded lifestyles. But evolution has tuned human systems further, turning us into consummate endurance athletes.

DOUBLE CIRCULATION

Ancestors of mammals pumped blood simultaneously to the lungs and the rest of the body with a four-chambered heart. This double circulation made exchange of gases and nutrients more efficient.

Blood pressure stays high in tissues around body, improving exchange

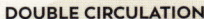

BREATHING AIR

Some fish that had evolved gas bladders (used in buoyancy) adapted their bladders to draw in air. They did this to get more oxygen when in oxygen-poor water. They pumped air into their newly acquired lungs.

Swim bladder received air pumped in by muscles in the mouth

220 MYA–PRESENT

M A M M A L S

375–220 MYA

DIAPHRAGM

Air-breathing animals draw in air by raising the ribcage with muscles between the ribs. Mammals add greater inflating power with their muscular diaphragm.

Diaphragm, a sheet of muscle, draws air in as it contracts and flattens

B I R D S

160 MYA–PRESENT

Heart has two upper chambers and one large, lower chamber

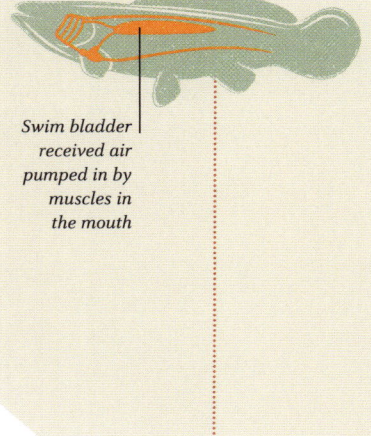

Accessory air sac acts as a reservoir that permits one-way air flow through the lung

UNIDIRECTIONAL AIR FLOW

Birds generate a one-way flow of air over their lung lining using air sacs filled before and after the air passes through the lungs.

Septum (wall) more effectively separates circulation into two

THREE-CHAMBERED HEART

Amphibians made circulation more efficient by separating the circulation with a three-chambered heart that sent blood separately to the lungs and the rest of the body.

LUNG-HEART HERITAGE

Mammals and birds have inherited their superior capacity to use oxygen from evolutionary ancestors that were the first aquatic animals. As animals diversified over millions of years – first in the seas, then on land – gills and simple tubular pumps gave way to lungs and multi-chambered hearts.

FOUR-CHAMBERED HEART

Birds evolved a four-chambered heart separately from mammals, and for similar reasons – to support their high-energy, warm-blooded lifestyle.

LUNGS AND HEART

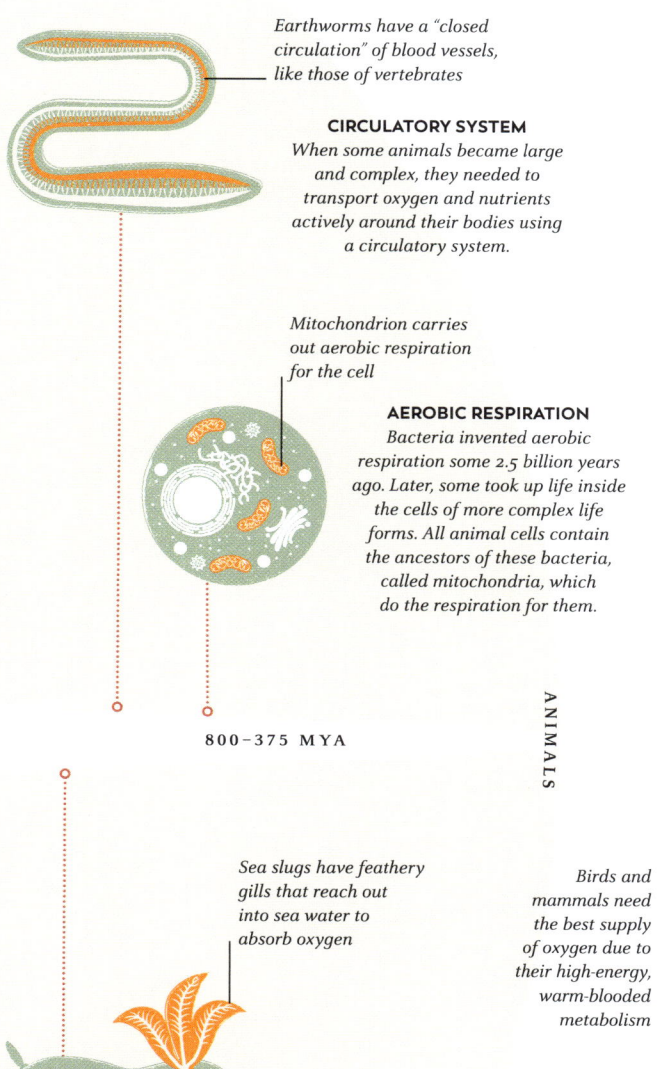

Earthworms have a "closed circulation" of blood vessels, like those of vertebrates

CIRCULATORY SYSTEM

When some animals became large and complex, they needed to transport oxygen and nutrients actively around their bodies using a circulatory system.

Mitochondrion carries out aerobic respiration for the cell

AEROBIC RESPIRATION

Bacteria invented aerobic respiration some 2.5 billion years ago. Later, some took up life inside the cells of more complex life forms. All animal cells contain the ancestors of these bacteria, called mitochondria, which do the respiration for them.

800–375 MYA

ANIMALS

LAND VERTEBRATES

Sea slugs have feathery gills that reach out into sea water to absorb oxygen

Birds and mammals need the best supply of oxygen due to their high-energy, warm-blooded metabolism

GILLS

Larger animals needed specialized structures with a high surface area to absorb enough oxygen for their respiration. When these structures were invented, all animals still lived in the ocean.

Oxygen is a vital commodity for most life. Life forms use it in a chemical process called aerobic respiration that generates usable energy. The story of animal evolution has oxygen as a key player. The first animals were sea-dwelling invertebrates that absorbed oxygen directly into their small bodies. But as invertebrates – including burrowing worms and crawling slugs – became larger, they evolved hearts that pumped oxygenated blood to deeper tissues, and gills that captured more oxygen from the water.

BACKBONED LAND ANIMALS

The first land animals were also invertebrates – worms, insects, and millipedes that absorbed oxygen from air. Backboned animals made the move out of water later, when amphibians evolved from fish that refashioned their swim bladders into lungs. Their blood circulations changed too: two-chambered hearts of fish evolved into four-chambered hearts of mammals and birds. These new hearts pumped oxygenated blood more efficiently, while rhythmic inflation of new lung systems increased oxygen uptake to satisfy an active warm-blooded lifestyle.

When it came to breathing, birds and mammals evolved along different trajectories. Birds rely on bellow-like sacs that flush air through their lungs, whereas mammals evolved bigger, spongier lungs packed with microscopic alveoli. Combined with strong hearts, the superior oxygen delivery helped to make these animals some of the fastest, most spectacular, and innovative animals on the planet.

The history of oxygen delivery

Animals have always needed oxygen, but as larger, more energetic animals evolved, their demand increased. Birds and mammals are the hungriest for oxygen and have the most efficient hearts and lungs.

The heart is 99 per cent muscle and does more physical work than any other organ. Its relentless exertions – pumping life-giving blood around the rest of the body – are testament to how much oxygen an animal needs to sustain its lifestyle. The heart of a 60-kg (132-lb) mako shark – a fast ocean predator – can pump 3 litres (6 pints) of blood every minute, about the same average as a chimpanzee. However, a human heart can pump 6 litres (13 pints), and around

six times this amount when running long-distance. Humans have evolved into athletes hungry for oxygen in the extreme.

The story of this human craving begins more than 2 billion years ago. Before then, the earliest organisms, such as certain groups of bacteria, were anaerobic – they did not rely on oxygen for energy – and were even poisoned by oxygen, being vulnerable to its destructive effects in the same way that metal can rust. But by evolving a

A human heart can pump 6 litres of blood around the body every minute, and around 35 litres when exercising

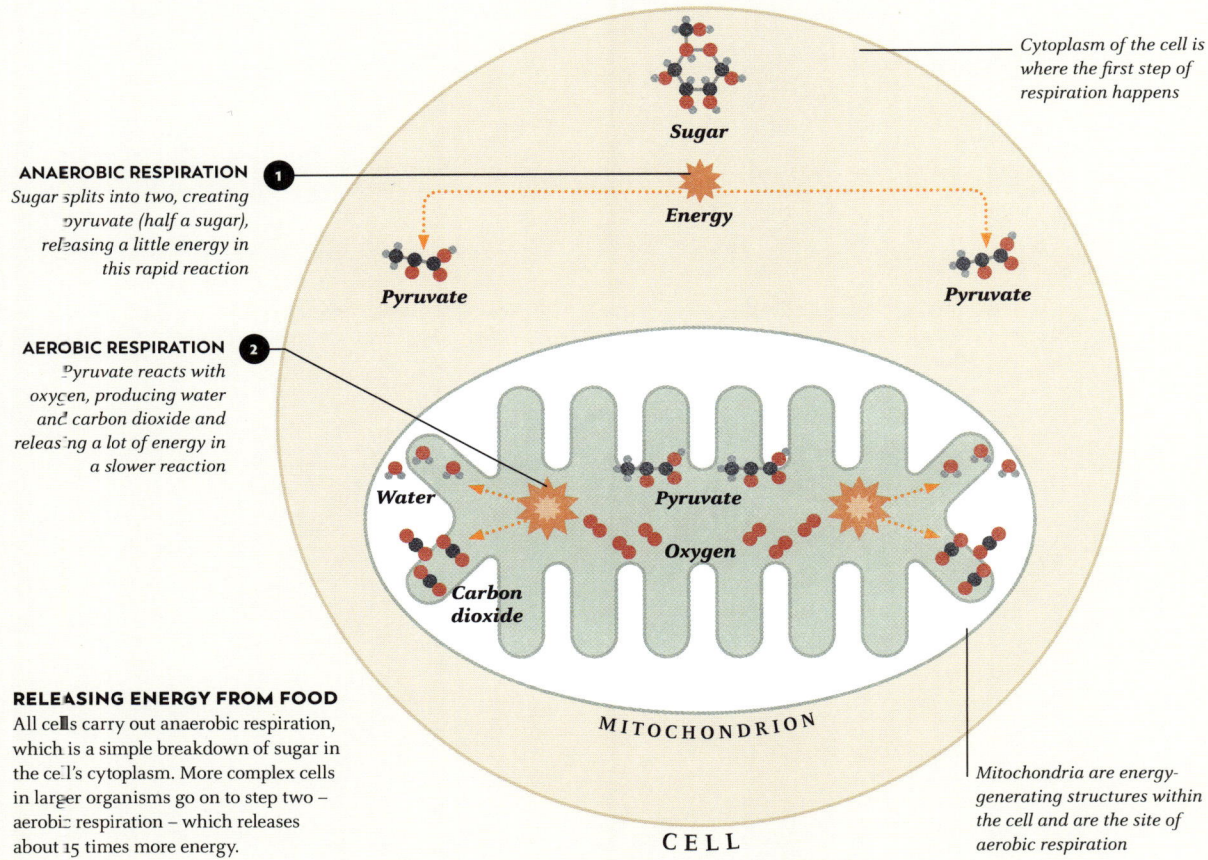

Cytoplasm of the cell is where the first step of respiration happens

ANAEROBIC RESPIRATION ❶
Sugar splits into two, creating pyruvate (half a sugar), releasing a little energy in this rapid reaction

Sugar

Energy

Pyruvate

Pyruvate

AEROBIC RESPIRATION ❷
Pyruvate reacts with oxygen, producing water and carbon dioxide and releasing a lot of energy in a slower reaction

Water

Pyruvate

Oxygen

Carbon dioxide

MITOCHONDRION

CELL

RELEASING ENERGY FROM FOOD
All cells carry out anaerobic respiration, which is a simple breakdown of sugar in the cell's cytoplasm. More complex cells in larger organisms go on to step two – aerobic respiration – which releases about 15 times more energy.

Mitochondria are energy-generating structures within the cell and are the site of aerobic respiration

Hungry for oxygen

Most animals use oxygen, but in humans the demand is especially strong. Three human attributes are largely responsible – big brains, warm blood, and muscle-powered endurance, all of which consume considerable energy and demand superior heart-lung performance.

chemistry that could harness the power of oxygen in a safe way, some single-celled microbes repurposed oxygen's reactivity to release useable energy in a reaction called aerobic respiration. By using this reaction to oxidize carbohydrates or fats into carbon dioxide, living organisms gained enough energy to evolve into more complex multicellular bodies – including the first moving animals with muscles and nerves.

BIG BODIES, BIG SOLUTIONS

Our atmosphere is just over 20 per cent oxygen. The oxygen content of water is lower. Because oxygen molecules are so tiny, they can seep into the bodies of small, aquatic animals. Single-celled organisms and simple multicellular life forms, such as flatworms and sea sponges, can rely on oxygen absorption through simple passive diffusion alone – where oxygen passes through a partially permeable surface layer. However, this process only works well over short distances, and bigger bodies require ways of getting oxygen to cells deep inside their bodies. The evolutionary solution to this challenge was active pumping of air and fluids. In air breathers like humans, this meant a muscular pump that sucked in air, and another one – the heart – that pumped oxygen-rich fluid through the body; and in between, a vast respiratory surface – the lining of the lungs – transferred gases between air and body tissues.

Some simple animals rely on diffusion alone to satisfy all of their oxygen needs

TWO PUMPS

Large, air-breathing animals have a high demand for oxygen and need specialized organs to absorb it. They also need two muscular pumps – the first pushes air into the lungs, and once oxygen is absorbed into the blood another muscle – the heart – pumps it around the body.

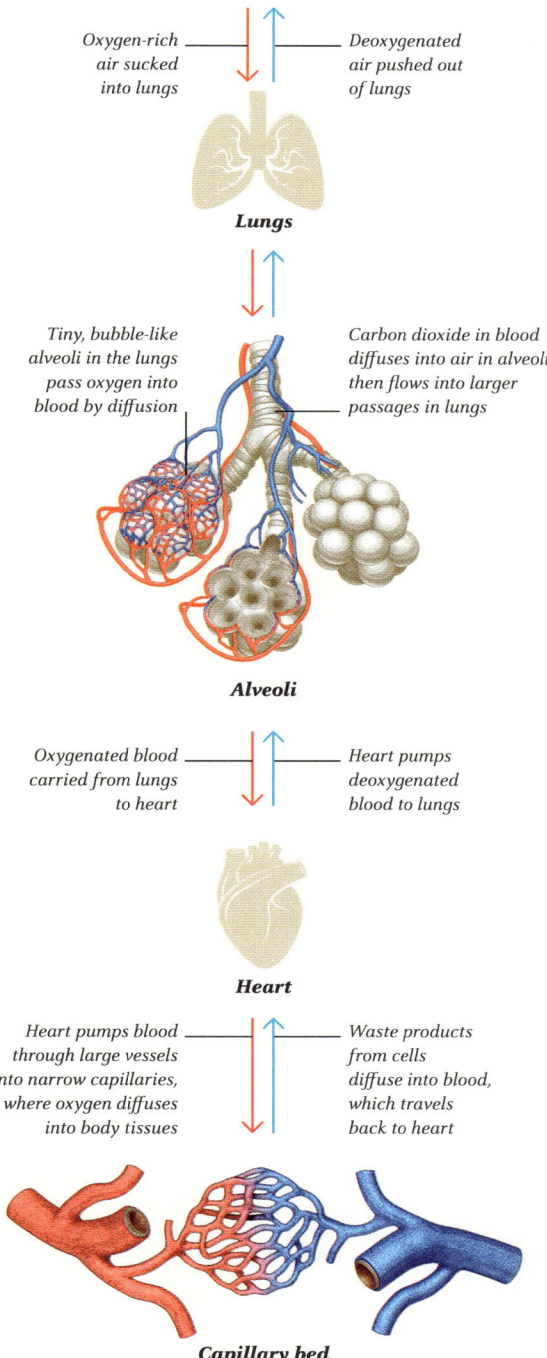

Oxygen-rich air sucked into lungs — Deoxygenated air pushed out of lungs

Lungs

Tiny, bubble-like alveoli in the lungs pass oxygen into blood by diffusion — Carbon dioxide in blood diffuses into air in alveoli, then flows into larger passages in lungs

Alveoli

Oxygenated blood carried from lungs to heart — Heart pumps deoxygenated blood to lungs

Heart

Heart pumps blood through large vessels into narrow capillaries, where oxygen diffuses into body tissues — Waste products from cells diffuse into blood, which travels back to heart

Capillary bed

Oxygen diffuses into the anemone's body from sea water — Carbon dioxide and wastes diffuse out of the anemone

Sea anemone

SEEPING IN AND OUT

If animals are small enough, they can get all the oxygen they need by the slow process of diffusion, which happens effectively over distances of a few millimetres. The waste products of respiration, such as carbon dioxide, can seep out by diffusion too.

HUNGRY FOR OXYGEN

SIMPLE CIRCULATION

In this system, blood passes through the heart only once per cycle. Commonly seen in fish, deoxygenated blood is first pumped to the gills, then the oxygenated blood flows to the rest of the body without returning to the heart.

DOUBLE CIRCULATION

Birds, mammals, and crocodiles have a double circulatory system, in which blood passes through the heart twice during one circuit of the body. A double circulation is composed of the pulmonary circuit (to the lungs) and the systemic circuit (around the body).

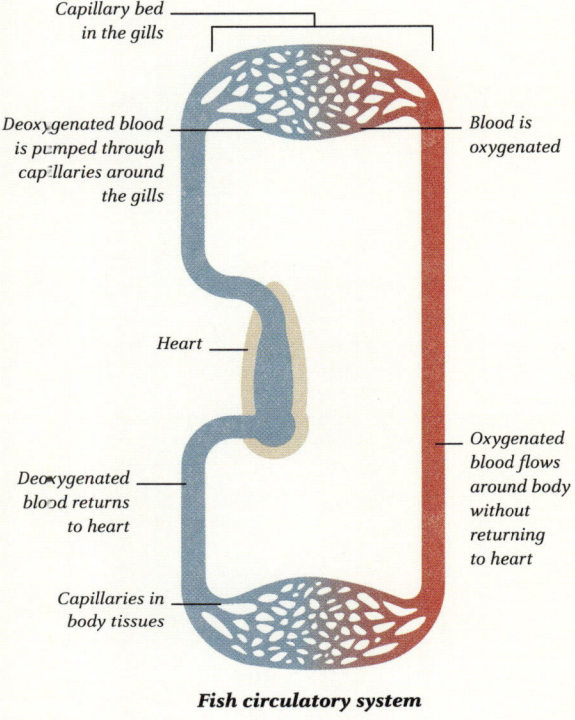

Capillary bed in the gills

Deoxygenated blood is pumped through capillaries around the gills

Blood is oxygenated

Heart

Deoxygenated blood returns to heart

Oxygenated blood flows around body without returning to heart

Capillaries in body tissues

Fish circulatory system

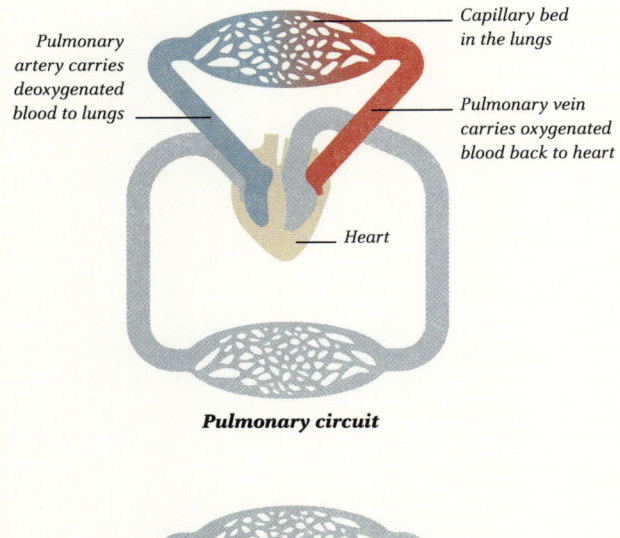

Pulmonary artery carries deoxygenated blood to lungs

Capillary bed in the lungs

Pulmonary vein carries oxygenated blood back to heart

Heart

Pulmonary circuit

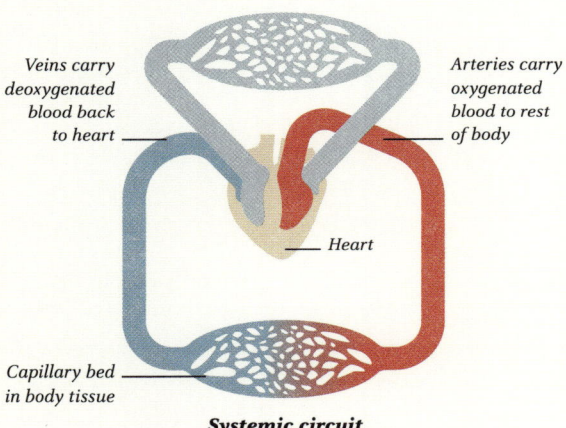

Veins carry deoxygenated blood back to heart

Arteries carry oxygenated blood to rest of body

Heart

Capillary bed in body tissue

Systemic circuit

Large animals need an expansive respiratory surface so they can absorb enough oxygen

Organs with an extra-high surface area have a long evolutionary heritage. Aquatic animals have a variety of feathery gills, while humans and other air breathers have finely divided lungs. In mammals, a huge surface area is achieved in the lungs through multiple branching air passages ending in microscopic chambers called alveoli, which greatly boost the quantity of oxygen absorbed by a body. Red blood cells packed with haemoglobin also increase the blood's carrying capacity. Mammal blood cells are so devoted to supplying their high-energy lifestyle that they lose their nucleus, along with their DNA, early in development, to pack in more oxygen-binding haemoglobin.

MUSCLES FOR MUSCLES

The heart and lungs are both muscular pumps that cause blood and air to flow from high to low pressure – but they work in opposite ways. The heart is a conventional pump: squeezing pushes blood out of its chambers; relaxation and expansion pull it in. The lungs are aspiration pumps: they suck air in when their contracting muscles expand the chest, then push air out as the lungs passively deflate.

As blood flows through the body, friction within the blood vessels leads to a decrease in blood pressure, particularly as it passes through the microscopically narrow capillaries that run through body tissues. However, the pressure is still high enough to push blood into tissues, delivering nutrients and oxygen to all areas. Maintaining pressure for this vital delivery has been critical in shaping the circulatory system of humans and our ancestors.

Higher blood pressure is essential to ensure blood reaches deep tissues

THE SHIFT TO A DOUBLE PUMP

Fish have the simplest kind of blood circuit, which has heart and oxygen-supply organs in series: blood from the heart passes to the gills to collect oxygen, then around the rest of the body for delivery. But this involves losing pressure in two networks of capillaries in quick succession. First the blood must force its way through the capillaries running through the gills, leaving it rich in oxygen, but much lower in pressure than when it left the heart. Then, each major artery must deliver oxygen to its target tissue through another bed of capillaries. Low pressure means less effective off-loading of oxygen in tissues. As a result, warm-blooded birds and mammals – with especially high energy demands – refashioned the heart into a double pump. A dividing wall kept the right side pumping blood to the lungs, but the left side received the returning oxygenated blood to pump it again around the rest of the body. This double circulatory system enabled blood to be pumped through the lungs at low pressure, preventing fluids being forced into the airways, then to respiring tissues at a higher pressure, effectively delivering oxygen to the body.

Oxygen-hungry humans have inherited this double circulation and several other major adaptations from earlier mammals that are crucial in delivering oxygen fast and efficiently. But humans have also evolved further refinements of the heart and lungs that make them superb endurance athletes (see pp.120–23).

Bigger bodies need oxygen to reach deep into their tissues. The evolutionary solution was the heart.

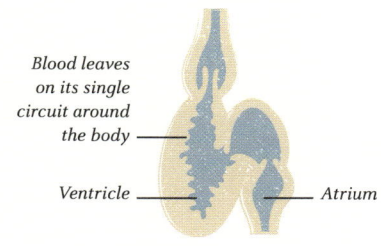

Blood leaves on its single circuit around the body

Ventricle — — Atrium

Fish

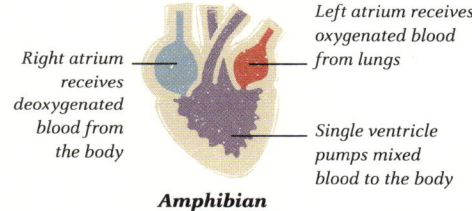

Right atrium receives deoxygenated blood from the body

Left atrium receives oxygenated blood from lungs

Single ventricle pumps mixed blood to the body

Amphibian

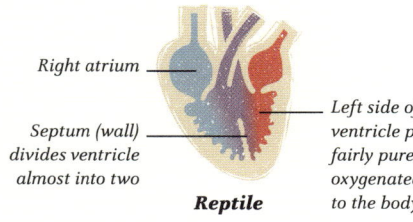

Right atrium

Septum (wall) divides ventricle almost into two

Left side of ventricle pumps fairly pure oxygenated blood to the body

Reptile

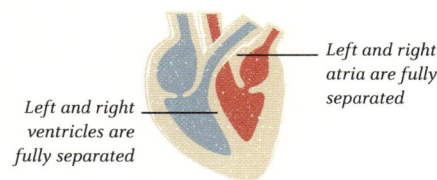

Left and right atria are fully separated

Left and right ventricles are fully separated

Mammal, bird, or crocodilian

HOW THE HEART DIVIDED INTO FOUR

The four-chambered heart evolved in steps. From ancestral fish, some amphibians had partially divided hearts, imperfectly separating oxygenated and deoxygenated blood – the heart pumped blood in separate directions, but the blood was mixed. Slowly, animals evolved better separation of the two circulations.

A breath of fresh air

A typical human cannot go for more than a minute or so without breathing. Every breath helps bring oxygen deep into the body and expels carbon dioxide. This gaseous exchange is made efficient by the lungs – each packed with millions of microscopic sacs.

When the first animals lumbered from the water onto land, they were exposed to a richer source of oxygen than previously experienced and needed to adapt to this new environment. Air contains at least 21 per cent more oxygen by volume than water. It is also thinner than water, so flows more easily – it takes less than a second for air to reach the deepest parts of human lungs with each inhalation, and a quarter of its oxygen is absorbed into the bloodstream before the air is exhaled. It therefore comes as no surprise that most energetic, warm-blooded vertebrates, including humans, are air-breathers. And while whales, penguins, and other marine mammals returned to the seas, no animal with lungs ever completely lost these organs and reverted to gills.

As animals moved onto land, their bodies had to adapt to the oxygen-rich atmosphere

IMPROVING GAS EXCHANGE

Lungs contain so much air that they are the only organs that float on water – most swimming mammals battle against their buoyancy to dive deep – and the biggest lungs usually power the most athletic

HOW TO CREATE SURFACE AREA

Large surface area maximizes the efficiency of gas absorption. As surface area increases, more oxygen is absorbed into the bloodstream. Animals increase the surface area of their lungs in a variety of ways, and birds have the additional advantage of one-way air flow.

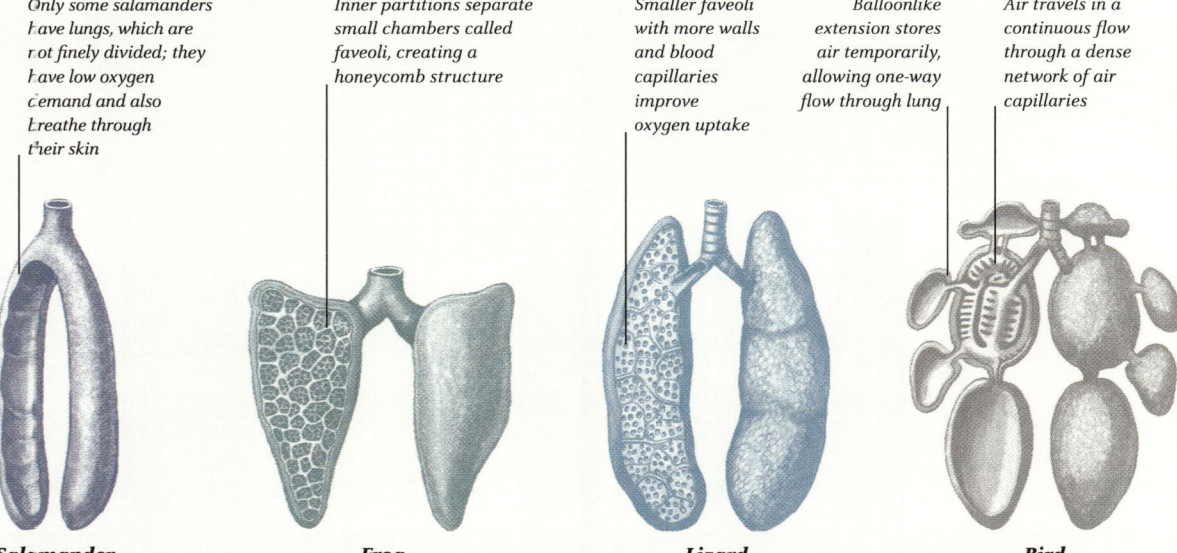

Only some salamanders have lungs, which are not finely divided; they have low oxygen demand and also breathe through their skin

Inner partitions separate small chambers called faveoli, creating a honeycomb structure

Smaller faveoli with more walls and blood capillaries improve oxygen uptake

Balloonlike extension stores air temporarily, allowing one-way flow through lung

Air travels in a continuous flow through a dense network of air capillaries

Salamander

Frog

Lizard

Bird

bodies. The first lungs of air-breathing vertebrates were like a honeycomb, with inhaled air being directed to the centre of the lungs, where oxygen then diffused outwards towards blood capillaries. This structure is retained in the lungs of today's amphibians and reptiles. As mammals evolved, lungs were restructured into a more efficient arrangement, delivering air via a network of branching tubes (bronchi and bronchioles), which were connected to tiny bubblelike sacs called alveoli, each one smaller than a grain of salt and enveloped by capillaries. This structure increased surface area and therefore made better use of lung capacity, allowing for greater output. An average adult human lung contains about half a billion alveoli – spread out, they would cover an area roughly the same size as a tennis court. Alveoli, and the blood capillaries that surround them, have incredibly thin walls, only one cell thick. The combination of thin walls and a large surface area makes gas exchange in the blood incredibly efficient.

Alveoli maximize lungs' absorption potential by increasing surface area

HOW DO HUMAN LUNGS DO IT?

Mammals increase their lung surface area through millions of tiny sacs called alveoli. Clustered at the ends of bronchioles, alveoli are surrounded by a dense network of blood capillaries and are the site of oxygen and carbon dioxide exchange.

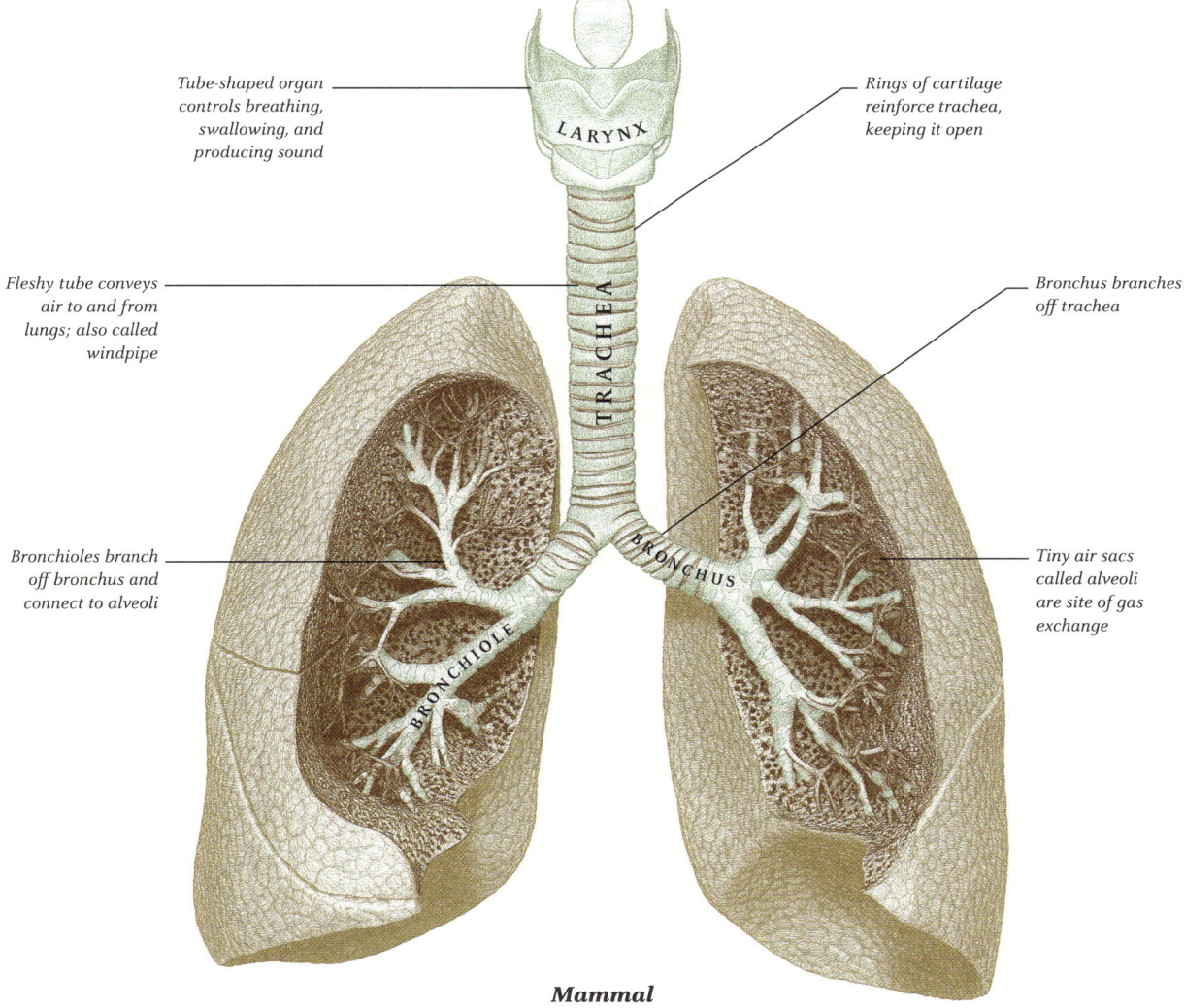

Tube-shaped organ controls breathing, swallowing, and producing sound

LARYNX

Rings of cartilage reinforce trachea, keeping it open

Fleshy tube conveys air to and from lungs; also called windpipe

TRACHEA

Bronchus branches off trachea

BRONCHUS

Bronchioles branch off bronchus and connect to alveoli

BRONCHIOLE

Tiny air sacs called alveoli are site of gas exchange

Mammal

A BREATH OF FRESH AIR

The structure of lungs alone is insufficient to serve the body. They must be ventilated, too, by breathing. Most active animals rely on breathing movements. Fish do it by gulping water and directing it over feathery, blood-filled gills on the sides of their heads. When certain fish became air breathers, they supplemented their gills by repurposing their swim bladder – a gas-filled organ that provides fish with buoyancy. While the bladder of some fish was sealed, in others it was connected to a tube at the back of the throat, allowing them to take fresh air in, making the bladder a primitive lung.

Having breathing organs deep in the body meant that air had to be moved from outside to the sites of gas exchange and then out again by the same route, like a tidal flow. This is much less efficient than simply allowing water to flow over gills in one direction, but fortunately, air is much thinner and easier to move than water and much richer in oxygen, so most air breathers, including humans and other mammals, are not severely limited by tidal-flow breathing. However, birds developed an even more efficient system, permitting unidirectional airflow in their lungs through an ingenious system of inflatable air sacs, which move air in and out of the lungs (see p.116).

Air breathers must draw air deep into their bodies then push it out the same way as it came in

BREATHING MUSCLES

In the first air-breathing amphibians, muscles in the throat and cheeks that originally ventilated gills in the heads of their ancestors became adapted to squeezing air into the chest – a process called buccal pumping. To this day, frogs still inflate their lungs by pumping from the mouth.

With the evolution of reptiles and mammals, new muscles in the chest made breathing more effective. Ribcage, or intercostal, muscles worked alongside a large, muscular diaphragm (in mammals and crocodilians) that formed the floor of the chest cavity. As they both contracted to raise the ribs and lower the chest floor, the chest cavity expanded and imitated a suction pump, drawing air into the lungs. When these muscles relax, the ribs drop, the chest floor raises up into a dome, and the chest cavity contracts, pushing air out.

FUELLING ENDURANCE

On average, humans breathe around 15 times per minute when resting and around 60 times per minute when running. The evolution of an upright posture was especially beneficial for

Oxygenated water flows over gill surface

Salamander gills

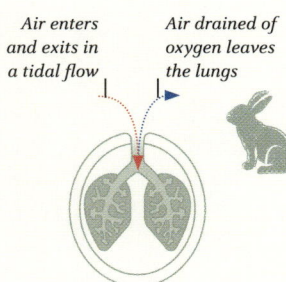

Air enters and exits in a tidal flow

Air drained of oxygen leaves the lungs

Rabbit lungs

IN WATER, ON LAND

Most aquatic animals extract oxygen from water as it passes over gills on the outside of their body. Air breathers have their respiratory surface (the lining of the lungs) inside their body, so air must flow all the way in, then out the same way.

MOUTH PUMP

Lacking a diaphragm and ribcage, amphibians rely on a method inherited from fish called buccal pumping, in which muscles in the mouth pump air into the lungs.

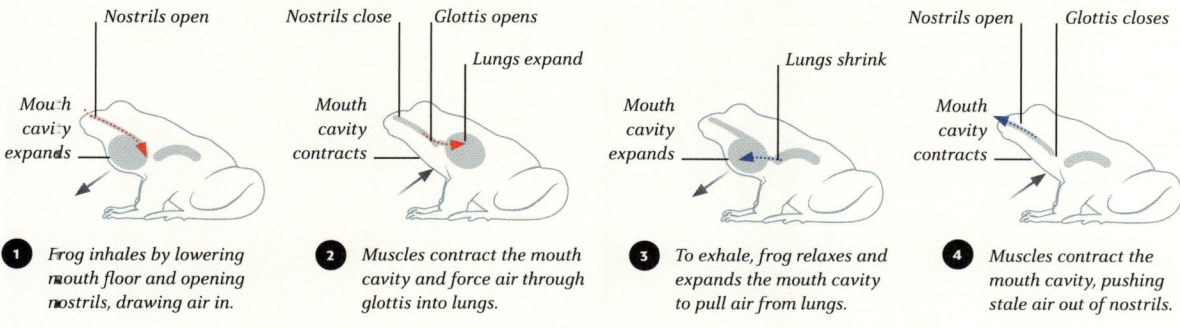

Nostrils open

Mouth cavity expands

1 Frog inhales by lowering mouth floor and opening nostrils, drawing air in.

Nostrils close *Glottis opens*

Lungs expand

Mouth cavity contracts

2 Muscles contract the mouth cavity and force air through glottis into lungs.

Lungs shrink

Mouth cavity expands

3 To exhale, frog relaxes and expands the mouth cavity to pull air from lungs.

Nostrils open *Glottis closes*

Mouth cavity contracts

4 Muscles contract the mouth cavity, pushing stale air out of nostrils.

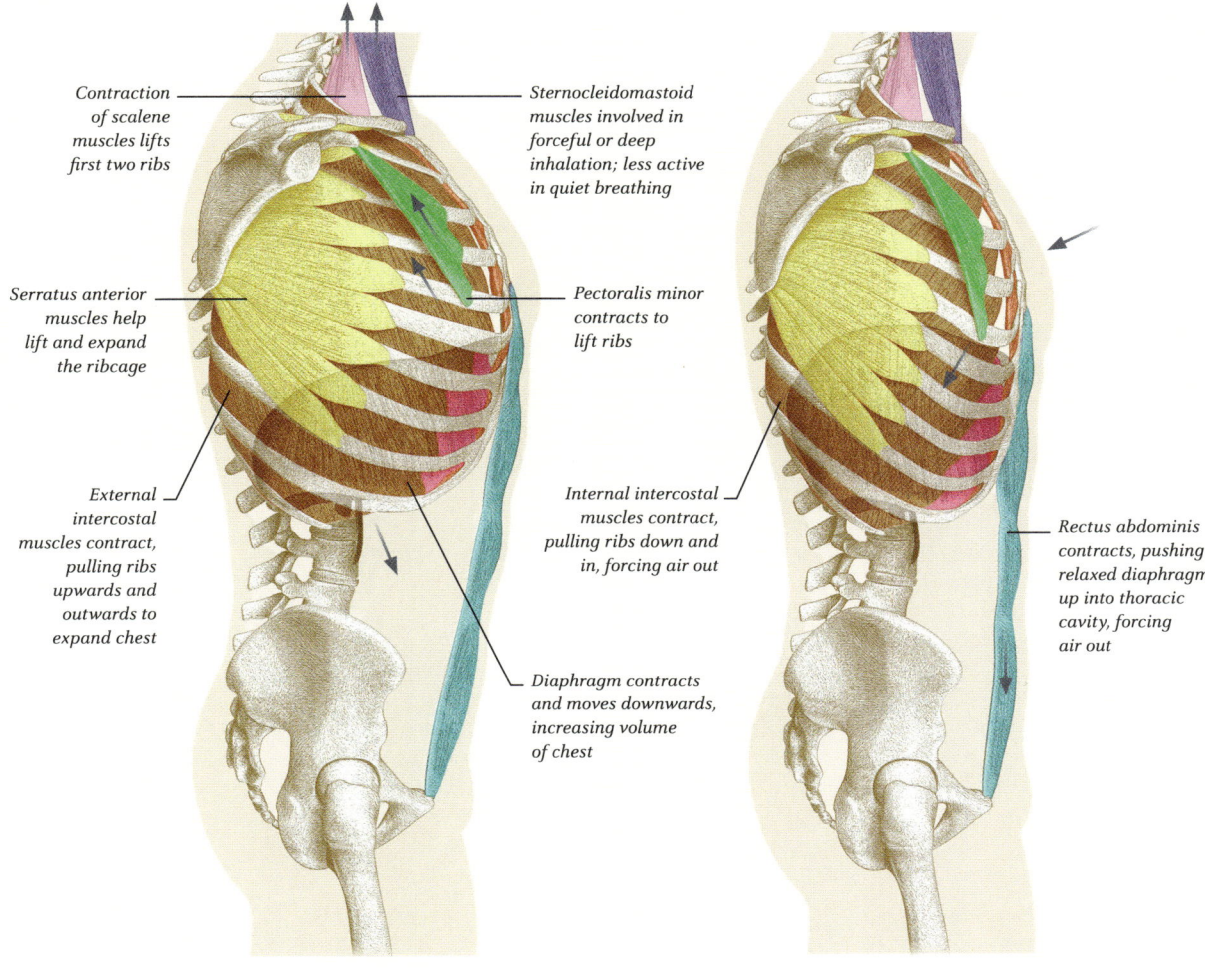

Contraction of scalene muscles lifts first two ribs

Sternocleidomastoid muscles involved in forceful or deep inhalation; less active in quiet breathing

Serratus anterior muscles help lift and expand the ribcage

Pectoralis minor contracts to lift ribs

External intercostal muscles contract, pulling ribs upwards and outwards to expand chest

Internal intercostal muscles contract, pulling ribs down and in, forcing air out

Diaphragm contracts and moves downwards, increasing volume of chest

Rectus abdominis contracts, pushing relaxed diaphragm up into thoracic cavity, forcing air out

Inhalation

Forced exhalation

HUMAN BELLOWS

Mammals, including humans, inflate their lungs using muscle power. Engaging the muscles between the ribs and the diaphragm allows the chest to expand. Exhalation can be passive and happens without effort when the breathing muscles relax, or humans can use alternative muscles to push the breath out with force.

allowing lungs to work at maximum efficiency. As human chests became wider and the ribcage more cylindrical, this increased lung capacity beyond that of tree-dwelling apes. Aided by gravity, this allowed more blood to reach the lower parts of the lungs to absorb more oxygen. The vertical alignment of the lungs, diaphragm, and pelvis even helped boost the breathing cycle – when running fast at a sustained pace, gait can be synchronized with breathing movements, and the force transmitted upwards through the limbs works like a piston to reinforce the push-pull action of the diaphragm. Already optimized for absorbing oxygen, when boosted by running human lungs become even more efficient – every step helping to bring a breath of fresh air.

An upright stance allows human lungs to work at maximum efficiency

A BREATH OF FRESH AIR

No other primates match the physical endurance of humans. At peak fitness, humans can run long distances for hours at a time before fatiguing – an ability that helped turn them into formidable predators on the open plains of Africa.

Endurance athletes

Walking on two legs set our ancestors on the path to becoming human

Africa became the birthplace of a new kind of primate when the first hominins developed a fully bipedal habit, walking upright on two legs, some 4–3 million years ago (MYA). Unlike other apes and monkeys – mostly climbers in forest and woodland, and eating a largely vegetarian diet – our ancestors became endurance athletes between 2.5 and 1.5 MYA, capable of chasing, exhausting, and catching prey on open terrain. For the first time in prehistory, primates had the potential to become top predators.

THE KEY TO ENDURANCE
Speed and strength are largely matters of mechanics; longer legs can swing through a wider stride, while thicker fibre-packed muscles

STEP COUNTS
This comparison of daily step counts shows that our species engages in much higher levels of physical activity than our closest ape relatives.

generate more force. But endurance is more about biochemistry and physiology. Running for longer periods required the production of energy over a longer time, as well as more efficient heart and lungs to deliver the necessary fuel and oxygen to where it is needed in the body.

Carbohydrate and fat have been the staple fuels for living things since their early origins. Oxygen can be used to oxidize or "burn" these organic compounds inside living cells, releasing energy to power activity. This process of aerobic respiration (see p.112) produces carbon dioxide and water as well as releasing energy, which is used to make a molecule called adenosine triphosphate (ATP) – the cell's energy "currency". In losing one of its phosphate groups to become adenosine diphosphate (ADP), ATP can in turn release energy to power reactions, such as those that cause muscles to contract. Reserves of carbohydrate and fat are stored throughout the body, ready to be called on when needed. However, oxygen must be delivered directly from the environment, via the lungs, gills, or

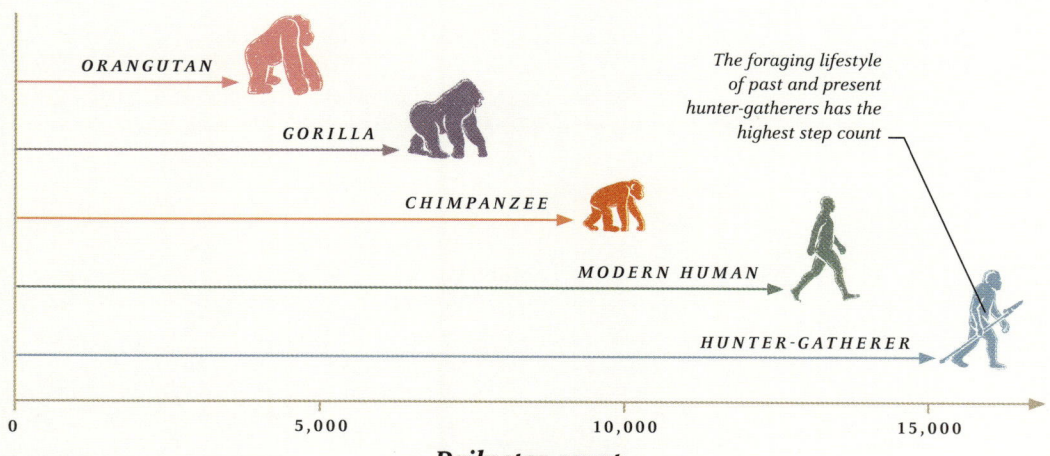

The foraging lifestyle of past and present hunter-gatherers has the highest step count

ORANGUTAN

GORILLA

CHIMPANZEE

MODERN HUMAN

HUNTER-GATHERER

0   5,000   10,000   15,000

Daily step count

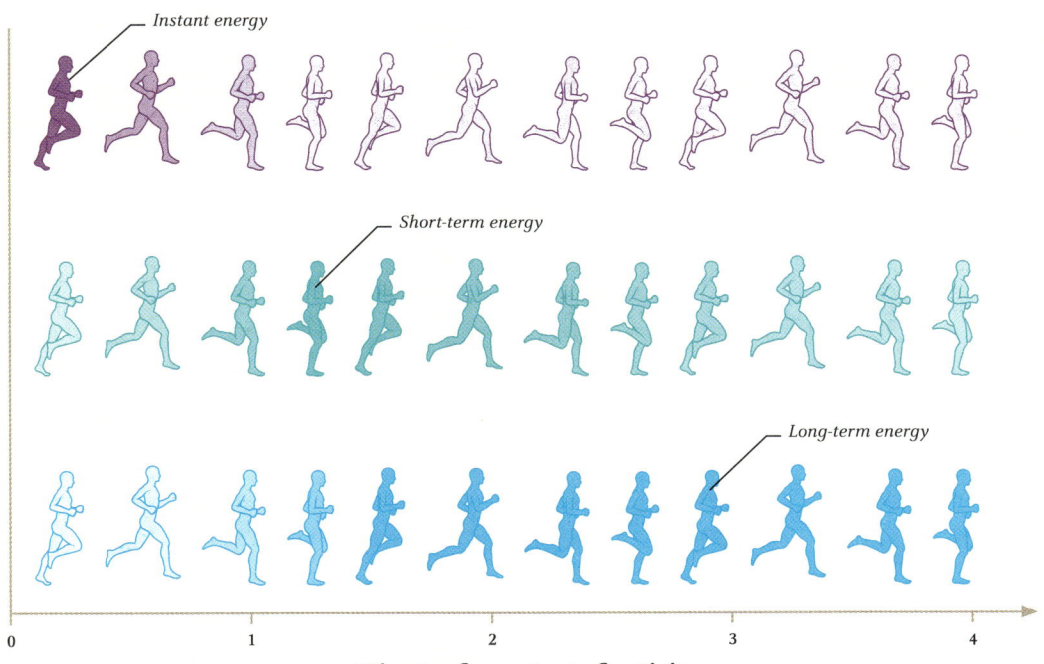

Instant energy

Short-term energy

Long-term energy

0   1   2   3   4

Minutes from start of activity

skin, and cannot be stored. For this reason, oxygen is a limiting factor when it comes to endurance running.

POWER WITHOUT OXYGEN?

There are metabolic pathways through which muscles can be powered without the need for oxygen or sugar (carbohydrate). Muscles contain the compound phosphocreatine – a molecule that regenerates ATP from ADP. Unfortunately, working muscles need so much energy that the phosphocreatine reserves are depleted within seconds of commencing exercise. Another way in which muscles can obtain energy without the need for oxygen involves breaking down sugars incompletely to produce energy in the form of ATP plus a waste product called lactic acid. This process is called anaerobic respiration. The accumulation of lactic acid in muscle during exercise produces an acidity that compromises muscle contraction, causing fatigue and cramp, and limiting endurance.

Powering muscles without using oxygen works well for animals that rely on bursts of movement – from small fish that dart away to escape predators to tarsiers, tiny tree-dwelling

Around 50 kg (110 lb) of ATP are processed by the human body every day

ENERGY SUPPLIES

Humans can draw on three energy systems to fuel their activity, but only aerobic respiration can support effort beyond a few minutes of duration.

● ATP–PHOSPHOCREATINE
● ANAEROBIC RESPIRATION
● AEROBIC RESPIRATION

primates that jump to catch insects. Reliance on bursts of movement also suited larger forest-dwelling primates, as well as human ancestors such as *Australopithecus*. But in a more open habitat, and with their growing taste for meat, bipedal humans developed muscles and adaptations of their circulation that could support aerobic respiration – a pathway that creates no lactic acid and generates 15 times more ATP than anaerobic respiration.

THE FAST AND THE SLOW

Muscles are made up of specialized cells called fibres, which contract when triggered by a nerve impulse. There are two main types of muscle

121

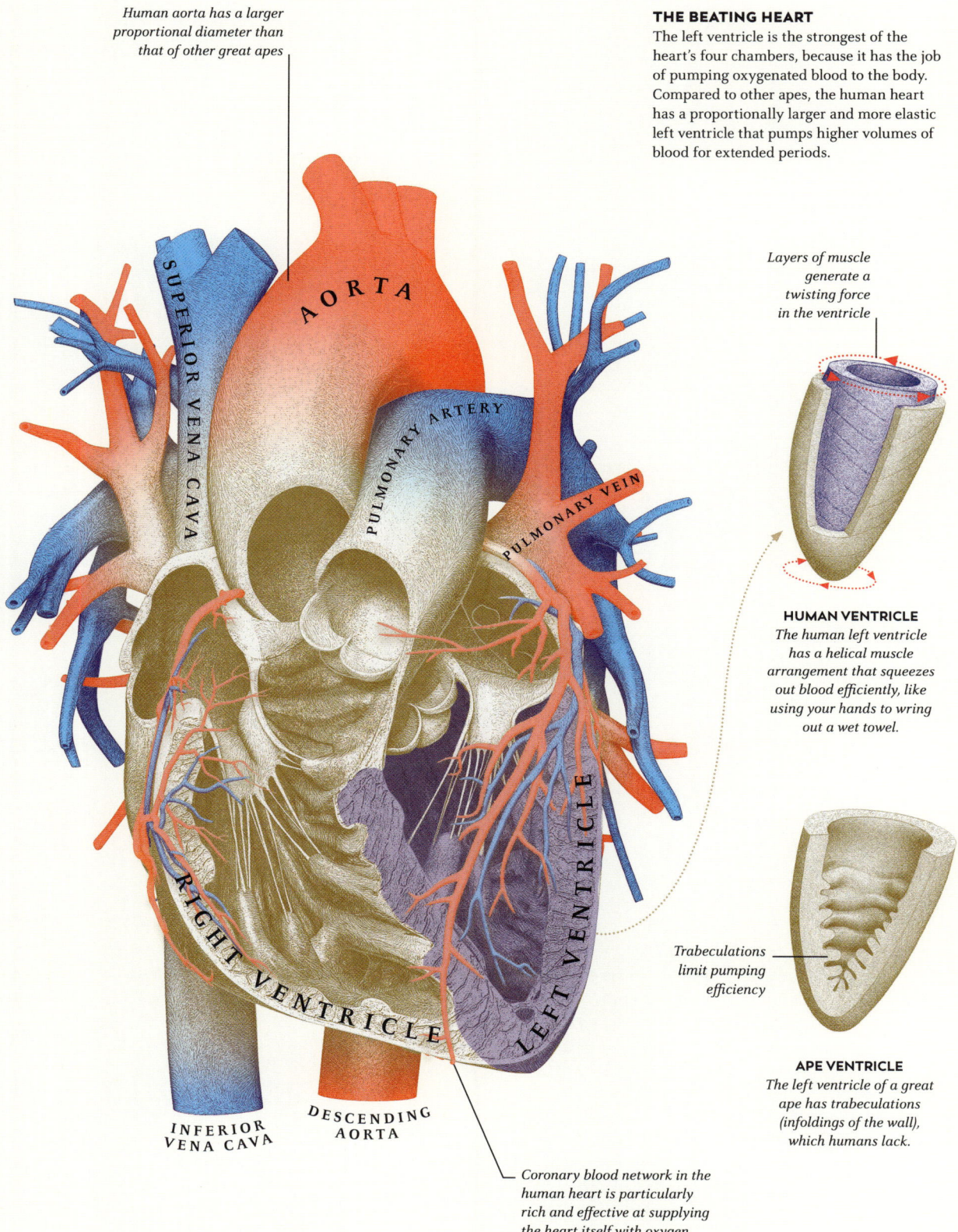

Human aorta has a larger proportional diameter than that of other great apes

THE BEATING HEART

The left ventricle is the strongest of the heart's four chambers, because it has the job of pumping oxygenated blood to the body. Compared to other apes, the human heart has a proportionally larger and more elastic left ventricle that pumps higher volumes of blood for extended periods.

Layers of muscle generate a twisting force in the ventricle

HUMAN VENTRICLE

The human left ventricle has a helical muscle arrangement that squeezes out blood efficiently, like using your hands to wring out a wet towel.

AORTA

SUPERIOR VENA CAVA

PULMONARY ARTERY

PULMONARY VEIN

RIGHT VENTRICLE

LEFT VENTRICLE

Trabeculations limit pumping efficiency

APE VENTRICLE

The left ventricle of a great ape has trabeculations (infoldings of the wall), which humans lack.

INFERIOR VENA CAVA

DESCENDING AORTA

Coronary blood network in the human heart is particularly rich and effective at supplying the heart itself with oxygen

MUSCLE TYPES

Endurance athletes, such as wolves and humans, have muscles with more slow-twitch fibres, which require oxygen. Such muscles are packed with more blood capillaries and the fibres contain more myoglobin – an oxygen-binding pigment similar to the blood's haemoglobin that acts as an oxygen reserve.

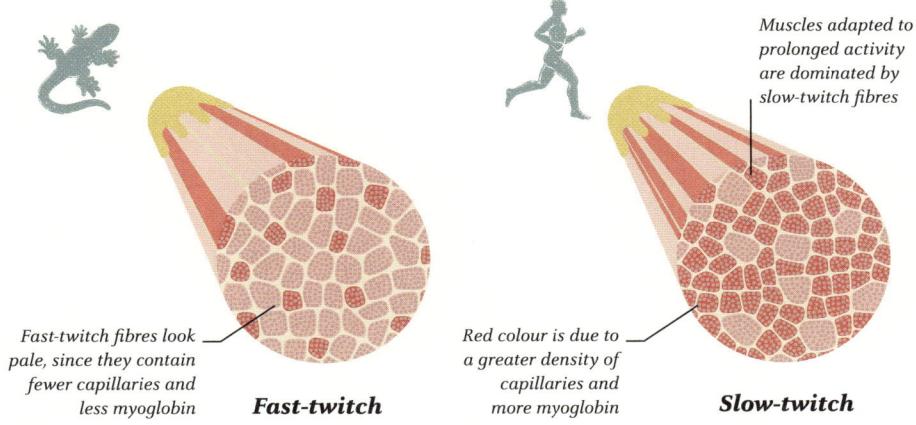

Fast-twitch fibres look pale, since they contain fewer capillaries and less myoglobin

Fast-twitch

Muscles adapted to prolonged activity are dominated by slow-twitch fibres

Red colour is due to a greater density of capillaries and more myoglobin

Slow-twitch

fibres – slow-twitch (Type I) and fast-twitch (Type II). Slow-twitch fibres can sustain significantly longer contractions and are suited to repeated, low-force movements over a long time, while fast-twitch fibres are best for producing short, intense bursts of force. Well-oxygenated slow-twitch muscle fibres have become the hallmark of endurance athletes in several vertebrate species, including wolves and humans.

THE HEART OF THE CHASE

No matter how well the limbs of endurance animals are adapted to their tasks (see pp.66–71), they depend on key organs to supply them with oxygen: big lungs that absorb more oxygen with every breath and a strong heart that pumps more oxygenated blood with every stroke. Human heart muscle packs a high density of contractile filaments, even though the smoother walls are proportionately thinner than in other apes. This means that the chambers of the heart can fill to greater capacity while their walls stay powerful for pumping, despite being more flexible than those of other apes. Muscle fibres in the heart are orientated spirally in the chamber walls. This results in a "wringing" action during each cycle that empties the heart with great power. The aorta – the main artery delivering blood around the body – is also proportionately wider in humans than in chimpanzees, reflecting greater blood flow and adding to the efficiency of our circulatory system.

Soft organs, muscles, and biochemicals rarely leave a trace in the fossil record, so it is hard to tell exactly when humans became endurance athletes. But comparing the anatomy and physiology of living apes with humans suggests the transition began with the genus *Homo*. The emergence of *Homo erectus* made sustained long-distance running possible: its legs were 50 per cent longer than the australopithecines that preceded it, and robust attachment points suggest that stronger tendons gave an extra spring in the step. These changes meant that early humans could potentially compete with other predators to steal meat from carcasses and – ultimately – bring down their own prey (see pp.142–47).

Homo erectus *was a persistence hunter, able to run prey to the ground even in hot conditions*

The heart's ventricle twists as it empties, pushing out blood like wringing out a towel.

123

The sound-making instrument

Few structures in our bodies are so intimately related with what it means to be human as the voice box. Our complex speech is the culmination of three million years of evolution that refashioned a part of the airway from a simple valve to a sound-making instrument.

Humans are the ultimate animal communicators. Speech delivers ideas, thoughts, and instructions far more precisely than grunts, barks, and whistles. And when humans talk to one another, sounds translate into phrases and stories that do more than communicate; they engage, inspire, and bind listeners into social groups. Speech is as much about social cohesion as it is about communicating specific ideas.

The ability to talk emerged from adaptations in two very different parts of the body – the respiratory system and the brain. Changes in the airway, and particularly the voice box, or larynx, allowed a range of sounds to be generated, while the emergence of specialized regions in the brain allowed both for the fine control over the airway needed to produce subtle sounds at different volumes, and enabled the sounds made to be understood by others. While the brain's speech centres are distinctly human, the origins of the voice box go right back to the evolution of air breathing in fish.

A NEW KIND OF AIRWAY

Fish breathe by taking water into their mouths and passing it over their gills – feathery structures on the sides of the head that contain a rich network of blood vessels. These absorb oxygen from the water and give up carbon dioxide (see pp.112–13).

Around 400 million years ago, some fish species that had adapted to live in water with a low oxygen content began to explore life on mud and land. Gradually, they transitioned from breathing through their gills to using lungs, which developed from structures similar to swim bladders in ancestral species. Swim bladders are air-filled sacs that are used by fish to control their buoyancy in the water. In some modern fish, the swim bladder remains connected to the airway via a tube. It is easy to see how such a structure, developing its own capillary network, would allow a fish to exchange gases and breathe air. When animals started to breathe air rather than

Limblike fins with internal bones and air-breathing lungs were key in the colonization of land

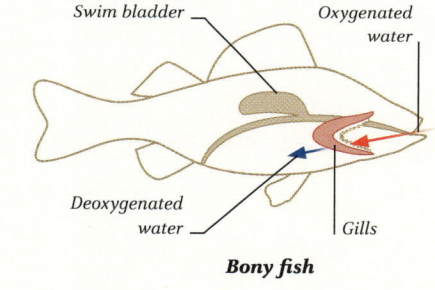

Swim bladder

Oxygenated water

Deoxygenated water

Gills

Bony fish

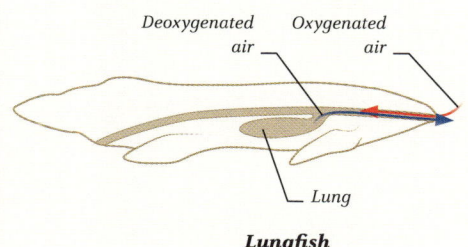

Deoxygenated air

Oxygenated air

Lung

Lungfish

FISH BREATH
Today's lungfish are similar to the first land vertebrates in that they breathe air through a modified swim bladder, as well as having gills typical of other fish that only breathe under water.

water, they required a tube that would not collapse when air was drawn through it. Both the larynx and trachea (windpipe) developed with cartilages in their walls to hold them open and to protect them from damage. Some of the cartilaginous tissue developed into a hinged lid – the epiglottis. This can be raised or lowered, allowing the airway to stay open while an animal is drinking, or closed while a lump of food is being swallowed. The larynx is capable of other functions too. Notably, it can regulate airflow in the course of breathing, widening the air passage during inhalation and funnelling air to the trachea and lungs. And crucially for human communication, it can also act as a type of wind instrument.

The larynx is key to regulating the flow of air into and out of the lungs

MAKING SOUNDS

Animals make sounds by using their muscles to vibrate parts of their body. These vibrations are transmitted to the surrounding water or air in the form of sound waves. Crickets create vibrations by rubbing their wings together; some fish do it by grinding fin rays or bones; and most mammals do it by forcing air through their larynx to vibrate special folds – the vocal cords housed within. The higher the speed, or frequency, of a vibration, the higher the pitch of the sound produced, while bigger vibrations – those with greater amplitude – result in louder sounds. Crickets sing at a higher pitch by rubbing their wings faster, but humans use a more sophisticated system of laryngeal muscles to alter the tension of the folds, allowing for changes in both frequency and amplitude that are under the control of the nervous system. The human larynx is a true voice box, and the sound emerging is a true voice, or vocalization.

Birds have a voice box that is unique to them – a structure, called a syrinx, at the base of the windpipe just where it splits to deliver air to the two lungs. Its two sides are controlled separately, so birds can make two sounds at once – an ability that helps explain the complexity of their songs. But their larynx remains the simple valve of their distant ancestors. Among amphibians and reptiles, vocal cords of the larynx are well developed in some species, such as chirping frogs and geckos, but not in the voiceless salamanders and snakes. Shifts toward vocalization happened in groups that relied heavily on sound for communication.

Most mammals, with their ancestry rooted in a nocturnal lifestyle, relied on smell and sound and many have well-developed vocal cords too.

HUMAN VS APE

Our vocal tracts are capable of finer, more rapid motions of the tongue than those of other apes. The larynx is also much lower in the throat than it is in other apes, allowing for a wider range of sounds to be produced.

- 🟢 LARYNX
- 🟡 HYOID
- 🟠 EPIGLOTTIS
- 🔴 TONGUE

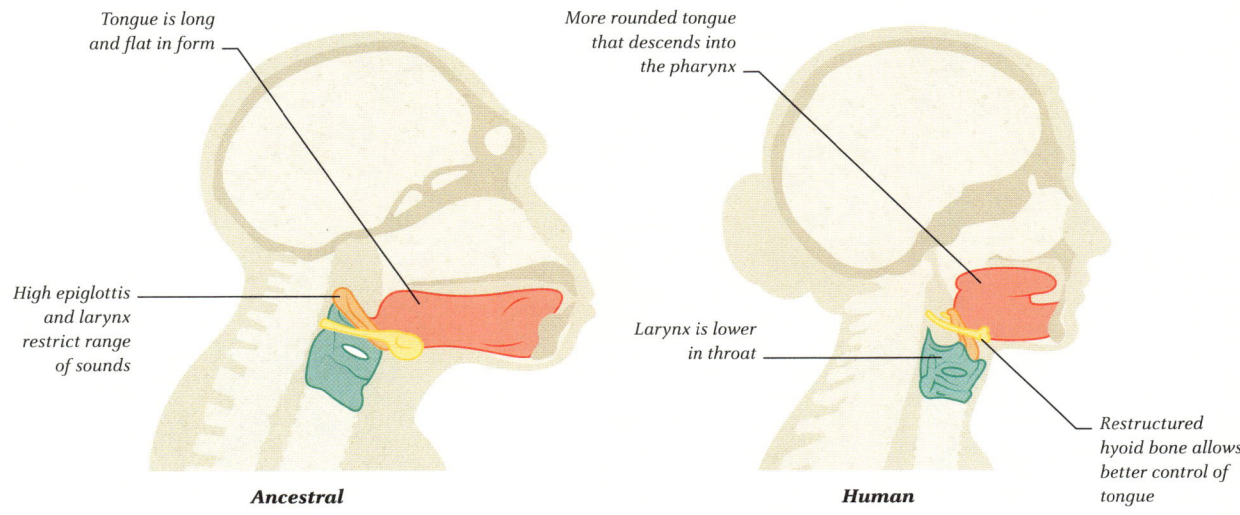

Tongue is long and flat in form

High epiglottis and larynx restrict range of sounds

More rounded tongue that descends into the pharynx

Larynx is lower in throat

Restructured hyoid bone allows better control of tongue

Ancestral

Human

THE SOUND-MAKING INSTRUMENT

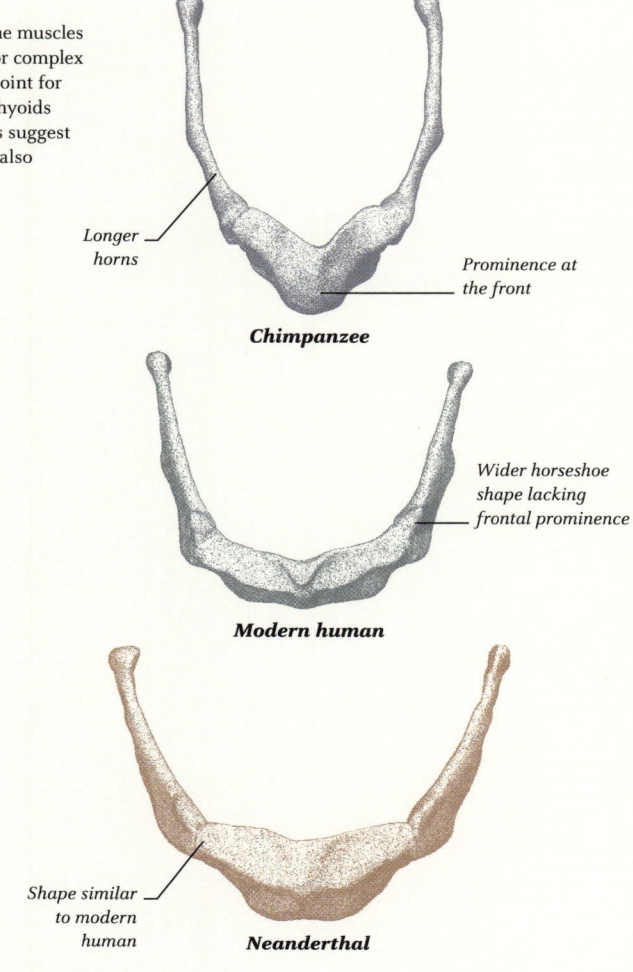

Longer horns

Prominence at the front

Chimpanzee

Wider horseshoe shape lacking frontal prominence

Modern human

Shape similar to modern human

Neanderthal

Primates – from crying bushbabies to whooping monkeys – are especially vocal, and the path to the human voice was set with their origins around 60 million years ago.

THE HOMININ VOICE

No-one can be sure when the cries and whoops of primates evolved into something more like a human voice, but influential factors would have included standing upright on two legs, flatter faces, and a lower larynx. In a bipedal posture, with the head balanced on top of the spine, there is less room to accommodate the larynx behind the jaw, and so it moved lower down in the neck. This created a longer tube for resonating the voice – especially to produce softer vowel sounds – making vocalization richer. But sound production is only the start of the story – other anatomical structures then shape the sound.

Frogs and howler monkeys amplify their calls with the help of a resonating throat sac, while monkeys and apes have flexible lips that can smack together or protrude for better resonating whooping. These kinds of anatomical modifiers became especially important in humans. The root of the tongue enlarged to affect the shape of the upper pharynx, introducing new ways to modify vowel and consonant sounds. At the same time, the soft palate – the flap of tissue behind the roof of the mouth – rose higher, which helped the switch between oral and nasal sounds. Collectively, all these anatomical attributes helped humans to produce a wide diversity of sounds, or phonemes. But they came at a cost: the longer throat that lowered the larynx also prevented its valve from guarding the windpipe as effectively, so humans became more vulnerable to choking when eating food, and the more exposed larynx was at greater risk of injury.

Despite these disadvantages, the benefits of a sophisticated vocal apparatus outweighed the costs. The ability to produce varied phonemes provided the raw material for complex language, which would help reinforce social bonds and make humans formidable cooperative hunters. But a capacity for vocalization is nothing without the brain to control and understand it.

THE VOCAL BRAIN

Speaking and understanding language are products of a big brain. Like other "higher" human functions, their control centres are located in the grey matter of the cerebral cortex that forms the outer walls of the brain's hemispheres. The cerebral cortex is responsible for the brain's principal processing functions. Cortex found around the side and back of the brain collects sensory information from around the body, while cortex in the frontal lobes controls muscular – or motor – responses. In nonhuman primates, part of the left frontal cortex controls

Complex thought and decision-making are human abilities rooted in the structures of the cortex

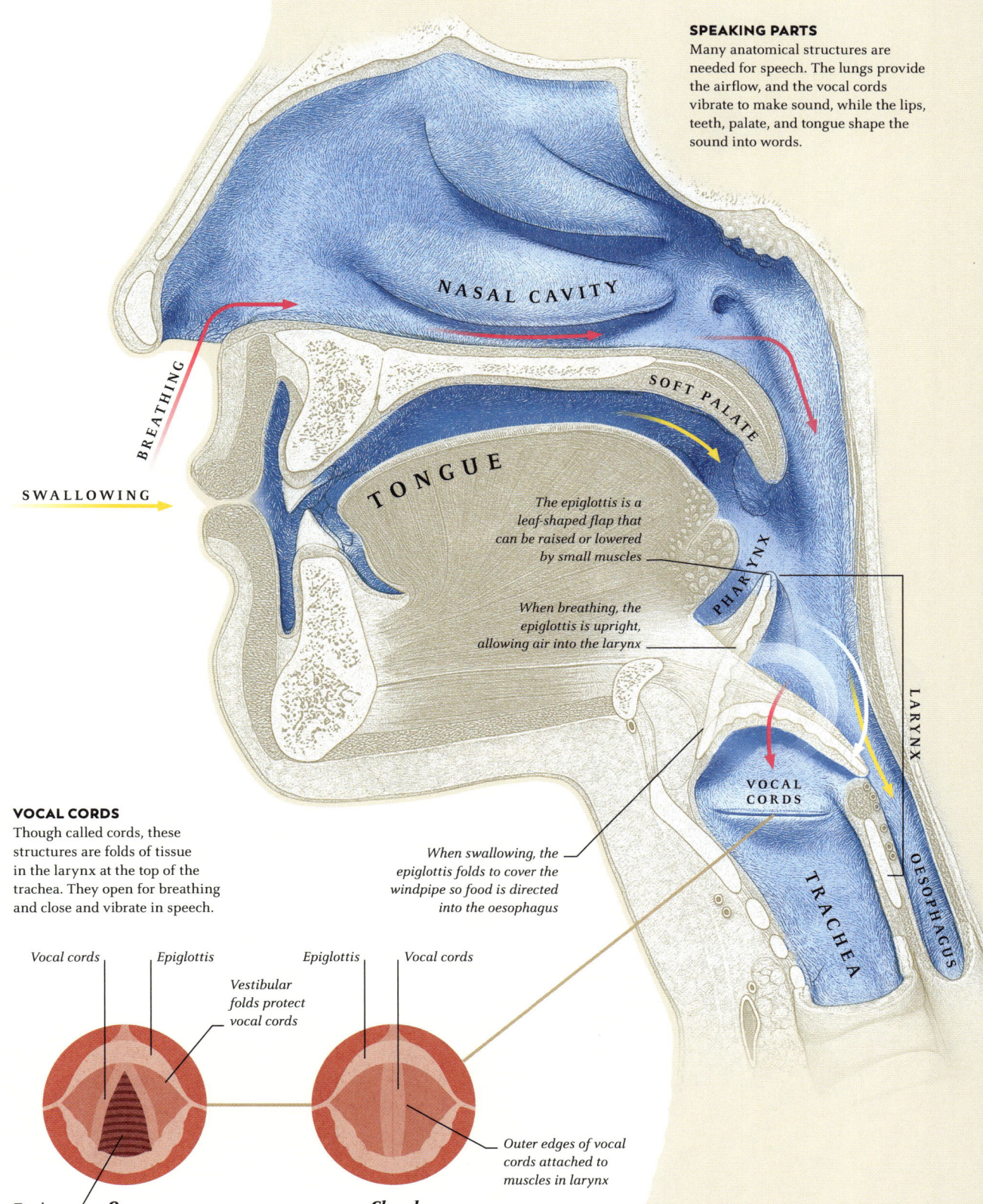

Many anatomical structures are
needed for speech. The lungs provide
the airflow, and the vocal cords
vibrate to make sound, while the lips,
teeth, palate, and tongue shape the
sound into words.

NASAL CAVITY

SOFT PALATE

BREATHING

SWALLOWING

TONGUE

The epiglottis is a
leaf-shaped flap that
can be raised or lowered
by small muscles

PHARYNX

When breathing, the
epiglottis is upright,
allowing air into the larynx

LARYNX

VOCAL
CORDS

OESOPHAGUS

VOCAL CORDS
Though called cords, these
structures are folds of tissue
in the larynx at the top of the
trachea. They open for breathing
and close and vibrate in speech.

When swallowing, the
epiglottis folds to cover the
windpipe so food is directed
into the oesophagus

TRACHEA

Vocal cords

Epiglottis

Epiglottis

Vocal cords

Vestibular
folds protect
vocal cords

Outer edges of vocal
cords attached to
muscles in larynx

Trachea

Open

Closed

127

THE SOUND-MAKING INSTRUMENT

the muscles involved in hand gestures and is also responsible for controlling sounds made in the vocal tract. A comparable region is active when humans talk. This is Broca's area (see pp.48, 249), named after a French physician who identified it after a post-mortem examination of a patient with a speech defect revealed damage in this region. Another region of cortex in the left temporal lobe (low down on the side of the brain, see p.249) is involved in comprehending speech. Named Wernicke's area after a German physician, it is functionally similar to an area in monkey brains involved in the interpretation of species-specific calls.

Despite the separation of these speech areas, a genetic factor apparently unites them. A gene called FOXP2 is involved in regulating the development of neuronal circuitry across both

regions. Apes and humans share this gene – but humans have a mutated version that is thought to be critical for circuits needed for complex speech. This mutation has also been found to be present in the Neanderthal genome, suggesting that the common ancestor of Neanderthals and humans and may have had a modern, humanlike voice more than 300,000 years ago.

The cerebral cortex is divided into four lobes, each of which has a specific role

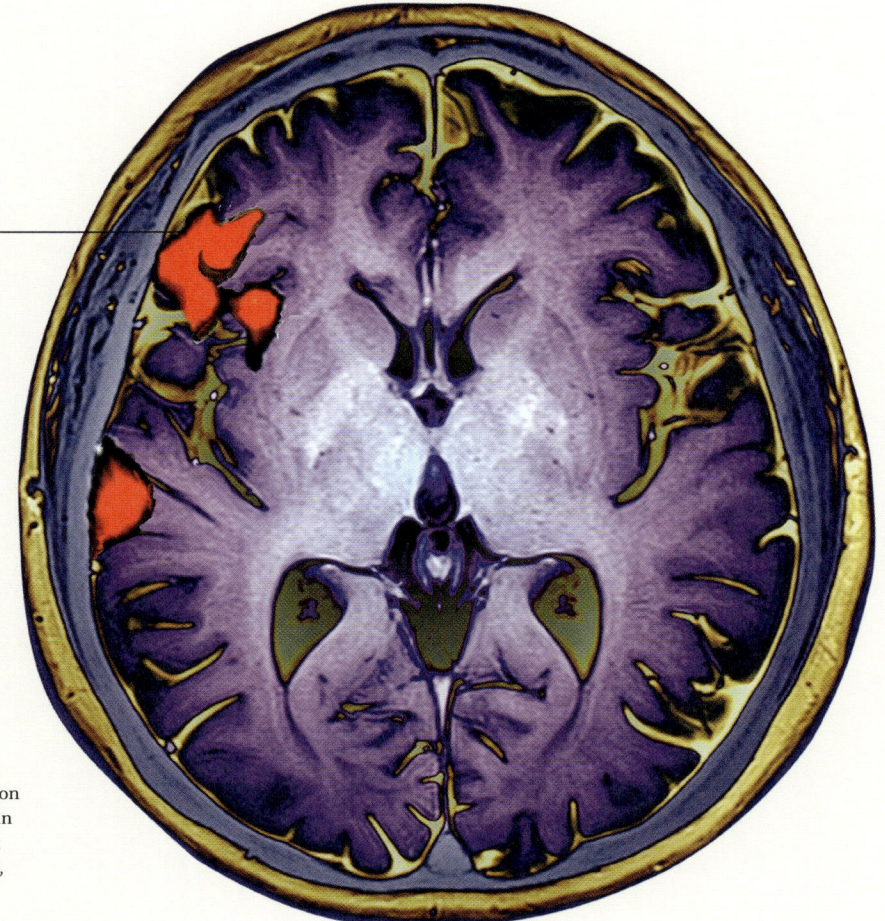

Broca's area is located in the left frontal lobe; it has a key role in language processing.

SPEECH CENTRES
This coloured scan of a section through the brain of a human test subject shows activation of Broca's area (red, top) and, Wernicke's region (red, centre) during conversation.

Speaking and understanding language are products of a big brain. Their control centres are located in the outer wall of the brain, the cerebral cortex.

Two important changes occurred in the evolution of our ancestors' cardiovascular and respiratory systems. One boosted our athletic performance, with a wider diet providing more energy. The other gave us a flexible way of making more sounds.

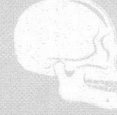

An upright posture, with the head balanced on top of the spine and a longer neck, lengthened the back of the throat.

Several features of the face, mouth, and throat freed up the tongue and larynx to make a greater range of sounds

Flattened face

Upright stance

Longer throat and lower voicebox

U-shaped hyoid bone

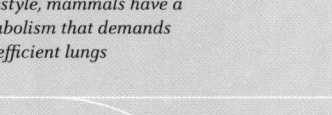

To maintain their body heat and active lifestyle, mammals have a high metabolism that demands the most efficient lungs

Lungs with alveoli

Fast metabolism

Diaphragm

Heart of the matter

Upgrades to the human respiratory and cardiovascular systems made us marathon specialists, with knock-on effects for nutrition and brains. Meanwhile, the upper respiratory tract took on a new role in complex communication.

Four-chambered heart

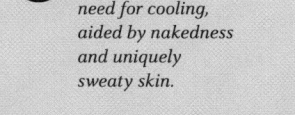

Increased activity lead to a great need for cooling, aided by nakedness and uniquely sweaty skin.

Cooling by sweating

Cooling by vasodilation

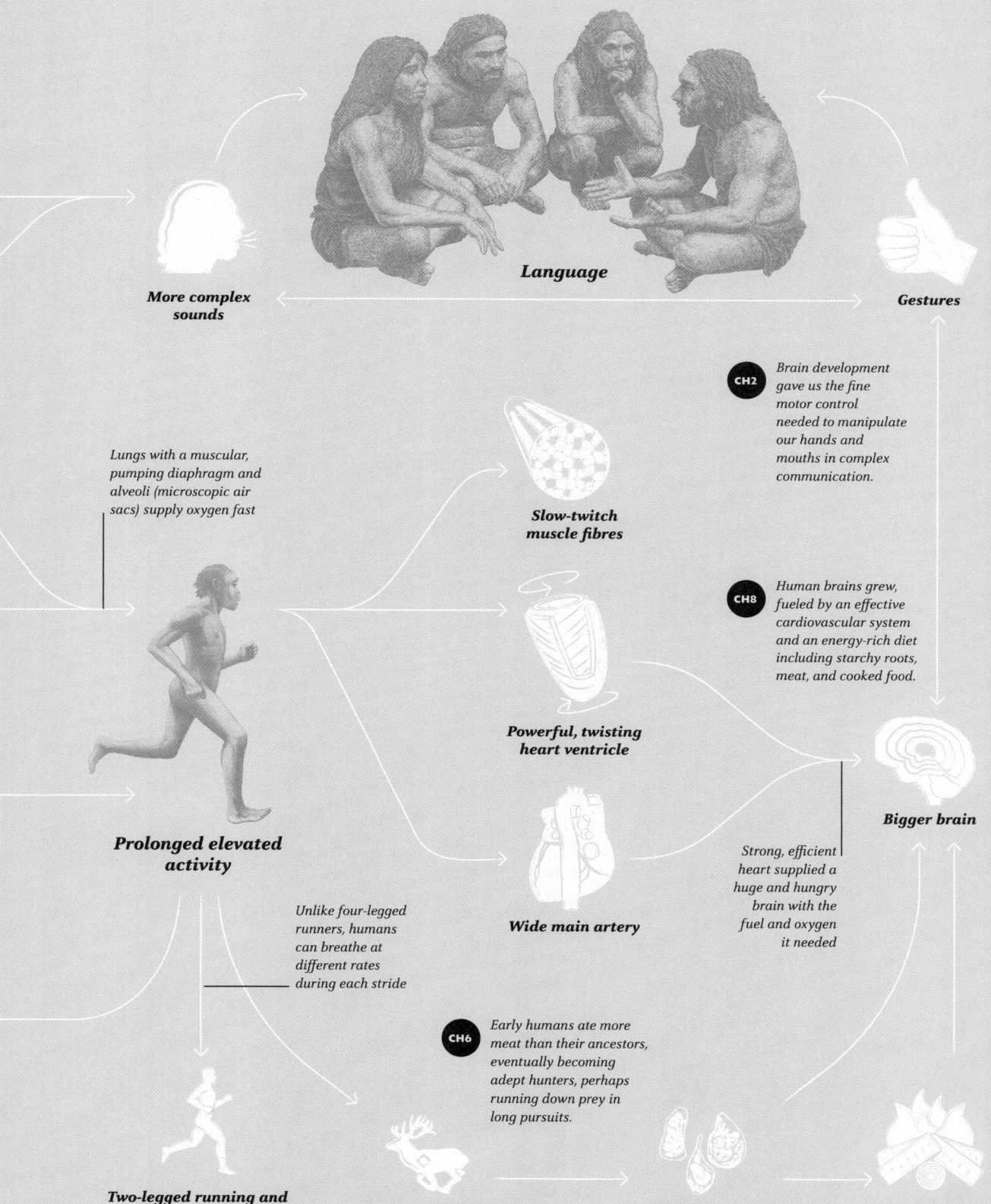

Language

More complex sounds

Gestures

Lungs with a muscular, pumping diaphragm and alveoli (microscopic air sacs) supply oxygen fast

Slow-twitch muscle fibres

CH2 Brain development gave us the fine motor control needed to manipulate our hands and mouths in complex communication.

CH8 Human brains grew, fueled by an effective cardiovascular system and an energy-rich diet including starchy roots, meat, and cooked food.

Powerful, twisting heart ventricle

Prolonged elevated activity

Bigger brain

Strong, efficient heart supplied a huge and hungry brain with the fuel and oxygen it needed

Unlike four-legged runners, humans can breathe at different rates during each stride

Wide main artery

CH6 Early humans ate more meat than their ancestors, eventually becoming adept hunters, perhaps running down prey in long pursuits.

Two-legged running and breathing decoupled

Hunting

More meat

Cooking

131

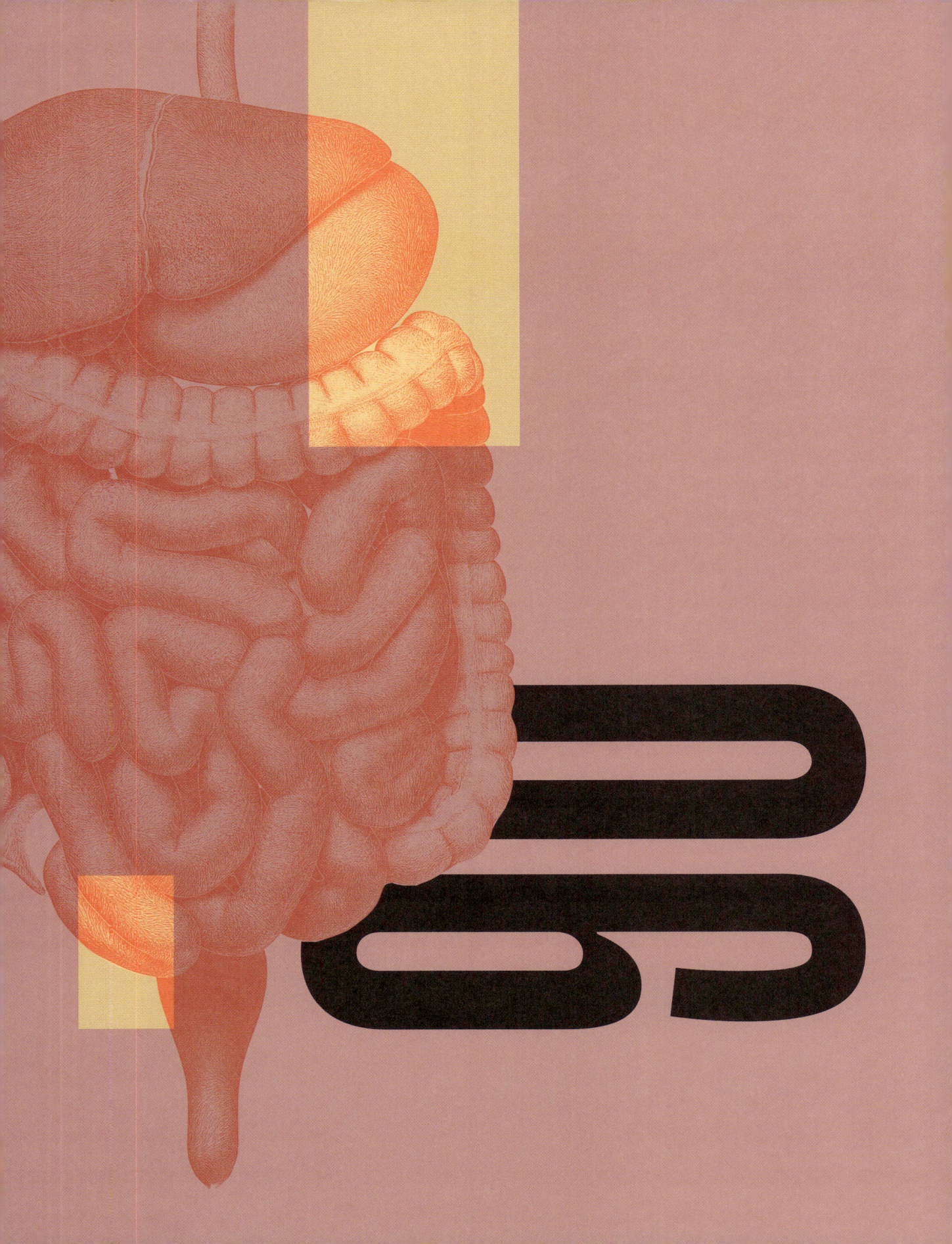

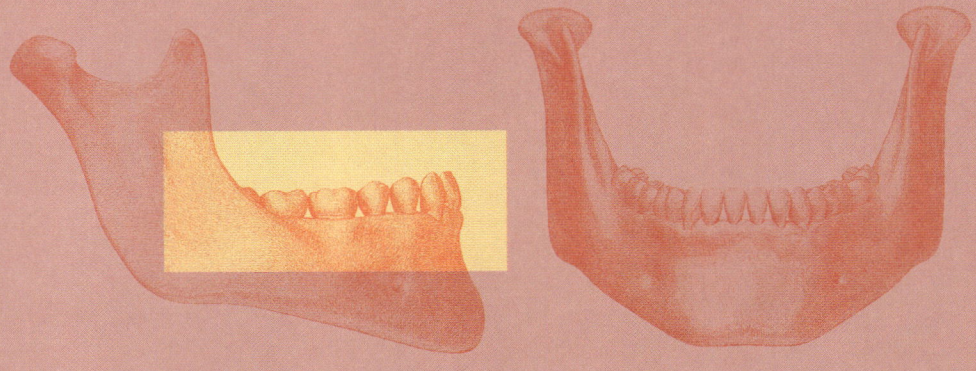

Mouth and gut

Humans have the broadest diet of any animal, although natural selection has given us modest teeth and an average gut. Our secret is our ingenuity in finding new resources, and processing and cooking those raw materials into digestible foods.

Everyone needs to eat. Over time, evolution has come up with ingenious ways to help animals digest their food. One of the first inventions was the gut. The earliest animals, including jellyfish, had a blind gut. This means they took in food and ejected waste through the same hole. However, most animals have a through gut: from their mouth, food passes through a series of tubes and chambers, before exiting through the anus. A through gut means the animal can put the food through a series of processes: it uses enzymes to digest the food into simple constituent nutrients such as glucose, it absorbs those nutrients, and recovers water.

DIGESTING BETTER AND FASTER

It helps to break up food before it even reaches the gut. That's why certain fish evolved teeth, and some toothed fish also evolved hinged jaws. Jaws allowed early fish to bite food – particularly useful with food that tries to get away.

Eating plants may seem easier, but plant tissue is full of indigestible cellulose. Plant-specialist animals therefore must evolve a partnership with cellulose-digesting microbes. These inhabit the gut, often in dedicated fermentation chambers, making herbivore guts complex, with twists, turns, and blind alleys.

Humans have some gut microbes, but we have also developed technologies for tenderizing food, like cutting tools and cooking fires. Our diet is unlike that of any other living animal.

Teeth, jaws, and gut bacteria are all innovations that have made digestion better

A billion years in the making

The human digestive system owes its features to a billion years of animal evolution, during which natural selection came up with ever-better ways of breaking up food and getting at its nutrients faster, and more completely.

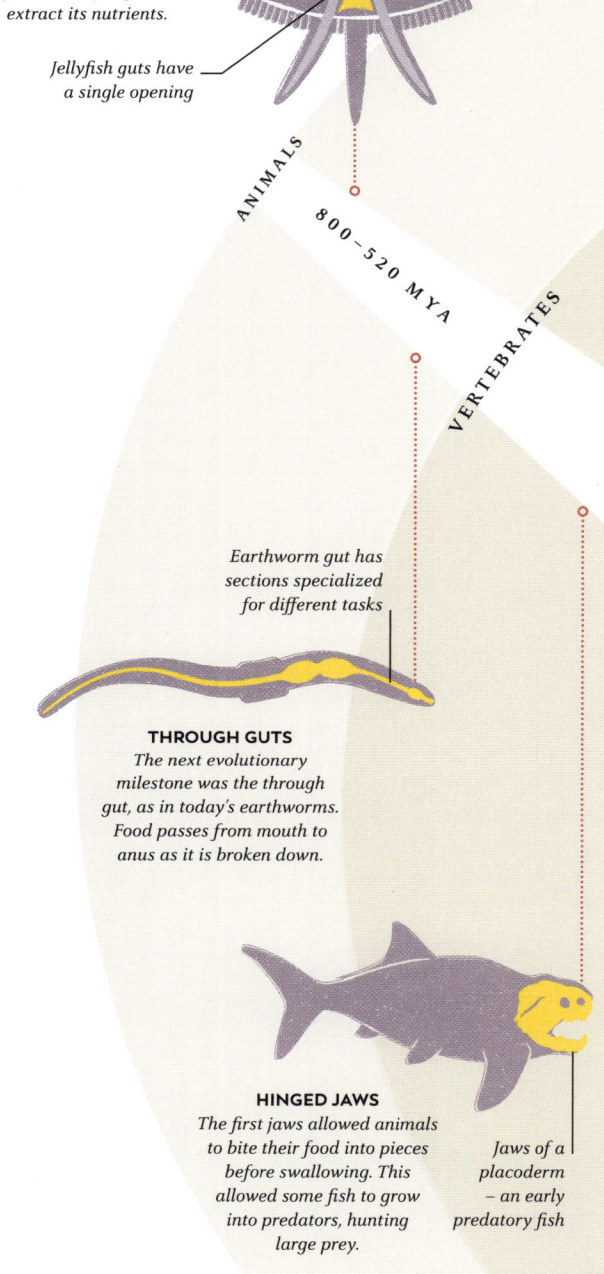

BLIND-ENDED GUTS
Early animals developed a cavity in their bodies where they held their food as they digested it to extract its nutrients.

Jellyfish guts have a single opening

ANIMALS

800–520 MYA

VERTEBRATES

Earthworm gut has sections specialized for different tasks

THROUGH GUTS
The next evolutionary milestone was the through gut, as in today's earthworms. Food passes from mouth to anus as it is broken down.

HINGED JAWS
The first jaws allowed animals to bite their food into pieces before swallowing. This allowed some fish to grow into predators, hunting large prey.

Jaws of a placoderm – an early predatory fish

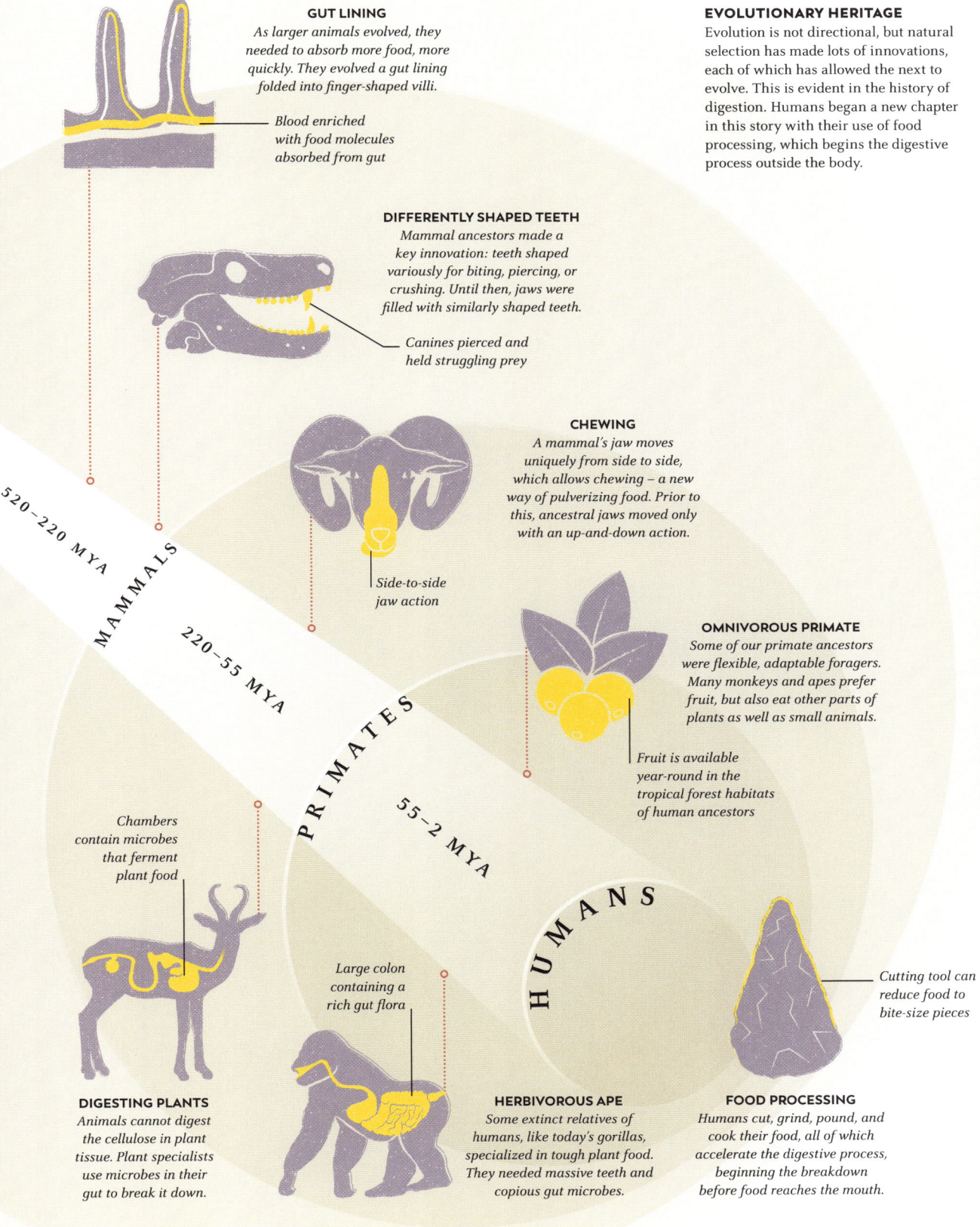

GUT LINING
As larger animals evolved, they needed to absorb more food, more quickly. They evolved a gut lining folded into finger-shaped villi.

Blood enriched with food molecules absorbed from gut

EVOLUTIONARY HERITAGE
Evolution is not directional, but natural selection has made lots of innovations, each of which has allowed the next to evolve. This is evident in the history of digestion. Humans began a new chapter in this story with their use of food processing, which begins the digestive process outside the body.

DIFFERENTLY SHAPED TEETH
Mammal ancestors made a key innovation: teeth shaped variously for biting, piercing, or crushing. Until then, jaws were filled with similarly shaped teeth.

Canines pierced and held struggling prey

CHEWING
A mammal's jaw moves uniquely from side to side, which allows chewing – a new way of pulverizing food. Prior to this, ancestral jaws moved only with an up-and-down action.

Side-to-side jaw action

520–220 MYA

MAMMALS

220–55 MYA

PRIMATES

55–2 MYA

HUMANS

OMNIVOROUS PRIMATE
Some of our primate ancestors were flexible, adaptable foragers. Many monkeys and apes prefer fruit, but also eat other parts of plants as well as small animals.

Fruit is available year-round in the tropical forest habitats of human ancestors

Chambers contain microbes that ferment plant food

Large colon containing a rich gut flora

Cutting tool can reduce food to bite-size pieces

DIGESTING PLANTS
Animals cannot digest the cellulose in plant tissue. Plant specialists use microbes in their gut to break it down.

HERBIVOROUS APE
Some extinct relatives of humans, like today's gorillas, specialized in tough plant food. They needed massive teeth and copious gut microbes.

FOOD PROCESSING
Humans cut, grind, pound, and cook their food, all of which accelerate the digestive process, beginning the breakdown before food reaches the mouth.

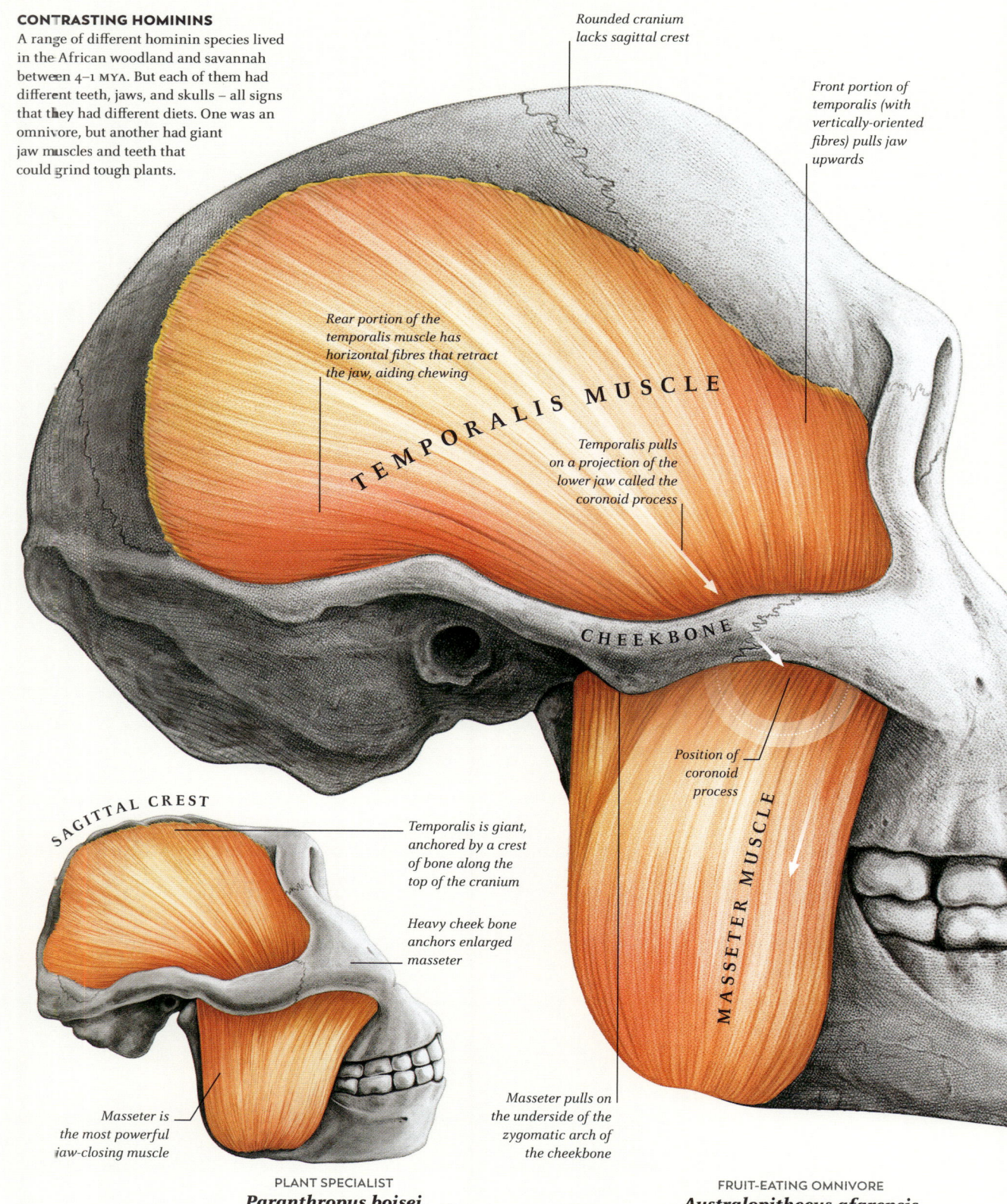

CONTRASTING HOMININS

A range of different hominin species lived in the African woodland and savannah between 4–1 MYA. But each of them had different teeth, jaws, and skulls – all signs that they had different diets. One was an omnivore, but another had giant jaw muscles and teeth that could grind tough plants.

Rounded cranium lacks sagittal crest

Front portion of temporalis (with vertically-oriented fibres) pulls jaw upwards

Rear portion of the temporalis muscle has horizontal fibres that retract the jaw, aiding chewing

TEMPORALIS MUSCLE

Temporalis pulls on a projection of the lower jaw called the coronoid process

CHEEKBONE

Position of coronoid process

MASSETER MUSCLE

SAGITTAL CREST

Temporalis is giant, anchored by a crest of bone along the top of the cranium

Heavy cheek bone anchors enlarged masseter

Masseter is the most powerful jaw-closing muscle

Masseter pulls on the underside of the zygomatic arch of the cheekbone

PLANT SPECIALIST
Paranthropus boisei

FRUIT-EATING OMNIVORE
Australopithecus afarensis

By 3 million years ago, humanlike apes had diversified into several species specializing in different diets and lifestyles. Those that would become our ancestors, however, were unspecialized, eating a wide diet.

All-devouring apes

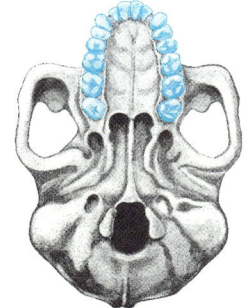

Australopithecus afarensis

Massive cheek teeth needed to grind down tough plants.

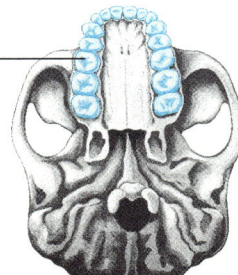

Paranthropus boisei

Paranthropus, boisei, misnamed "Nutcracker Man", was a specialist herbivore that could survive on grass if needed

Hominins ate many different diets, and their bodies evolved to match. Some were equipped to eat tough plants such as grass. Others had more varied diets, enjoying everything from fruit and vegetables to insects.

Paranthropus boisei was a hominin species with a specialized diet. It lived in east Africa, 2.3–1.2 million years ago (MYA), meaning it emerged at around the same time as our genus, *Homo*, but long before our own species evolved. The first fossil discovered was a skull, found in Tanzania in 1959. It had big back teeth, far bigger than any person alive today. This earned the species the nickname "Nutcracker Man".

We now suspect that *Paranthropus boisei* didn't eat hard foods like nuts, at least not often. Instead, those big teeth were used for grinding up tough plants, such as grasses and sedges. To do this, *Paranthropus boisei* had to evolve large

SPECIALIST TEETH

Among upright ape species, *Paranthropus* tended to be specialized in eating the toughest of fibrous plants and had large molars, while *Australopithecus* had a more general diet and modest cheek teeth.

and powerful chewing muscles. As a result, it had massive cheekbones – giving it a flat, dish-shaped face.

It probably ate softer and tastier foods as well, such as fruit and termites. However, being able to eat grasses would have helped it survive lean periods, when more nutritious foods were not available. That helps explain why it survived for over a million years – far longer than our species has managed so far.

ADAPTABLE OMNIVORE

Australopithecus afarensis was a very different kind of hominin. It also lived in east Africa, but earlier: about 3.9–2.9 MYA. And unlike *Paranthropus*, it ate a varied diet: that is, it was an omnivore. We can't be sure exactly what it ate. We can draw inferences from its teeth and jaws, which compared to *P. boisei*, were more modest in size, in terms of both dentition and chewing muscles, so we suspect it wasn't eating so much chewy food. As a result, *Au. afarensis* cheeks were smaller, giving it a more rounded face.

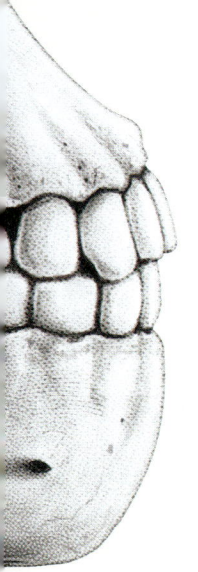

Australopithecus afarensis was a flexible hominin that ate a wide variety of foods

Apart from the shape and size of jaws and teeth, there are other clues to prehistoric diets. Scientists sample the fossil dental enamel, which, when being laid down in early life, preserves the ratio of carbon isotopes in food eaten. This gives us insight into the plant types in the diet, as different plants handle carbon in contrasting ways – forest trees and herbs contain a different ratio of carbon isotopes to the grasses and sedges of savannah and succulents in arid habitats. Nitrogen isotopes, meanwhile, indirectly tell us how much animal matter is in the diet.

EXPANDING ONTO THE SAVANNAH

Isotope studies of hominins suggest a widening of diets around 4–3 MYA, adding savannah plants. Forest plants remained on the menu, as they had been for earlier hominins, such as *Ardipithecus*, but *Au. afarensis* ate a greater variety, possibly due to living in a mosaic habitat of woodland and patches of grassland. Its diet of plant material would have included leaves and fruit from the forest and – on the savannah – seeds, and the underground starchy food stores of certain plants, including tubers and corms. Like modern chimpanzees, they probably also ate insects such as termites. They may also have eaten meat from large animals such as antelope, but nitrogen isotope data tells us it wasn't a major part of their diet. Grasslands had begun expanding at this time, and it seems the response of *Au. afarensis* was to become more adaptable. As the grasslands spread further after 2.6 MYA, however, its relative *Paranthropus boisei* seems

to have become a grassland specialist. Its carbon isotope profile is similar to modern grass eaters such as gelada baboons. In these ways, *Au. afarensis* was more similar to us and it is possible that it was our distant ancestor, while *P. boisei* was our more distant cousin.

THE END OF THE SNOUT

Our closest living relatives, the chimpanzees and bonobos, have faces similar to ours and yet different. That's the result of millions of years of gradual change – and some of the biggest changes have been to our mouths and jaws.

One obvious change is that humans have small teeth compared to apes. That's especially true of the pointed canine teeth, which have shrunk dramatically, and of the blocky molar teeth at the backs of our mouths. Early hominins such as *Australopithecus* still had relatively large, apelike teeth, but over the generations, in the lineage leading through *Homo* species to humans, hominins' teeth tended to shrink. That's probably because of eating fewer leaves and more fruit, but there were bigger changes to

DENTAL ARCHES

Today's nonhuman apes have U-shaped dental arches, matching their protruding jaws, and large canines. As the human face shortened, its dental arch developed a parabolic curve. *Australopithecus* shows an intermediate shape.

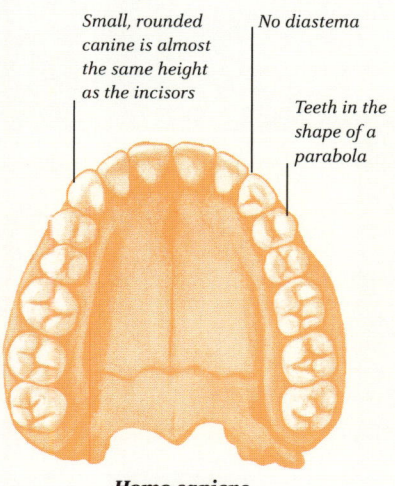

Small, rounded canine is almost the same height as the incisors

No diastema

Teeth in the shape of a parabola

Homo sapiens

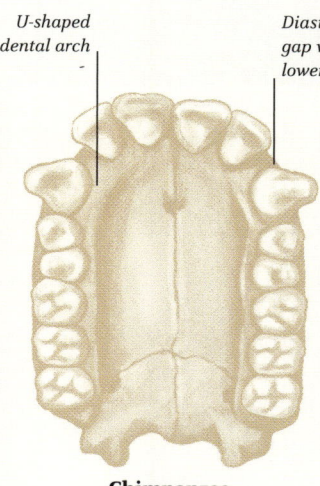

U-shaped dental arch

Chimpanzee

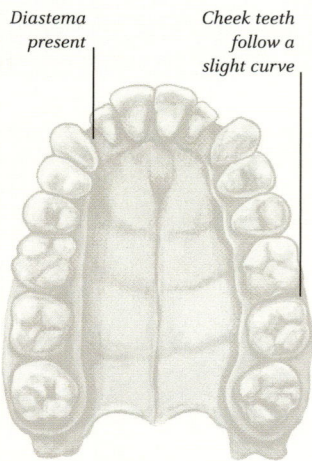

Diastema – the gap where the lower canine fits

Diastema present

Cheek teeth follow a slight curve

Australopithecus afarensis

SHORTENING FACE

Most apes have a prognathous face, meaning the jaw protrudes. *Australopithecus* retained this tendency, but it dwindled in the human lineage to a completely flat face in *Homo sapiens*.

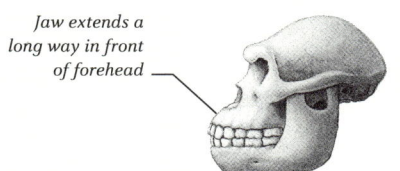

Jaw extends a long way in front of forehead

Australopithecus afarensis

come. Early members of the *Homo* genus began incorporating more meat into their diet (see p.142) and they used stone tools to process their food (see p.142). Later, hominins also worked out how to control fire (see p.148) and started cooking their food, making it easier to chew. Each of these steps reduced the pressure on the teeth and jaws to process our food.

Perhaps hominins also didn't need to bite their enemies quite so often, or bare their teeth in aggression. Canine teeth are good for threatening rivals and defending territories, but our ancestors also gradually evolved less aggression in their social behaviour (see p.186). So while today's nonhuman apes need a space in their dental arch, called a diastema, to slot in their opposing canines, humans have entirely lost this. As our teeth evolved to be smaller, so our jaws became shorter. A chimpanzee's jaw thrusts forward several centimetres ahead of their nose and eyes, but human jaws sit beneath their forehead.

Anyone who has worn orthodontic braces has experienced the consequences of the way our faces have evolved. Because our jaws have

In human evolution, teeth have tended to get smaller, especially canines

Smaller jaw, significantly less prognathous

Homo erectus

Middle of face still markedly prognathous

Homo neanderthalensis

Rounded cranium could be linked developmentally to the short jaw

Jaw extremely shortened

Homo sapiens

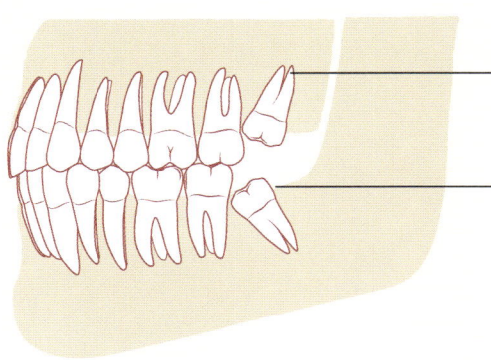

Wisdom tooth is tilted due to lack of space in jaw

Like the upper wisdom tooth, this lower one is impacted and needs dental intervention

NO ROOM FOR OUR TEETH

Many of today's humans still develop their third molars, or wisdom teeth – two in the upper jaw and two in the lower. However, they often don't have room in their jaws for them to develop properly, leading to dental problems.

Wisdom tooth problems

139

shortened, we don't always have room to fit in all our teeth, so they end up growing wonky or becoming damaged. In modern humans, rear molars, called "wisdom teeth" (see p.139), often don't fit properly and have to be taken out. It is not only evolution to blame, though. The problem has been exacerbated by the popularity of puréed baby foods. If, early in life, we do not exercise our jaws by gnawing on solids, we increase the likelihood of an underdeveloped jaw and wisdom tooth problems later in life.

Our jaws are now so short, they cannot comfortably accommodate all of our teeth

FLEXIBLE GUTS

If there's one thing that marks out the human digestive system, it is its adaptability. We can eat a huge variety of foods. Meat, insects, fish, fruit, vegetables, mushrooms: our guts can deal with all of it. Potatoes and tomatoes are native to the Americas, so no European, African, or Asian saw one until about 500 years ago. Chips and ketchup are recent inventions – but we all have got used to them. Our guts have evolved this way because variety has long been a feature of our diet. If you compare a human and

LARGE SMALL INTESTINE

The human small intestine, despite its name, is the largest part of our gut. In contrast, a chimp's color, which is full of cellulose-digesting microbes, is far larger than ours, both absolutely, and in proportion to the rest of its gut.

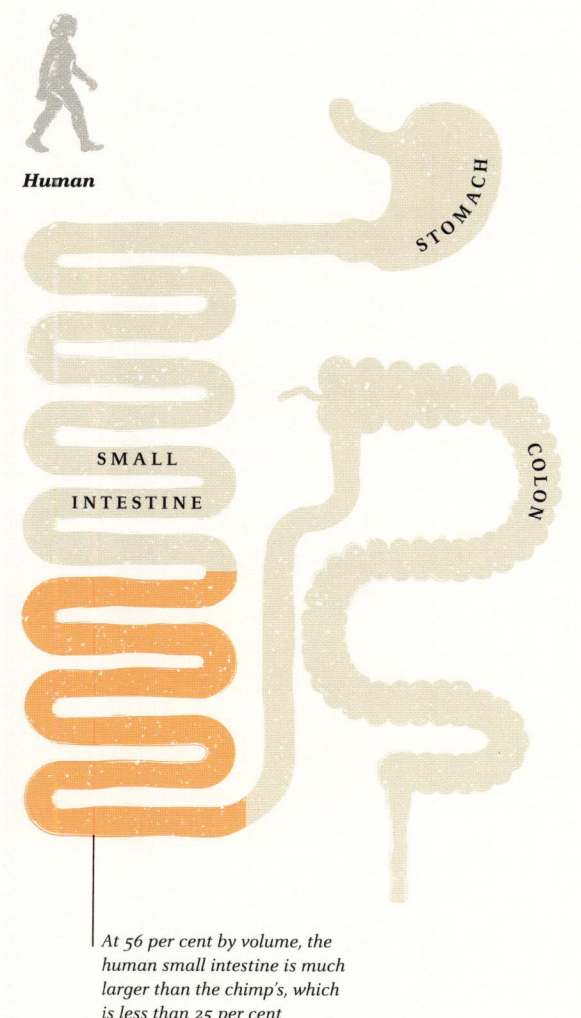

Human

STOMACH

SMALL INTESTINE

COLON

At 56 per cent by volume, the human small intestine is much larger than the chimp's, which is less than 25 per cent

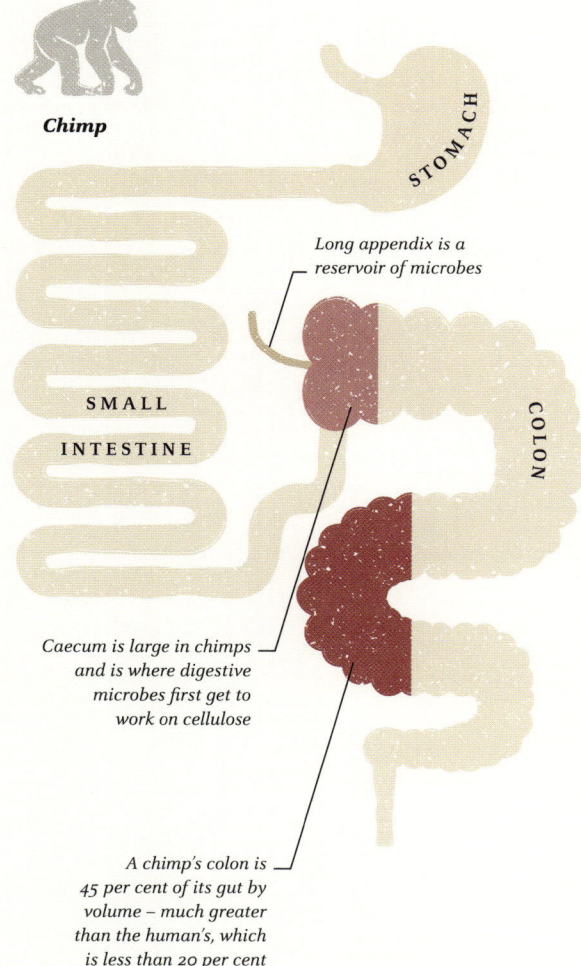

Chimp

STOMACH

Long appendix is a reservoir of microbes

SMALL INTESTINE

COLON

Caecum is large in chimps and is where digestive microbes first get to work on cellulose

A chimp's colon is 45 per cent of its gut by volume – much greater than the human's, which is less than 20 per cent

HOW HOMININ DIETS EVOLVED

Beginning as generic woodland apes, eating a variety of fruit, leaves, and possibly occasional animals, hominins changed to become more adaptable in a patchwork environment of forests and grasslands.

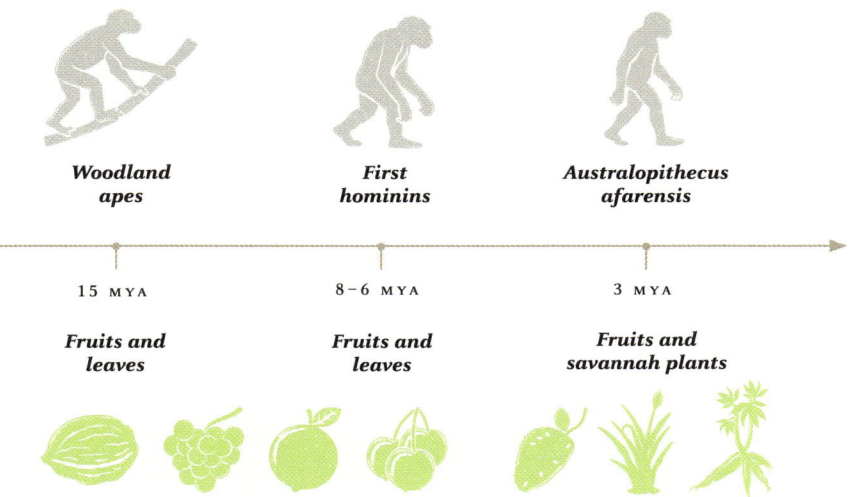

Woodland apes	First hominins	Australopithecus afarensis
15 MYA	8–6 MYA	3 MYA
Fruits and leaves	Fruits and leaves	Fruits and savannah plants

chimpanzee gut, there are obvious similarities but also stark differences. The overall structure is the same, reflecting our shared evolutionary heritage. First a modest stomach containing strong acid, then a small intestine, then another tube called the caecum, and finally another tube, the colon, which leads to the anus. But they're all different sizes. In humans, the small intestine makes up more than half the volume, and much more than half the length, of the gut. But in chimps, the colon is the biggest part, both in terms of volume and surface area. Chimps need a large colon because they eat so many leaves, which are indigestible. To break them down, chimps rely on millions of single-celled microorganisms living in their colon. Humans don't eat many leaves, so we can do without this. Instead we eat a lot of easily-digestible foods, which the small intestine handles.

The human colon is smaller than that of other apes

If there's one thing that marks out the human digestive system, it is adaptability.

All of these processes begin the digestive process so that our guts don't have to take the whole strain. And because we don't eat so many fibrous plants, we have no need of the large fermenting colon of other apes.

PROCESSED FOOD

Chimps are, in fact, quite typical of today's great apes, and it is humans that are unusual. Our gut size and proportions are in such contrast probably because of our higher quality diet. By the time our food reaches our gut, we have selected the most nutritious foods, cut them up with sharp tools, or ground, or pounded, or cooked them to tenderness.

EVOLUTIONARY JOURNEY

Humans are relatively unremarkable in terms of their teeth and guts, but that is because we are unspecialized. We seem to have descended first from forest apes that ate a variety of plant foods, followed by upright apes, such as *Australopithecus afarensis*, that took a flexible approach when their forest home became a mosaic of woodland and grassland.

A taste for meat

Early hominins were probably omnivorous, but some found ways of acquiring a new food full of easily digestible calories. These meat eaters eventually became technologically advanced top predators.

The record of early humans hunting animals does not stretch back millions of years, partly because hunting tools, such as spears, would have been made of perishable wood. Isolated evidence includes a single 400,000-year-old wooden spear tip preserved in England and an unmistakable spear wound in a 500,000-year-old horse's shoulderblade from another English site. But hominins started eating meat long before this.

It's thought that our ancestors more than 3 million years ago were omnivorous (see p.137). Most apes today eat a mainly plant-based diet, but also eat insects, eggs, and small animals when available. But meat would become a more important component of some hominin diets. The earliest evidence is marks in bones made by ancient butchers: cut marks suggest flesh removal, and impact marks suggest the breaking of bones to extract marrow. Butchered bones of antelopes and relatives of horses were found at Bouri, Ethiopia, dated to 2.58 MYA and those of hippopotamuses were unearthed at Nyayanga, Kenya, which could be up to 3.03 million years old. These dates correspond to age of the earliest Oldowan tools (see pp.36–37), found at Gona, near Bouri, and at Nyayanga itself. Who were these butchers? Several species of hominin could be implicated – not just early *Homo*, but also *Australopithecus garhi*, whose remains were found at Bouri, and

Butchery marks on bones tell us that hominins ate meat more than 2 million years ago

EARTH'S FIRST BUTCHERY
These marks match those made by stone tools slicing flesh off the bone and are by far the oldest known. Found in Dikika, Ethiopia, they are disputed, but if they are genuine butchery marks, they must have been made by a pre-*Homo* species, such as *Australopithecus*.

HOW HUMANS FORAGED FOR MEAT
Our relationship with meat has undergone some major transformations as humans gradually changed from scavengers to top predators.

HOMO HABILIS
Cutting tools gave hominins access to scavenging the meat of large animals.

HOMO ERECTUS
Technology and teamwork allowed early humans to hunt the largest animals.

HOMO NEANDERTHALENSIS
Neanderthals used projectile weapons to kill fast prey as the largest mammals declined.

HOMO SAPIENS
Humans, now equipped with bows and possibly dogs, hunted a broad range of quarry.

| 2,600,000 | 1,000,000 | 200,000 | 30,000 |

Years ago

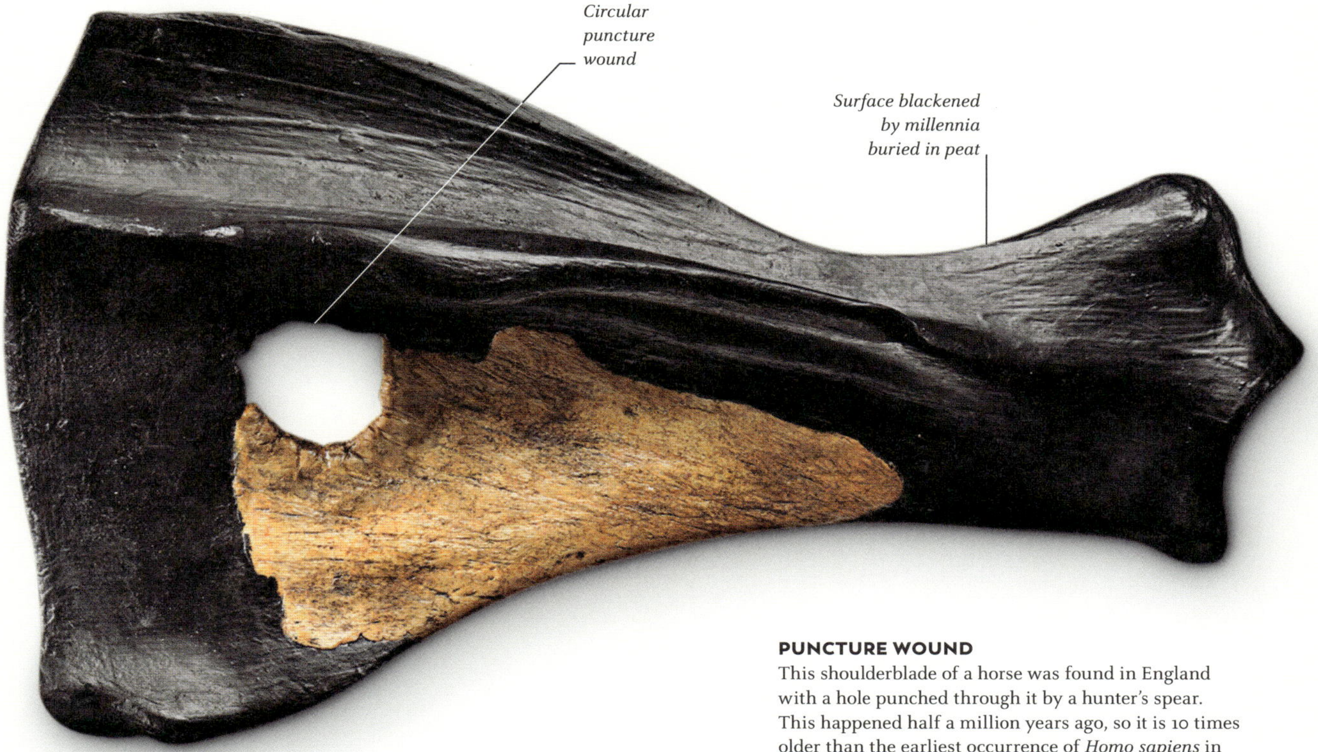

*Circular
puncture
wound*

*Surface blackened
by millennia
buried in peat*

PUNCTURE WOUND

This shoulderblade of a horse was found in England
with a hole punched through it by a hunter's spear.
This happened half a million years ago, so it is 10 times
older than the earliest occurrence of *Homo sapiens* in
the region. No remains of the hunter have been found.

OLDEST SPEARS

Wood does not usually endure in the
archaeological record for longer than a
few centuries. It is incredible, then, that
archaeologists have discovered wooden
spears in Schöningen, Germany, that have
recently been dated to 200,000 years old.

*Sharpened tip made
by splitting wood
chips from the shaft*

*Shaft made of
spruce or pine*

A TASTE FOR MEAT

possibly *Paranthropus*. An even earlier site of butchered bones, Dikika, is far older than *Homo*, at 3.39 MYA. This corresponds roughly in age to the oldest known flaked tools (see p.35).

SCAVENGERS TO HUNTERS

By 2.5 MYA, hominins were already doing something modern apes rarely, if ever, do: scavenging meat. This probably began furtively, but some hominins may have become more confrontational scavengers, perhaps even challenging formidable competitors, such as hyenas and big cats. This may have led to the evolution of the larger size and swifter running abilities of *Homo erectus* (often known as *H. ergaster* in Africa). This early human seems to have developed the persistence hunting that honed its hips, legs, heart, lungs, and cooling adaptations of the skin to become an endurance athlete capable of running prey to heat exhaustion. By the end of such a hunt, the prey was unable to defend itself from a cooperative team of early humans armed with stones and clubs – spears and arrows may not have been necessary.

As our ancestors challenged other scavengers and predators, they learned to be predators themselves

Exhausted prey could not defend itself from a team of armed early humans.

SUSTAINING THE BRAIN

From 2.0–1.5 MYA, there is growing evidence of eating meat, from sites such as Kanjera and Koobi Fora in Kenya and Oldupai Gorge in Tanzania. It is easy to focus here on meat-eating, but in fact, incorporating more animal protein was part of a broadening of the diet, and plants were still very important as foods. Cooking

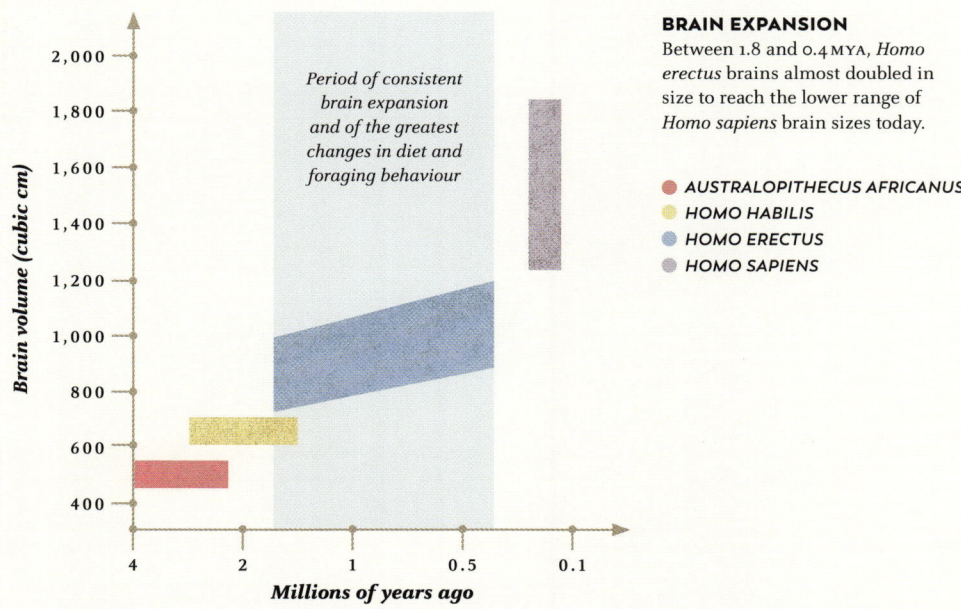

Period of consistent brain expansion and of the greatest changes in diet and foraging behaviour

Brain volume (cubic cm)

2,000
1,800
1,600
1,400
1,200
1,000
800
600
400

4   2   1   0.5   0.1

Millions of years ago

BRAIN EXPANSION

Between 1.8 and 0.4 MYA, *Homo erectus* brains almost doubled in size to reach the lower range of *Homo sapiens* brain sizes today.

● *AUSTRALOPITHECUS AFRICANUS*
● *HOMO HABILIS*
● *HOMO ERECTUS*
● *HOMO SAPIENS*

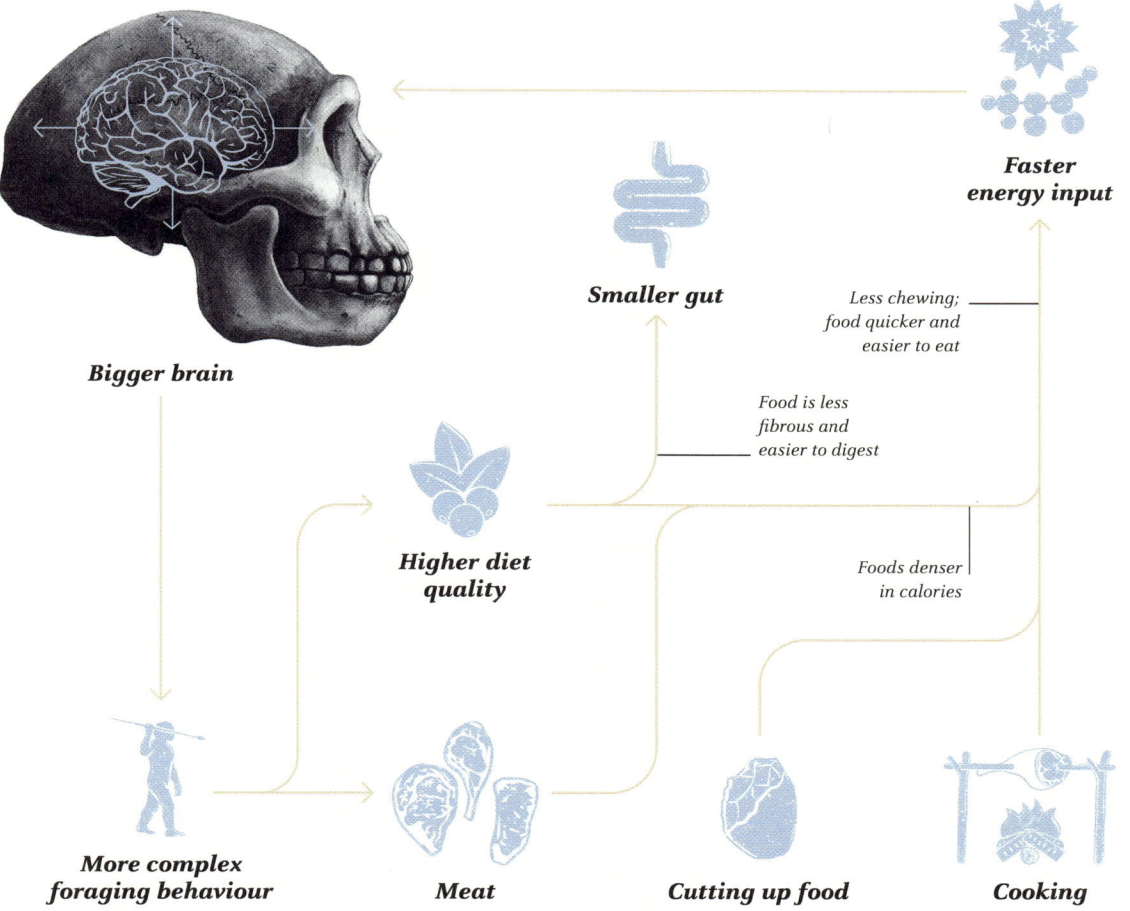

Bigger brain

Smaller gut

Faster energy input

Less chewing; food quicker and easier to eat

Food is less fibrous and easier to digest

Higher diet quality

Foods denser in calories

More complex foraging behaviour

Meat

Cutting up food

Cooking

technology developed between 1.5 and 0.8 MYA (see pp.148–50), and this made plants more digestible. The first food processing – simply cutting up food into bite-size morsels, or pounding it, made eating quicker and more efficient. All of these factors contributed to a higher calorie intake, which may have provided the energy for higher physical activity, increased reproduction – and perhaps, it has been suggested, for growing larger brains.

It's true that brains are energetically expensive: brain tissue consumes 10 times the energy, per unit mass, of other body tissue, but our metabolic rate is the same as any mammal of our size. The "Expensive Tissue Hypothesis" suggests that brain growth was possible because of a reduction in other energy-hungry tissue – in the guts. Humans don't carry bulky large intestines that break down plants as do gorillas and chimps (see p.140), but overall, our guts are the size expected for a fruit eater. In fact, it is low-cost tissue that balances the energy equation. We have larger fat deposits than other apes (an evolutionary buffer against starvation) and fat tissue consumes very little energy to maintain. Take it away, and our remaining metabolic rate per unit body weight is actually rather high, and explains how we fuel our brain.

Brain size was thought to be traded-off against gut size, but evidence does not support this

BRAIN FOOD

Foraging and hunting can be seen as part of an evolutionary feedback loop involving better food and quicker digestion, more energy delivered to the brain, and better cognitive abilities further improving foraging and diet.

A TASTE FOR MEAT

HOW TO USE AN ATLATL

An atlatl, or spear-thrower, uses leverage to amplify throwing speed. The spear is secured by a hook at one end. This end travels much faster than the fingertips, and its rapid motion is transferred to the spear as it is released.

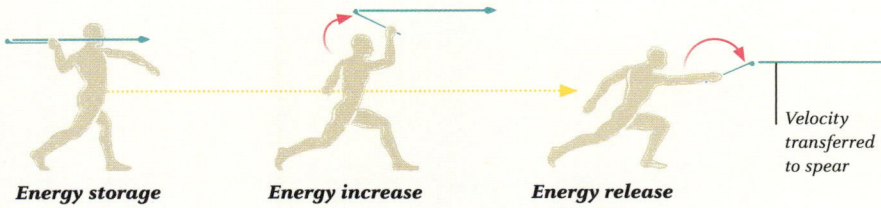

Energy storage

Energy increase

Energy release

Velocity transferred to spear

Sharp point could penetrate fast and dangerous prey at a safe distance

PROJECTILE POINT

This is not a hand tool, but a projectile point – the tip of a throwing spear – once hafted to a shaft by glue, twine, or both. This one is from Omo Kibish, Ethiopia, and is 104,000 years old.

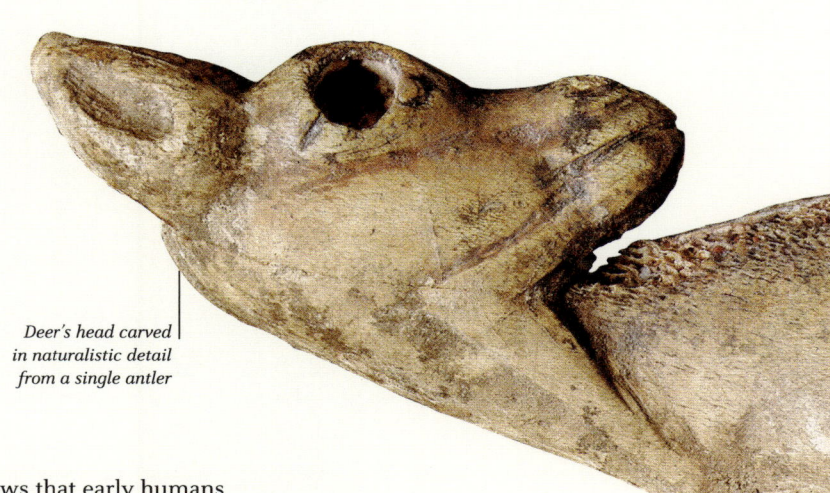

Deer's head carved in naturalistic detail from a single antler

PROJECTILES

From half a million years ago, evidence grows that early humans had developed specially crafted projectiles, such as spears, that could be thrown at fast and dangerous prey. Although few shafts survive, experts can identify stone spearheads without them, since many points bear the distinctive scars of high-velocity, point-first impacts. This is also evidence that the spears were thrown, rather than thrusted, and that their owners had found ways of lashing stone tools to shafts. Some of the oldest known are 500,000 years old and were found in Kathu Pan, South Africa.

Spear points can be identified by characteristic high-impact damage

Later, spear-throwers were developed that could propel spears or darts faster and further. The oldest remains of these are 30,000 years old and belonged to the *Homo sapiens* of Europe, but there were variations, such as the woomera in Australia and the atlatl in the Americas. One 45,000-year-old skeleton from Lake Mungo, Australia, had severe osteoarthritis in one elbow – a sign, perhaps, of a lifetime of throwing. With bows and arrows also added to human toolkits some time after 80,000 years ago, human hunters were well-equipped to become top predators wherever they went.

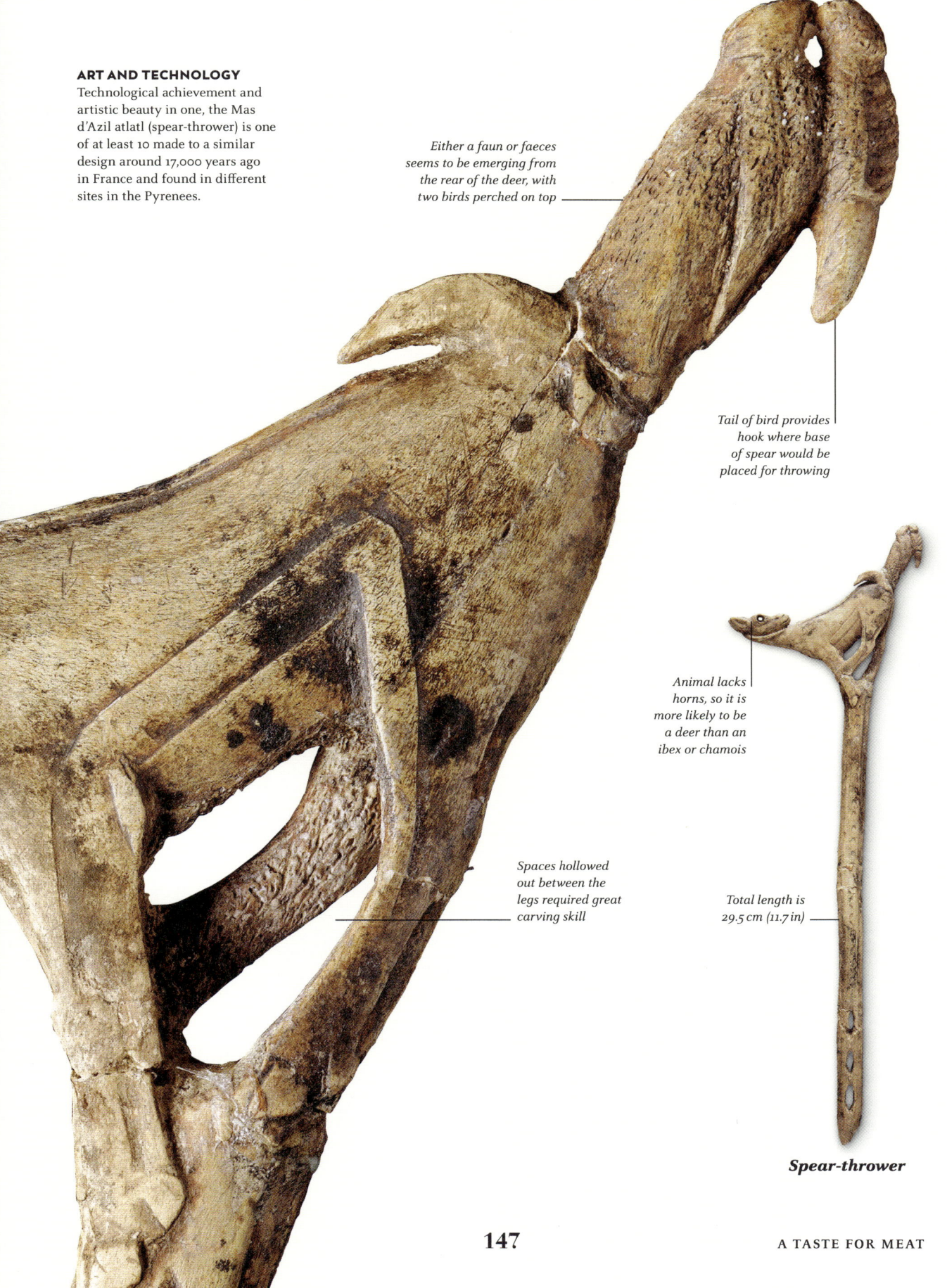

ART AND TECHNOLOGY
Technological achievement and
artistic beauty in one, the Mas
d'Azil atlatl (spear-thrower) is one
of at least 10 made to a similar
design around 17,000 years ago
in France and found in different
sites in the Pyrenees.

*Either a faun or faeces
seems to be emerging from
the rear of the deer, with
two birds perched on top*

*Tail of bird provides
hook where base
of spear would be
placed for throwing*

*Animal lacks
horns, so it is
more likely to be
a deer than an
ibex or chamois*

*Spaces hollowed
out between the
legs required great
carving skill*

*Total length is
29.5 cm (11.7 in)*

Spear-thrower

A TASTE FOR MEAT

Cooking and the control of fire

When human hunter-gatherers first used fire to cook their food, they hit upon something very unusual indeed in the animal kingdom. Cooked food tasted better, was easier to digest, and safer to eat. And cooking had consequences for further human evolution, too.

Some evidence suggests that early humans turned to cooking their food not long after the emergence of *Homo erectus*. Sites in Africa contain fossils of early humans alongside sediment containing ash made from burnt plants and bones dating from around 1.5 million years ago (MYA). At sites in Europe from 400,000 years ago, this type of ash was so abundant that experts concluded that cooking had become routine by that time.

Cooking required more than just hunting and foraging skills. It involved the deliberate control of fire. Perhaps the only other example of this in nature is that of Australian black kites and brown falcons, also known as "firehawks", which spread wildfires by transporting lighted sticks. But with early humans, possessing both the brains to problem-solve and the manual dexterity to manipulate tools, control of fire reached another level.

FOLLOWING FIRE

The African landscape inhabited by early humans was shaped by a drying climate, with forests transforming into bushland. Increasingly common wildfires were fanned by winds over exposed savannahs, just like in fire-prone parts of Australia today. Early humans would have witnessed fire frequently, observing its power as well as its potential use – fire flushed out prey, or cleared forests to make way for the spread of food plants. Fire also generated light and warmth and deterred predators.

Humans probably learned that wildfires could be controlled and maintained long before they created their own. Some fuels, such as animal dung, could burn slowly enough that fire could be kept alight and transported. Over time, humans learned how to ignite their own fires. Early evidence includes pieces of iron pyrite, tools, and burnt clay at a 400,000-year-old site in England, suggesting a method of fire-starting still used to this day. Archaeologists studying a similar, younger, collection of tools and iron pyrite in France decided to recreate the process. When they did so, the wear created on their flint tools when they struck them against pyrite matched that on the ancient tools.

This replica was made using Palaeolithic glue-manufacture techniques

STICKY ARROW

Natural adhesives such as pitch were made by heating bark from certain trees, in this case birch. The resultant tarlike substance could be used to make compound tools, for example, sealing arrow barbs to wooden shafts.

Heating birch bark produced tar, which early humans used as an adhesive

When controlled, fire could be used to fashion better tools – heat-treated stone and wood produce sharper flints and sticks. Fire could also be used to prepare adhesives, such as sticky pitch, which was made from heated birch bark.

COOKING CHEMISTRY

It is uncertain why humans began cooking their food. However, experts suggest that early humans had to witness the effect of burnt or heated foodstuffs before they adopted it as part of their lifestyle. Foods would have been unintentionally "cooked" in wildfires triggered by lightning strikes or other natural phenomena, leaving charred seeds and animals in their wake. Human foragers may have taken advantage of this bounty, noticing how cooked food tasted better and was softer and easier to eat. Eventually, it may have become clear that cooked food was also safer – the sanitizing effect of fire resulting in fewer bouts of illness after eating certain foods.

NEW FLAVOURS

Cooking drives chemical reactions in food that generate new flavours – sugars turn to caramel and react with amino acids in proteins to create more savoury flavours. Tasting delicious foods stimulates pleasure centres in the brain, which would have encouraged early humans to recreate these flavours by cooking deliberately. Fire softens food too. Heat breaks down plant pectin, the gel that binds plant cells together, which causes the cells to separate and softens tough, fibrous vegetables. At the same time, solid grains of starch, such as wheat, swell when heated in water, weakening and absorbing liquid, which turns them into a gel form that is

Striker is an old, worn flint tool

Flint impact breaks off fragments of iron pyrite (a form of iron sulfide)

Dried fungal tinder ignites easily and burns slowly

Sulfur particles oxidize quickly in air, getting hot enough to form glowing sparks

IGNITION

Humans could ignite fire by striking a stone tool against iron pyrite to produce sparks. When these sparks come into contact with a dry tinder material, the tinder could set alight and build into a flame.

COOKING AND THE CONTROL OF FIRE

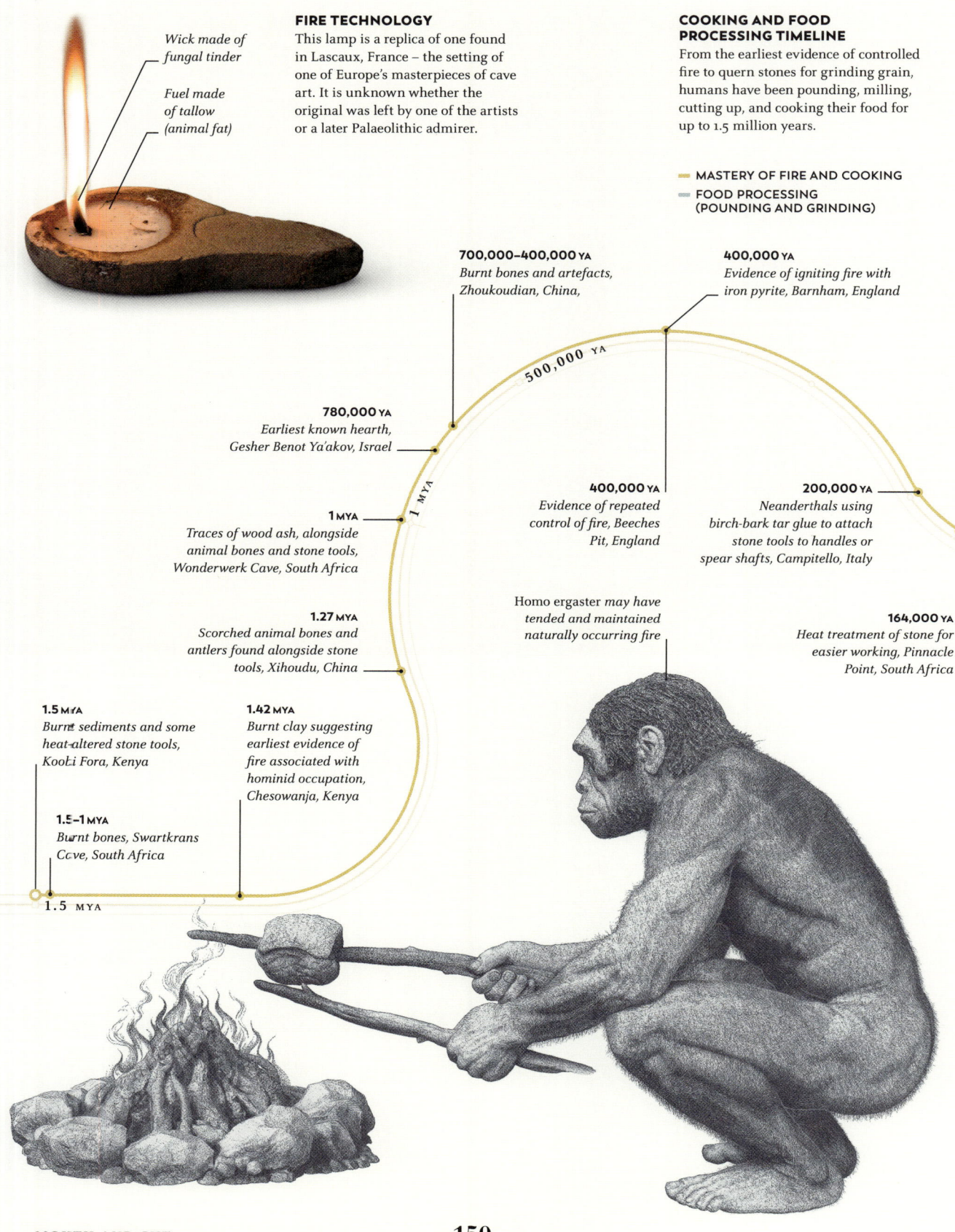

FIRE TECHNOLOGY

Wick made of
fungal tinder

Fuel made
of tallow
(animal fat)

This lamp is a replica of one found
in Lascaux, France – the setting of
one of Europe's masterpieces of cave
art. It is unknown whether the
original was left by one of the artists
or a later Palaeolithic admirer.

COOKING AND FOOD PROCESSING TIMELINE

From the earliest evidence of controlled
fire to quern stones for grinding grain,
humans have been pounding, milling,
cutting up, and cooking their food for
up to 1.5 million years.

━━ MASTERY OF FIRE AND COOKING
━━ FOOD PROCESSING
(POUNDING AND GRINDING)

700,000–400,000 YA
*Burnt bones and artefacts,
Zhoukoudian, China,*

400,000 YA
*Evidence of igniting fire with
iron pyrite, Barnham, England*

500,000 YA

780,000 YA
*Earliest known hearth,
Gesher Benot Ya'akov, Israel*

1 MYA

1 MYA
*Traces of wood ash, alongside
animal bones and stone tools,
Wonderwerk Cave, South Africa*

400,000 YA
*Evidence of repeated
control of fire, Beeches
Pit, England*

200,000 YA
*Neanderthals using
birch-bark tar glue to attach
stone tools to handles or
spear shafts, Campitello, Italy*

1.27 MYA
*Scorched animal bones and
antlers found alongside stone
tools, Xihoudu, China*

*Homo ergaster may have
tended and maintained
naturally occurring fire*

164,000 YA
*Heat treatment of stone for
easier working, Pinnacle
Point, South Africa*

1.5 MYA
*Burnt sediments and some
heat-altered stone tools,
Koobi Fora, Kenya*

1.42 MYA
*Burnt clay suggesting
earliest evidence of
fire associated with
hominid occupation,
Chesowanja, Kenya*

1.5–1 MYA
*Burnt bones, Swartkrans
Cave, South Africa*

1.5 MYA

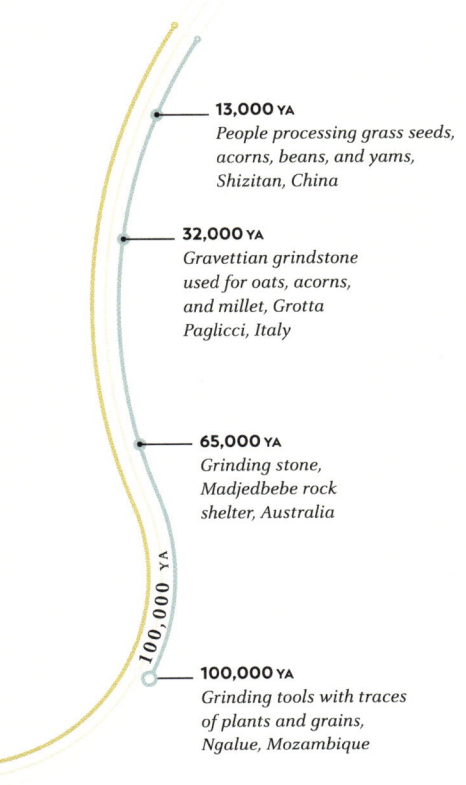

13,000 YA
*People processing grass seeds,
acorns, beans, and yams,
Shizitan, China*

32,000 YA
*Gravettian grindstone
used for oats, acorns,
and millet, Grotta
Paglicci, Italy*

65,000 YA
*Grinding stone,
Madjedbebe rock
shelter, Australia*

100,000 YA

100,000 YA
*Grinding tools with traces
of plants and grains,
Ngalue, Mozambique*

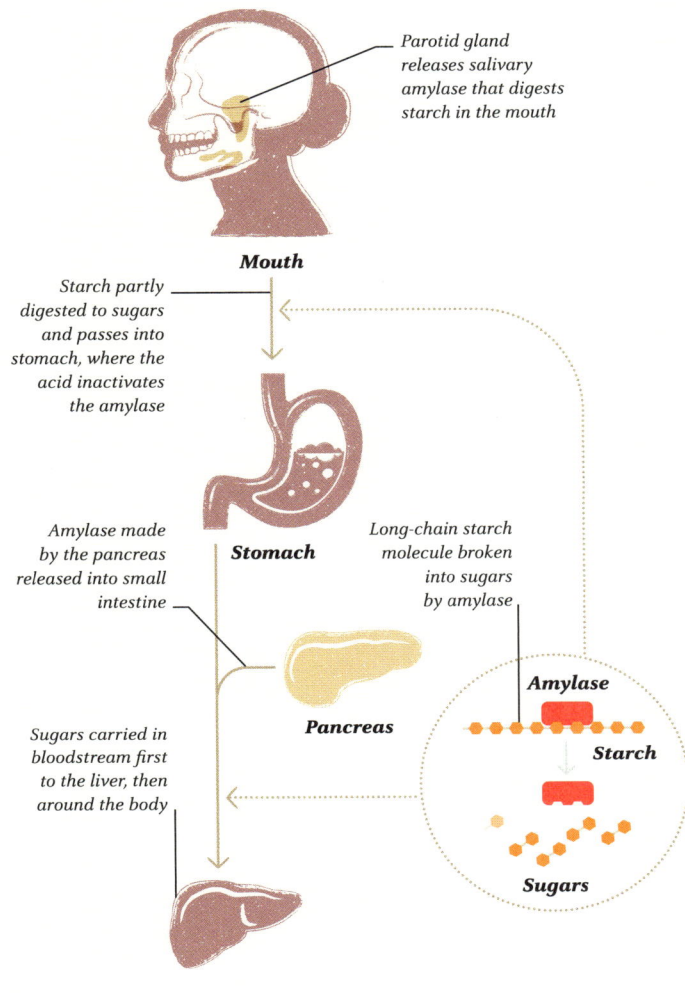

*Parotid gland
releases salivary
amylase that digests
starch in the mouth*

Mouth

*Starch partly
digested to sugars
and passes into
stomach, where the
acid inactivates
the amylase*

Stomach

*Amylase made
by the pancreas
released into small
intestine*

*Long-chain starch
molecule broken
into sugars
by amylase*

Amylase

Starch

Pancreas

*Sugars carried in
bloodstream first
to the liver, then
around the body*

Sugars

Liver

more easily digested. Cooking also breaks down collagen fibres in muscle, reducing it to tender meat. In these ways, cooking makes more foods palatable and digestible and allows the human diet to be broader than that of any other animal.

UNLOCKING NUTRIENTS

This breakdown of complex foods into more digestible materials is thought to have provided the fuel for a further increase in human brain size. While cooking made more foods available, it also reduced the energy cost of digestion, making more energy available for a large, expensive brain.

Dietary shifts in early humans can be tracked by examining fossil teeth and cooking tools, which retain residues of meals eaten hundreds of thousands of years ago. Additionally, genetic research of digestive enzymes reveals what foods humans were ingesting on a regular basis.

Each nutritional component in food needs a different type of digestive enzyme to break it down so the resulting nutrients can be absorbed into the blood. For example, protein molecules are broken down into simple amino acids by protease

DIGESTING STARCHY FOODS

Humans' starch-digesting enzymes have multiplied over the last million years, spreading from the pancreas to our salivary glands. This could be linked to eating more starchy foods, such as roots, made more edible by cooking.

enzymes. Long chains of starch are digested by an enzyme called amylase, which converts starch to absorbable simple sugars. In vertebrates, amylase is made mainly in the pancreas and secreted into the small intestine.

Analysis of Neanderthal and Denisovan genomes shows that both species of early human possessed multiple copies of the gene that codes for amylase, including one version that triggers the production of the enzyme in the salivary glands. Mixing salivary amylase with food

*Some studies
reveal multiple
amylase genes
in Neanderthals
and Denisovans
as well as in
modern humans*

GROUND FOODS

Grinding is another form of food processing which, like cooking, allowed early humans to expand their diet, rendering a wider range of tough grains and vegetables digestible and safe to eat. Study of trace residues on grindstones may reveal which plants were processed.

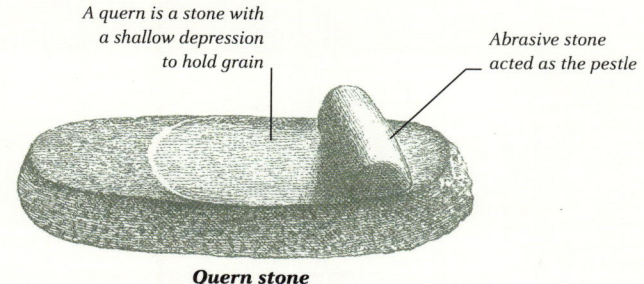

A quern is a stone with a shallow depression to hold grain

Abrasive stone acted as the pestle

Quern stone

Sorghum

Pigeon peas

False banana

Acorns

Millet

Oats

MOZAMBIQUE, 100,000 YA
Dozens of stone tools, including grindstones, found deep in a cave, bore traces of sorghum, false banana, and pigeon peas.

ITALY, 32,000 YA
Gravettian people ground oats and other grains to make flour. We don't know if they used it to make porridge, bread, or patties.

through chewing begins the breakdown of starchy food before it reaches the stomach, increasing the efficiency of digestion. The number of amylase-producing genes correlates to the amount of starch in an animal's diet. Therefore, some experts hypothesize that some humans were upping their intake of starchy foods before Neanderthals and Denisovans split from a common ancestor around 800,000 years ago.

Researchers suggest this increase in amylase production may have been encouraged by food processing, such as grinding and cooking, which would have made these foods available in the diet. A feedback loop may have been created, in which the evolving digestive system adapted to the increasingly starchy diet by producing better starch-digesting juices.

If an animal has a high-starch diet it tends to have a greater number of amylase genes

ACCESS TO STARCHES

Although systematic agriculture first emerged around 12,000 years ago, the discovery of much older Palaeolithic grindstones reveals that some early humans were processing certain foods way before the rise of organized farming.

In northern Australia, researchers unearthed one of the oldest known grindstones, dating from around 65,000 years ago. Neanderthal sites

in Europe have also yielded rudimentary grindstones linked to the late Mousterian period, around 40,000 years ago. Some archaeological sites associated with early *Homo sapiens* that migrated into Europe and north Africa contain grindstones of various shapes – elongated grinders and circular wheels – suggesting that our early ancestors may have been experimenting with and adapting their technology to suit different habitats.

Analysis of residues preserved on grinders support the idea that these tools were used to grind starchy plants for food, including cereal grains and acorns. Grinding would have softened plants and grains for better digestion. Quern stones found in Wadi Kubbaniya, Egypt, are 21,000 years old and bear residues of *Bolboschoenus maritimus*, a plant with a fibrous, indigestible texture that could be broken down by grinding. Grinding may also have been used to make food more palatable – the same querns also had traces of purple nutsedge, a tuberous plant containing bitter chemicals that can be leeched out through grinding.

Cooking opened up a diet wider than that of any other animal, meaning humans are not only omnivores, they are cucinivores – eaters of prepared food.

Domesticating animals

*Humans' relationship with the animals around them began
in the African bushland, but shifted closer to settlements
when hunters turned to keeping livestock. The new
ecological interactions that followed changed the
course of evolution for both animals and humans.*

No species lives in isolation: every animal is exposed to predators, prey, competitors, pathogens, and parasites. Early humans succumbed to leopards, hunted gazelles, faced up to scavenging hyenas, and survived disease-causing organisms. But when people began keeping animals they formed new types of partnerships with species as varied as dogs, cattle, horses, chickens, and honeybees. Today, domesticated animals differ so much from their wild ancestors and are so dependent on their human caretakers that most can no longer survive in the wild. They appeared relatively quickly, most within the last 10,000 years – testament to the power of human intervention over natural selection in breeding animals with traits helpful to humans.

In purely genetic terms, domestication benefits both humans and animals

Humans who domesticated animals benefited from not only a ready supply of food, but also clothing, labour, and companionship. In an evolutionary sense, the animals benefited too: livestock proliferated with a continual supply of food and protection from other predators. It is little wonder that domesticated livestock came to outnumber large wild animals: today more than 60 per cent of the biomass of all mammals – humans included – comes from domesticated cattle, pigs, and pets.

DOGS AND HUMANS

The earliest undisputed evidence for animal domestication comes from an archaeological site in Germany, 14,000 years old, where the partial skeleton of a dog was found interred alongside two human bodies. Pathological evidence suggests that this animal survived canine distemper – a recovery that would not have been possible without human help – meaning the dog was probably a pet. More uncertain records of domesticated dogs in Siberia date back even further to 30,000 years ago. All these records are from a time that pre-dates the emergence of farming and agriculture, indicating that dogs were domesticated long before cattle and similar livestock. Dog domestication was practically unique: dogs were not kept for food, but rather as companions and extended members of the human tribe.

WORKING DOGS

Dogs have been used as working animals for millennia, providing humans with services such as guarding camp, hunting, and transport by pulling sleds.

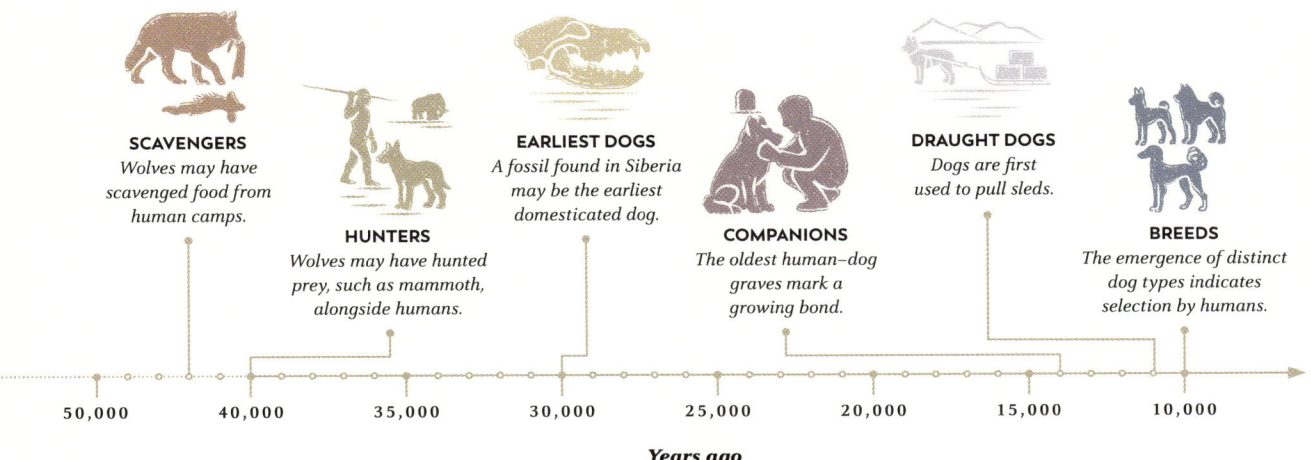

SCAVENGERS
Wolves may have scavenged food from human camps.

HUNTERS
Wolves may have hunted prey, such as mammoth, alongside humans.

EARLIEST DOGS
A fossil found in Siberia may be the earliest domesticated dog.

COMPANIONS
The oldest human–dog graves mark a growing bond.

DRAUGHT DOGS
Dogs are first used to pull sleds.

BREEDS
The emergence of distinct dog types indicates selection by humans.

50,000   40,000   35,000   30,000   25,000   20,000   15,000   10,000

Years ago

DOMESTICATING ANIMALS

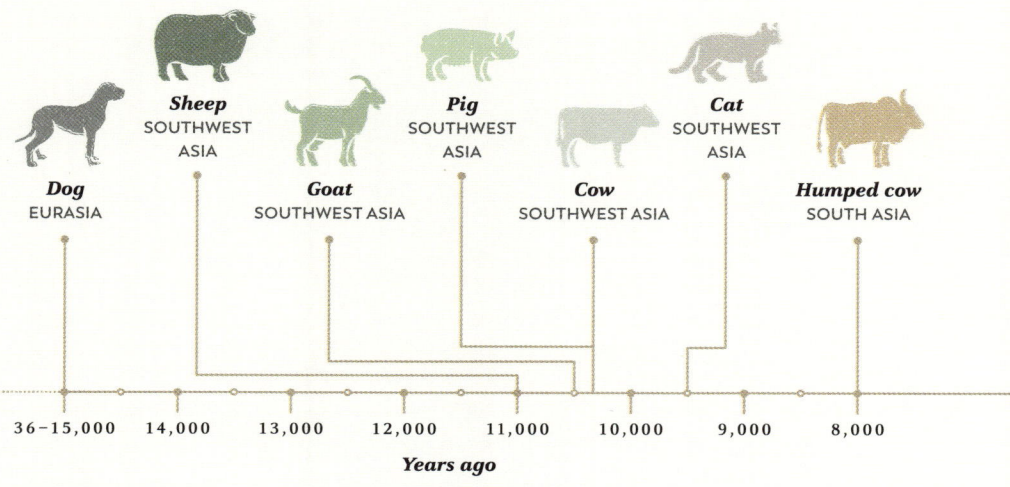

CENTRES OF DOMESTICATION

Archaeological and genetic studies have shown that the domestication of livestock animals took place in early agricultural communities in centres around the world.

Dog
EURASIA

Sheep
SOUTHWEST ASIA

Goat
SOUTHWEST ASIA

Pig
SOUTHWEST ASIA

Cow
SOUTHWEST ASIA

Cat
SOUTHWEST ASIA

Humped cow
SOUTH ASIA

36–15,000   14,000   13,000   12,000   11,000   10,000   9,000   8,000

Years ago

The dog's wild ancestor is the pack-hunting grey wolf (*Canis lupus*), which was widespread across Eurasia 14,000 years ago. The process of domestication most likely began when bold individuals ventured close to human camps to steal food. Gradually, they were tolerated and even encouraged. But why would humans share their food with animals? A leading theory cites a key difference in the nutrition of the two potentially competing species: wolves eat lean meat and little else, while humans, despite their emerging hunting lifestyle, retained a more omnivorous capability inherited from herbivorous ancestors. Too much protein-rich lean meat – especially abundant during Ice Age winters when prey have exhausted their fat reserves – is harmful to humans. The protein-partitioning theory suggests that surplus meat was discarded by Palaeolithic hunters and attracted wolves. Over time food was deliberately shared – perhaps as the benefits of wolves as hunter–companions or guards was recognized. And hormonal biology suggests parallel paths in terms of temperament; modern humans and domesticated dogs both have high levels of oxytocin, a hormone that promotes social bonding.

> # Sheep and goats, rather than gazelles and antelopes, were the first to be domesticated.

The early domestication of dogs may have depended on the differing protein requirements of dogs and humans

CONTROLLED BREEDING

The earliest step in the domestication of livestock was bringing prey animals under management – much like the practice of reindeer herders today who drive animals to areas of local vegetation where they can graze or browse safely. This developed into confining livestock in paddocks, a practice seen at archaeological sites in Syria dating back 12,000 years, where sheep and goat dung was found within settlements. This timing is backed up by DNA analysis, which indicates that domestic sheep diverged from their wild ancestors – the Anatolian mouflon

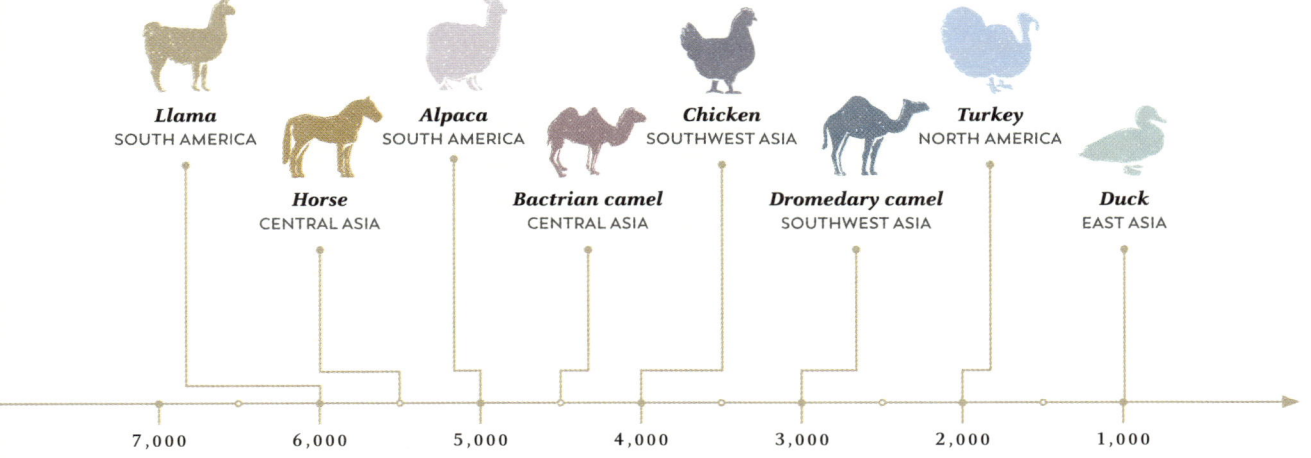

Llama SOUTH AMERICA		**Alpaca** SOUTH AMERICA		**Chicken** SOUTHWEST ASIA		**Turkey** NORTH AMERICA
	Horse CENTRAL ASIA		**Bactrian camel** CENTRAL ASIA		**Dromedary camel** SOUTHWEST ASIA	**Duck** EAST ASIA

7,000   6,000   5,000   4,000   3,000   2,000   1,000

– at around the same time. Docile, sociable animals of manageable size were easiest to control, and it is no surprise that sheep and goats, rather than gazelles and antelopes, were the first to be domesticated. Living in close contact with their animals, herders soon recognized the power of selective breeding – choosing individuals with desirable traits to mate and produce the next generation. They bred out characteristics such as aggression and weapons (pointed horns) and used their growing insight into breeding to enhance characteristics such as meat yield.

Genes for docility were advantageous for certain species because they then proliferated in great numbers

Across Eurasia and north Africa, just four species of ungulates dominated global livestock for the next 5,000 years – the sheep, goat, cow, and pig. Evidence from skeletal anatomy and DNA shows that they descended from wild species – Anatolian

ANCESTRAL SPECIES
Domesticated animals are genetically and functionally distinct from their wild ancestors, though they may belong to the same species and occasionally interbreed.

DOMESTIC ANIMALS
Desirable traits are enhanced by selective breeding. For example, wild sheep typically have smaller litter sizes than domesticated breeds.

Cow   Goat   Cat   Dog

Pig   Sheep   Horse   Chicken

WILD SPECIES
The wild animals selected for domestication had traits that were useful to humans – they had warm coats or produced nutritious milk, for example.

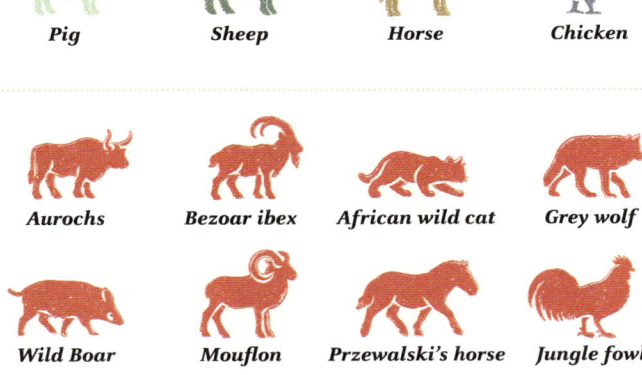

Aurochs   Bezoar ibex   African wild cat   Grey wolf

Wild Boar   Mouflon   Przewalski's horse   Jungle fowl

157

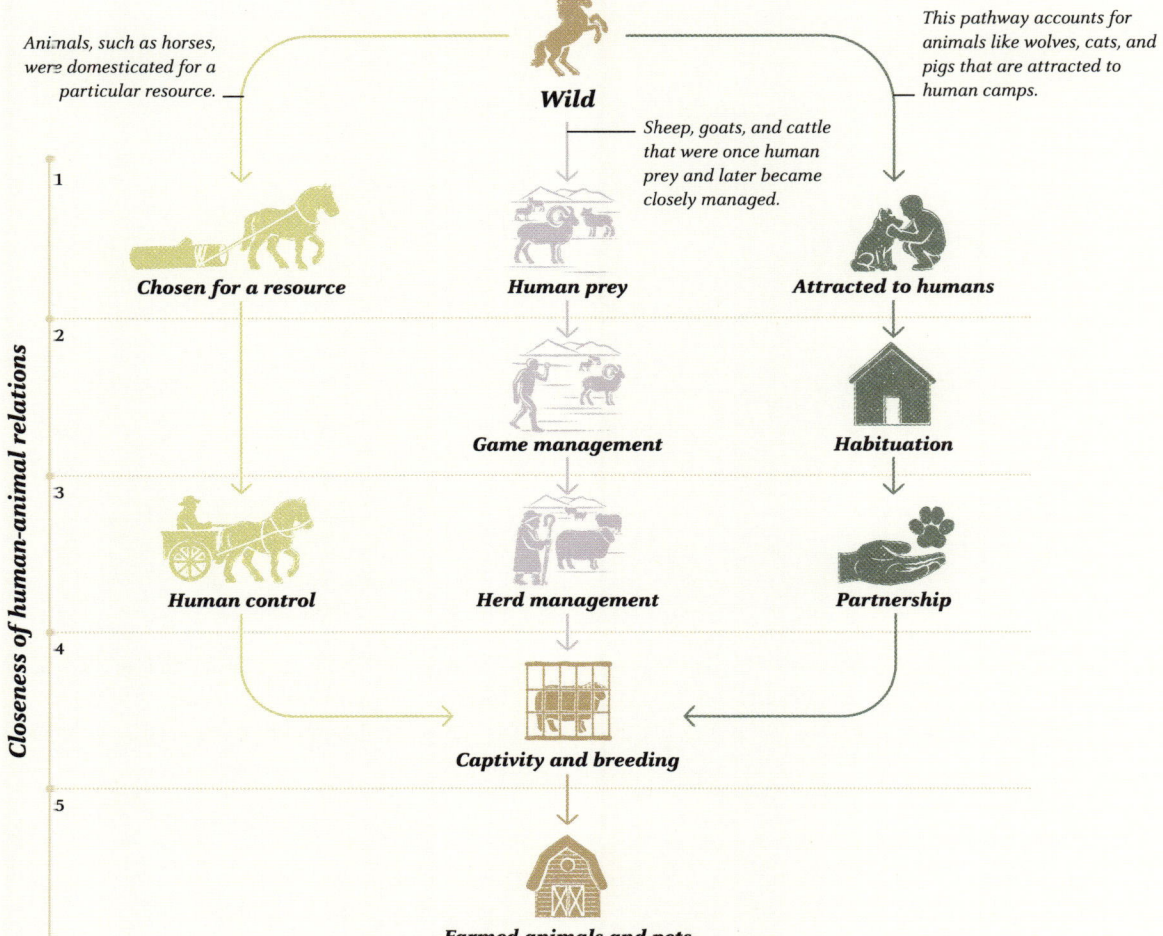

Animals, such as horses, were domesticated for a particular resource.

This pathway accounts for animals like wolves, cats, and pigs that are attracted to human camps.

Wild

Sheep, goats, and cattle that were once human prey and later became closely managed.

Closeness of human-animal relations

1

Chosen for a resource **Human prey** **Attracted to humans**

2

Game management **Habituation**

3

Human control **Herd management** **Partnership**

4

Captivity and breeding

5

Farmed animals and pets

PATHS TO DOMESTICATION

There are broadly three pathways to domestication, depending on the animal, how easy it is to control, and the resources that the animal provides to humans.

mouflon, Bezoar ibex, aurochs, and wild boar, respectively. Datable archaeological deposits show where and when these animals were bred, but only DNA analysis reveals their true origin. Goats emerged with sheep about 11,000 years ago in Mesopotamia, cattle about the same time in Africa and Asia, and pigs about 1,000 years later in southwest Asia. Horses – initially used for meat and milk, later for riding – were bred from Mongolian wild horses on the steppes of Central Asia around 6,000 years ago. As humans settled elsewhere in the world within the last two millennia, local species added to the inventory: chickens from jungle fowl in Southeast Asia, for example.

EFFECTS ON HUMANS

New breeds have provided humans with specific benefits: dogs like the Saluki are excellent hunters, while sheep have been bred to provide thick wool and cattle to provide plentiful milk. Domestication has changed the course of evolution on both sides. With the rise of dairy farming, more humans retained an

Aurochs are now extinct; they were larger than modern cattle, bulls reaching over 1.8 m (6 ft) at the withers

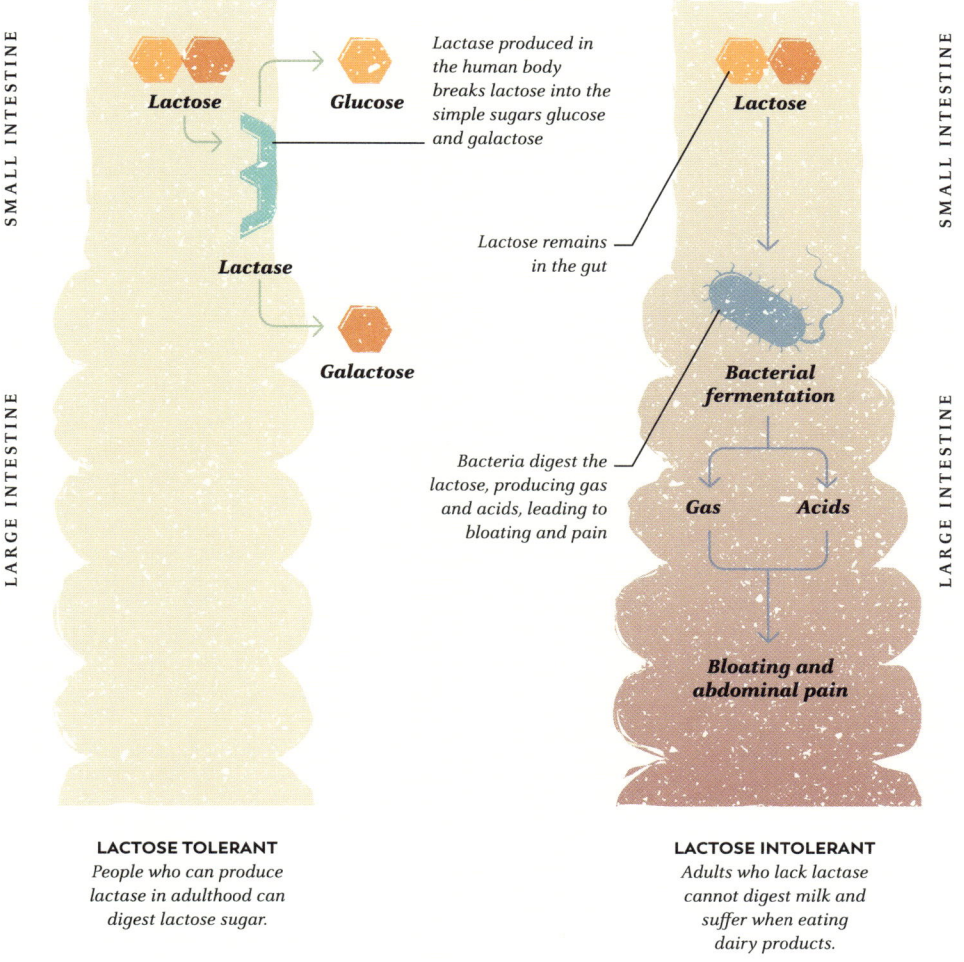

LACTOSE TOLERANT
People who can produce lactase in adulthood can digest lactose sugar.

LACTOSE INTOLERANT
Adults who lack lactase cannot digest milk and suffer when eating dairy products.

Lactase produced in the human body breaks lactose into the simple sugars glucose and galactose

Lactose remains in the gut

Bacteria digest the lactose, producing gas and acids, leading to bloating and pain

ability to digest milk into adulthood – the result of a mutation for lactose tolerance that originated in southwest Asia and spread around the world. Domestication also made infectious diseases spread more easily. Brucellosis – a bacterial disease transmitted from hoofed animals to humans in milk – became endemic with the rise of dairy farming. And in some cases, domestication has triggered the emergence of new diseases that have crossed species barriers. Similarities in the genetic material of viruses suggests that measles may have originated from rinderpest, an infection of cattle. With this, the human genome responded by evolving new variants of genes for infection-fighting antibodies.

Animal domestication has altered the course of biodiversity through the emergence of new varieties of animals as well as new diseases, but it has also changed the biology of humans and changed their relationships with species around them.

Brucellosis comes from eating dairy produc ts. It causes fever and joint pain but can be treated with antibiotics

LACTOSE TOLERANCE
Human infants produce the enzyme lactase, which helps them to digest the sugar lactose that is present in milk. A mutation that occurred around 8,000 years ago extended that ability to some adults, who could then benefit from drinking the milk of domesticated animals.

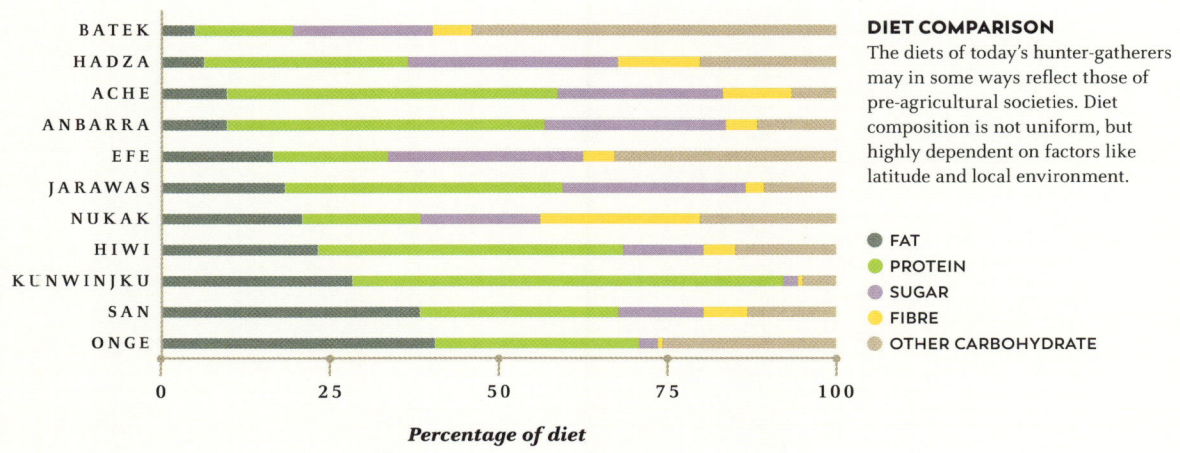

Percentage of diet

The move to an agricultural way of life represented a seismic shift in the ecology of prehistoric humans. Time and energy once channelled into foraging over large distances was instead diverted to cultivating crops closer to home.

Cultivating plants

The practice of farming gave humans reliable access to food throughout the growing season and was able to feed many more mouths than the long-distance foraging of wild plants. Over thousands of years, it became the dominant ecology for practically all people on Earth, but not without costs to human biology.

SHIFT AWAY FROM A NOMADIC DIET

During the last Ice Age, which ended some 11,700 years ago, it paid to be nomadic. Food was highly seasonal and often spread over a wide area, so the hunter-foragers of temperate Eurasia had to move from place to place to survive. Much of what we know about their diets comes from isotope analysis of their bones and teeth: percentages of certain nitrogen and carbon isotopes in

Our ancestors moved camps up to several times a year to exploit seasonal resources

Hafts were made from wood or sometimes bone

Flint blades could be resharpened or replaced after they became worn

NEOLITHIC TOOLS
Settled life and agriculture led to the development of heavier tools for farming, such as hoes and sickles. This modern replica of a Neolithic sickle was used in much the same way as the modern tool.

their remains show the range of plant and animal proteins they consumed and where these foods came from. Isotope studies show that hunter-gatherers enjoyed a varied diet of uncultivated tubers, seeds, and fruits, as well as meat and fish, and that there was considerable seasonal variation in what was eaten – nuts and berries in autumn, more meat in summer.

Permanent settlements became possible with climate change, warmer summers and milder winters replacing the retreating ice. Some locations – often those near waterways with good fishing – offered sufficient resources to stay put all year round. Foraging and hunting could be done within walking distance of the camp, satisfying the needs of the whole community. Camps turned into stronger buildings and fortified settlements and group size

Permanent encampments provided central hubs for the activities of hunter-gatherers

161

SHIFTING DIETS

The diets of prehistoric hunter-gatherers were higher in protein and lower in carbohydrates than modern Western diets, reflecting a growth in consumption of cereal.

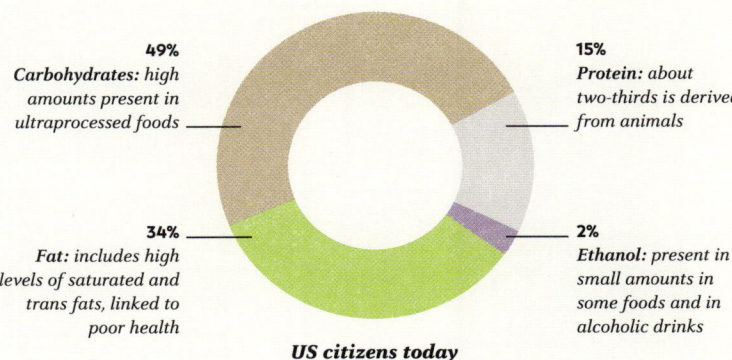

19–35%
Protein: highly variable (can be as little as 19%), but generally higher than in industrialized societies

22–40%
Carbohydrates: higher in tropical or grassland regions; can be as low as 22% in northern areas

28–47%
Fat: high levels of unsaturated fats from fish, nuts, and meat; can sometimes fall to 28%

Palaeolithic hunter-gatherers

49%
Carbohydrates: high amounts present in ultraprocessed foods

15%
Protein: about two-thirds is derived from animals

34%
Fat: includes high levels of saturated and trans fats, linked to poor health

2%
Ethanol: present in small amounts in some foods and in alcoholic drinks

US citizens today

increased. In some places, people began to clear plots of undesirable vegetation and introduce edible plants instead. Some soils, such as wind-blown loess, proved fertile and easy to till, making them most suitable for early farming, but the development of heavier tools, such as hoes and sickles with flint blades, made agriculture possible at an even wider range of sites. The reward was a high yield of food produced on the doorstep, though the range of plants grown was far smaller than that available in the forest and bush, compromising nutritional balance.

THE FIRST CROPS

Routine agriculture began some 12,000 years ago, in the highly fertile floodplains of Mesopotamia – the "land between the rivers" of the Euphrates and Tigris, in modern-day Iraq. From here, it spread west and east into Europe and Asia, while independent centres of agriculture later emerged in similarly fertile floodplains – 9,000 years ago in China, and from 5,000 years ago in Africa and the Americas (see pp.282–83).

Quick-growing, high-yielding herbaceous plants have the best crop potential, and early farmers soon focused on grass species that had seeds rich in starch and protein. These form the basis of staple crops to this day: wheat in Mesopotamia, rice in China, maize in North America, and sorghum in Africa. Elsewhere, in the tropics,

Early forms of wheat, maize, and sorghum became the staple crops that feed the world

tubers and legumes were the mainstay, yams in Asia, and beans and potatoes in tropical America. The evidence for this profound dietary shift comes from the isotopic signatures of Neolithic bones – specifically the ratio of the "heavy" isotope of nitrogen compared to the "normal" isotope in bones and teeth. Heavy nitrogen is incorporated into flesh and so is at higher levels in top predators than in herbivores. Comparing the bones of Palaeolithic and Neolithic peoples reveals a drop of up to 50 per cent in the heavy isotope because the first farmers relied so much more on plants.

A diet dominated by starchy crops left humans lacking in the nutrients that were more abundant in a meatier diet.

CHANGING BIOLOGY

Pre-agricultural diets may have resulted in low incidence of diabetes and cardiovascular disease

These dietary changes – together with the more sedentary lifestyle permitted by an agricultural way of life – seem to have had significant effects on the human body. Some studies have shown that Neolithic farmers tended to be smaller in stature than their hunter-gatherer ancestors, with lighter, less-dense bones. Human jaws became less powerful and smaller, though teeth remained around the same size, leading to dental problems. The farmers also suffered more from tooth decay and calculus resulting from a diet rich in starchy foods.. Studies of Neolithic bones point to even more damaging effects. A diet dominated by starchy crops could leave humans lacking some vital nutrients that were more abundant in a meatier diet: deficiencies of vitamins B12 and D are indicated by deformities in Neolithic skeletons such as spongy porosities and rickets.

CEREAL ADOPTION

This timeline follows the stages in domestication and exploitation of cereal crops in the Levant – one of the pioneering places where agriculture developed in the Fertile Crescent.

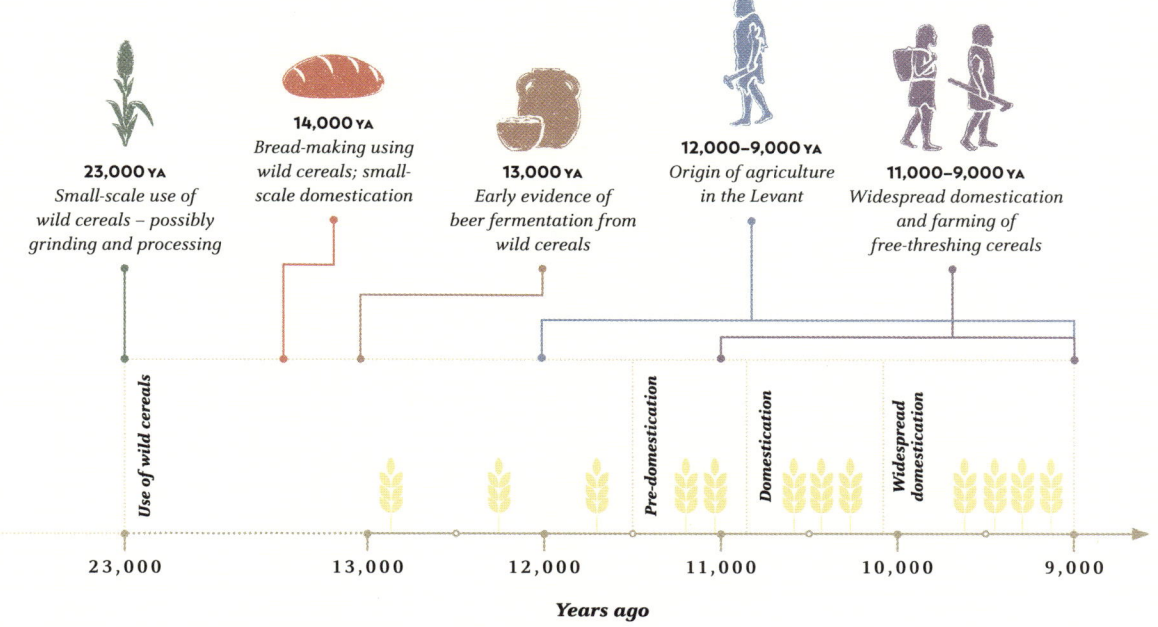

23,000 YA
Small-scale use of wild cereals – possibly grinding and processing

14,000 YA
Bread-making using wild cereals; small-scale domestication

13,000 YA
Early evidence of beer fermentation from wild cereals

12,000–9,000 YA
Origin of agriculture in the Levant

11,000–9,000 YA
Widespread domestication and farming of free-threshing cereals

Use of wild cereals

Pre-domestication

Domestication

Widespread domestication

23,000   13,000   12,000   11,000   10,000   9,000

Years ago

CULTIVATING PLANTS

GIGANTIC IMPROVEMENT

Teosinte, a wild form of maize, carries very few kernels on a head just 2.5 cm (1 in) tall. The head of its modern counterpart may be more than 30 cm (12 in) in length.

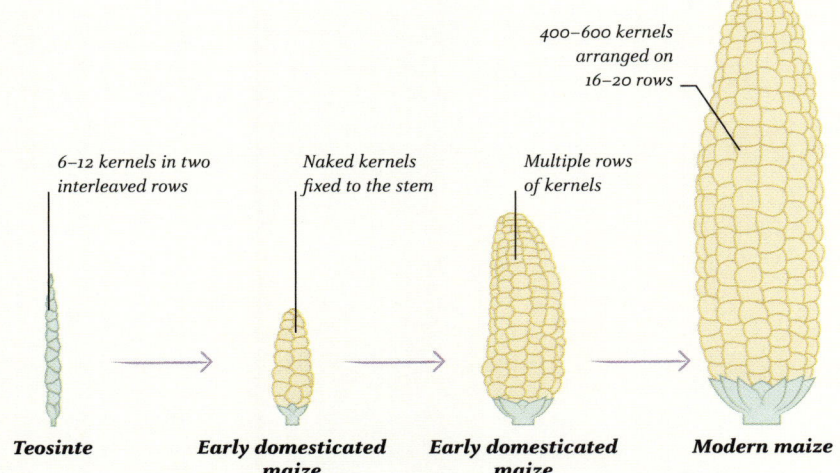

6–12 kernels in two interleaved rows

Naked kernels fixed to the stem

Multiple rows of kernels

400–600 kernels arranged on 16–20 rows

Teosinte

Early domesticated maize

Early domesticated maize

Modern maize

GATHERABLE GRAINS

Generations of breeding have produced wheat with fatter nutrient-rich grains. Human selection took wheat from wild, shattering varieties (those where the seeds detach and fall from the plant when ripe) into nonshattering varieties.

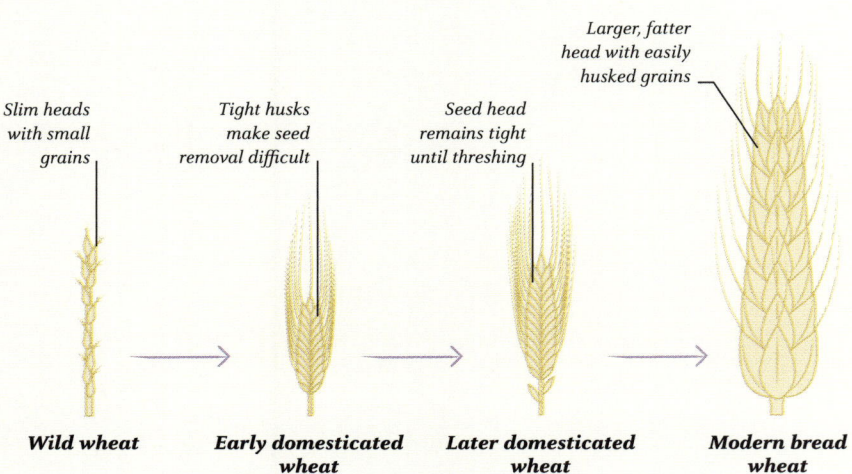

Slim heads with small grains

Tight husks make seed removal difficult

Seed head remains tight until threshing

Larger, fatter head with easily husked grains

Wild wheat

Early domesticated wheat

Later domesticated wheat

Modern bread wheat

BIGGER BEANS

Wild beans – a staple food in Mesoamerica – have small pods that release their seeds when ripe. Domesticated plants produce bigger seeds that remain in their pods until harvested.

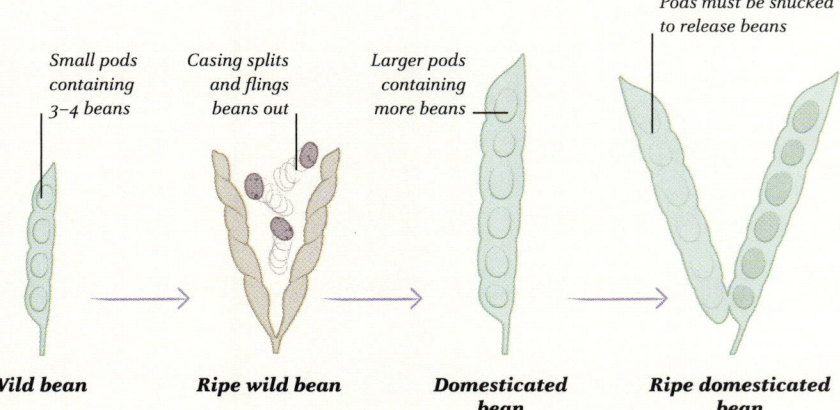

Small pods containing 3–4 beans

Casing splits and flings beans out

Larger pods containing more beans

Pods must be shucked to release beans

Wild bean

Ripe wild bean

Domesticated bean

Ripe domesticated bean

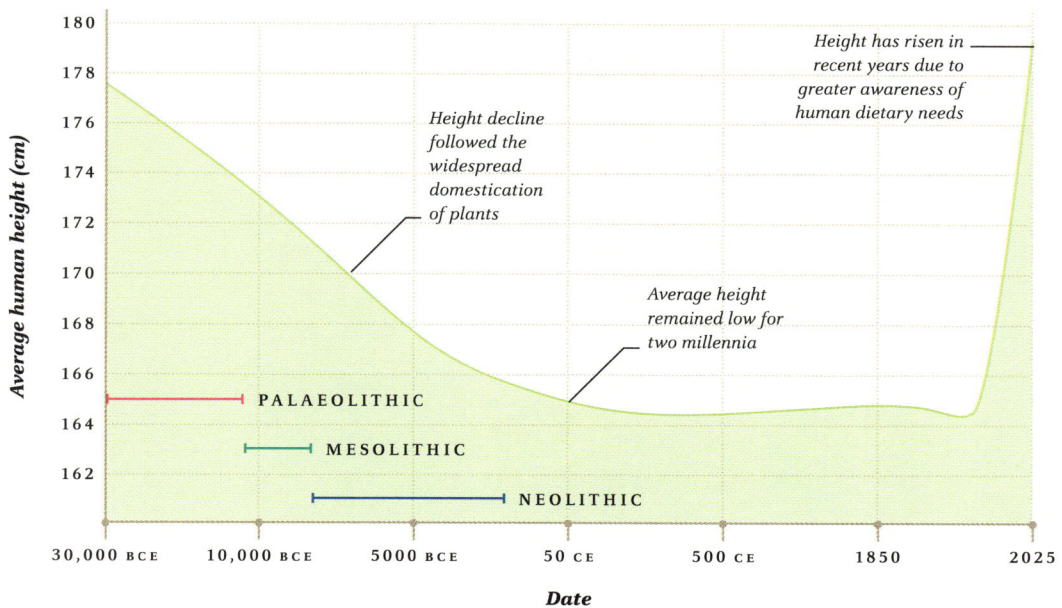

Average human height (cm)

180
178
176
174
172
170
168
166
164
162

Height decline followed the widespread domestication of plants

Height has risen in recent years due to greater awareness of human dietary needs

Average height remained low for two millennia

PALAEOLITHIC

MESOLITHIC

NEOLITHIC

30,000 BCE   10,000 BCE   5000 BCE   50 CE   500 CE   1850   2025

Date

HEIGHT AND NUTRITION
Hunter-gatherers living at the end of the Palaeolithic were taller than Neolithic farmers, whose height decreased significantly due to a less varied starch-rich diet.

PLANT-HUMAN COEVOLUTION

Just like the farmers who bred livestock, early horticulturalists learned that plants could pass on useful characteristics to their offspring. And as the best characteristics were selected each time – initially by accident, later by design – they became amplified. Grains, beans, and tubers got bigger and plants became easier to grow and harvest, with new varieties emerging from the wild species. Crop improvement began with wheat. Around five wild wheat species of the genus *Triticum* are found across warm temperate Eurasia from the Mediterranean to India. Our bread wheat today has a complicated evolutionary history, involving the domestication of emmer wheat and einkorn, as well as interbreeding with other wild species.

Farming caused genetic changes not only in crop plants, but in their human stewards. Natural selection favoured an increase in genes for the production of amylase – an enzyme that digests starch – as well as genes that help trigger immunity in denser populations. The rapid shift in diet also affected the human internal microbiome (see pp.166–69). A mismatch between Neolithic immunity genes and a Palaeolithic microbiome may have been responsible for a rise in autoimmune disorder in modern humans, including bowel diseases.

The agricultural revolution moulded plants and humans together – as wild cereals turned into cultivated crops, and Palaeolithic hunter-gatherers became Neolithic farmers. Within a few thousand years, agriculture would feed a global human population that had swelled into billions.

Humans have coevolved with bacteria in their digestive systems

165

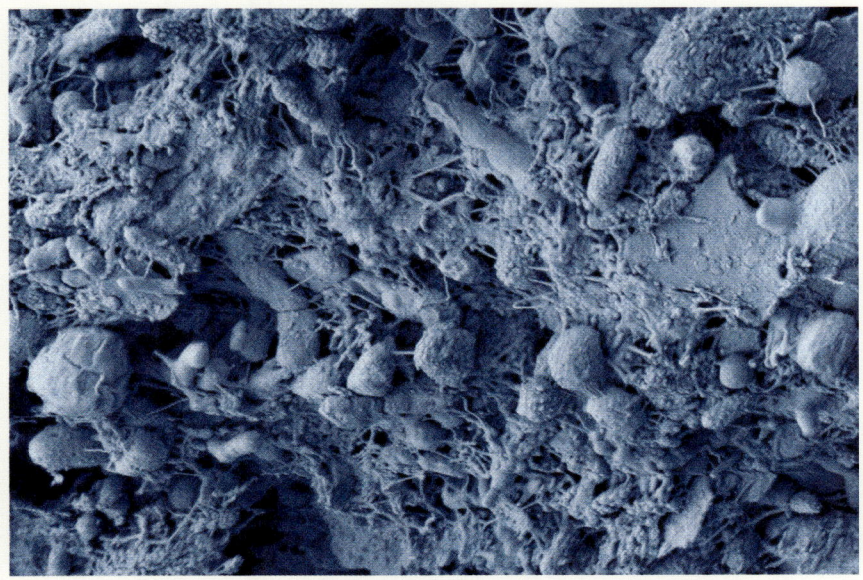

Humans and microbes

The coevolving partnerships between humans and other organisms reaches its closest intimacy with microbes that live within our bodies. More than 3,000 species thrive inside the human gut, and many of them are beneficial.

All vertebrates have some kind of gut microbiome – a community of microbes that lives permanently inside the digestive system. Most belong to groups of the smallest, simplest organisms – bacteria and archaea. These microbes are single-celled organisms that lack a nucleus and contain a single loop of DNA. Other gut inhabitants include more complex single celled organisms, such as yeasts.

These communities are products of evolution and ecology. They have changed and adapted to the evolving bodies of their hosts – just as their hosts have responded to their unseen lodgers. The microbes thrive in an environment packed with food. Many of them are commensal (from the Latin meaning "sharing a table") and these organisms enjoy the benefits of being surrounded by nutrients but give nothing back in return. Others are mutualistic or symbiotic, benefiting from living in the gut but also assisting digestion and promoting health in the host. The human gut microbiome was inherited from leaf-eating ancestors, but is now moulded by a host that eats more meat and grain.

SEEDING THE GUT

The developing gut of a mammal embryo, like all embryonic organs, is largely sterile – so microbes must colonize the body's interior from outside. Much of this happens during birth

Babies must acquire gut microbes from their mother or the environment

when microbes enter the gut from the birth canal. They include surface bacteria of the skin microbiome – such as *Lactobacillus* that use oxygen but can survive without it inside the gut. There are also true anaerobic gut bacteria called *Bifidobacterium* that come from the mother's intestine via faecal contamination. Both types of bacteria are linked to a defining feature of mammals – breastfeeding. They digest milk carbohydrates, helping microbe and infant alike. Their presence also serves to protect the infant from harmful invaders because these benign bacteria expose the baby's immune system to "foreign" particles, or antigens on cell surfaces. This triggers the infant's white blood cells and other pathogen-fighting machinery to be primed for future infection. The gut microbiomes of all mammals are broadly similar at first, but significant changes occur after weaning when the maturing digestive system begins to process more specialized diets.

The genera of gut bacteria and their relative abundances vary greatly, depending on the host

EARLY HOMININ GUT MICROBES

Just as herbivores and carnivores differ in their metabolism and nutritional requirements, so do their gut microbiomes. Many microbes in the guts of herbivores are adapted to extract nourishment by digesting fibre, which is

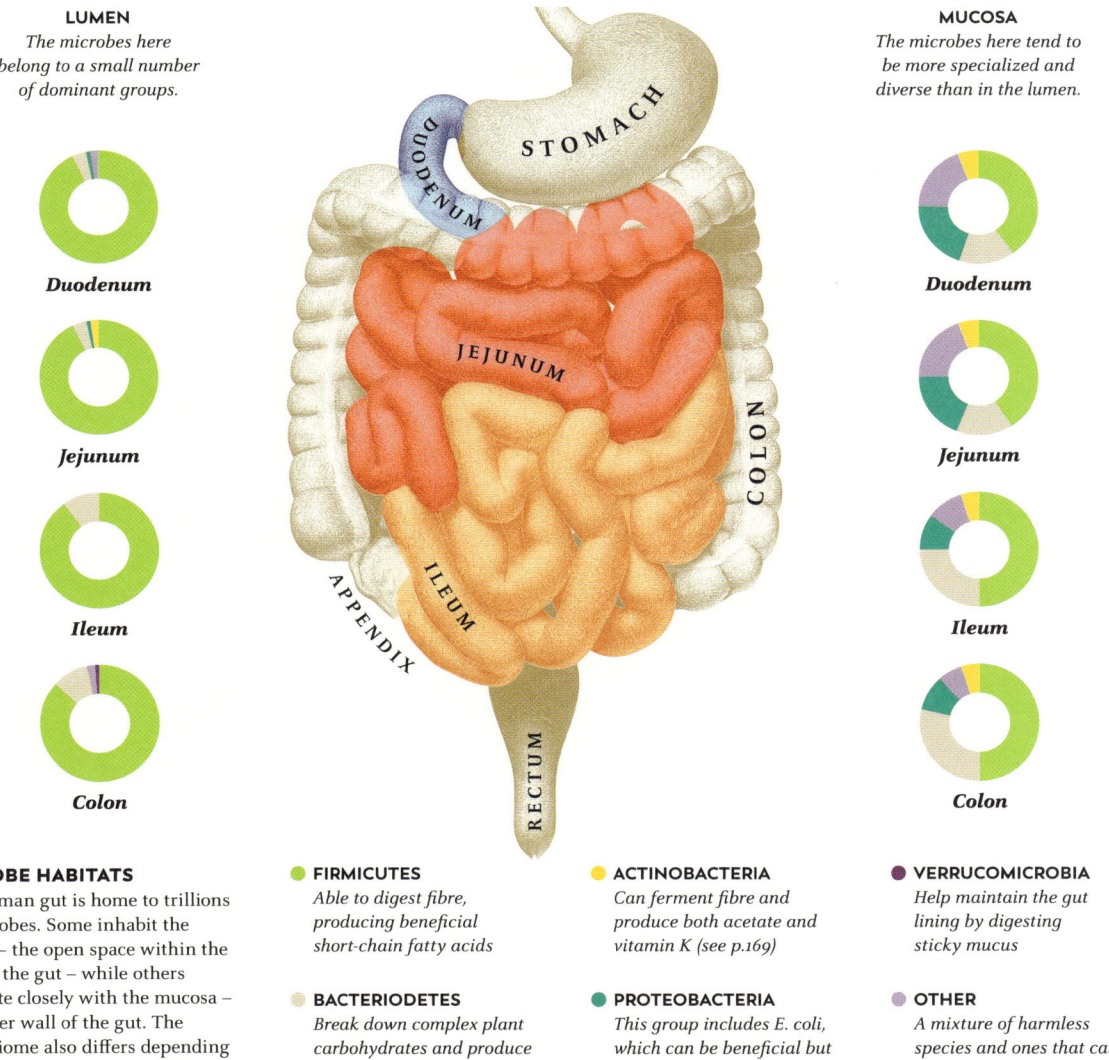

LUMEN
The microbes here belong to a small number of dominant groups.

Duodenum

Jejunum

Ileum

Colon

MUCOSA
The microbes here tend to be more specialized and diverse than in the lumen.

Duodenum

Jejunum

Ileum

Colon

MICROBE HABITATS
The human gut is home to trillions of microbes. Some inhabit the lumen – the open space within the tube of the gut – while others associate closely with the mucosa – the inner wall of the gut. The microbiome also differs depending on the region of the digestive tract.

● FIRMICUTES
Able to digest fibre, producing beneficial short-chain fatty acids

● BACTERIODETES
Break down complex plant carbohydrates and produce propionate (see p.169)

● ACTINOBACTERIA
Can ferment fibre and produce both acetate and vitamin K (see p.169)

● PROTEOBACTERIA
This group includes E. coli, which can be beneficial but also pathogenic

● VERRUCOMICROBIA
Help maintain the gut lining by digesting sticky mucus

● OTHER
A mixture of harmless species and ones that can become pathogenic

abundant in plants. The walls of plant cells are composed of cellulose, a polymer made up of glucose molecules linked together in long chains. Cellulose digestion demands an enzyme called cellulase, which can be manufactured by some insects and earthworms, but not by vertebrates, which lack genes for its production. Herbivorous mammals instead rely on microbes in their gut to ferment (break down) plant matter. Some, such as cattle, are foregut fermenters, housing these microbes in their stomachs. Others. including horses and apes, are hindgut fermenters, with fermentation taking place in the large intestine. It has been suggested that hominins such as *Australopithecus* were probably hindgut fermenters too.

DIMINISHING CELLULASE

The shift towards a meat-based diet in the more predatory, upright *Homo* (see pp.144–47) was accompanied by less reliance on fibre-digesting microbes. Today's, humans have a proportionally smaller colon and caecum (the saclike expansion of the hindgut, see p.140) than do living apes,

DIGESTIVE DECLINE

Many species of bacteria can digest cellulose. They are widespread among nonhuman primates and in hunter-gatherers, but are much less diverse in populations of industrialized societies.

and faecal analysis has revealed a corresponding loss of cellulase-producing microbes in humans. In our species, the small intestine – the principal site of nutrient absorption – has become more important for digesting an omnivorous diet.

Some cellulose-digesting bacteria, such as *Ruminococcus* and *Fibrobacter*, remain in the human colon and are most abundant in populations with a higher-fibre diet, such as the Hadza people of Tanzania who eat a large array of plant foods in far larger quantities than those in industrialized nations.

GUT BACTERIA FOR A MODERN AGE

The agricultural revolution was the most likely trigger for further changes to the human gut microbiome. Beginning some 12,000 years ago, diets came to be dominated by wheat, yams, and other starchy crops. The production of amylase – the enzyme that splits starch into simple sugars (see p.151) – increased in the pancreas and salivary glands of our ancestors and a new set of bacteria colonized their guts. Among these bacteria were members of the genus *Bacteroides*, which get their energy by fermenting a range of grain-specific fibres that the human gut cannot break down. Like many other organisms of the microbiome, their benefits go beyond augmenting digestion. For example, some produce the compounds acetate and propionate during fermentation: acetate helps to prevent

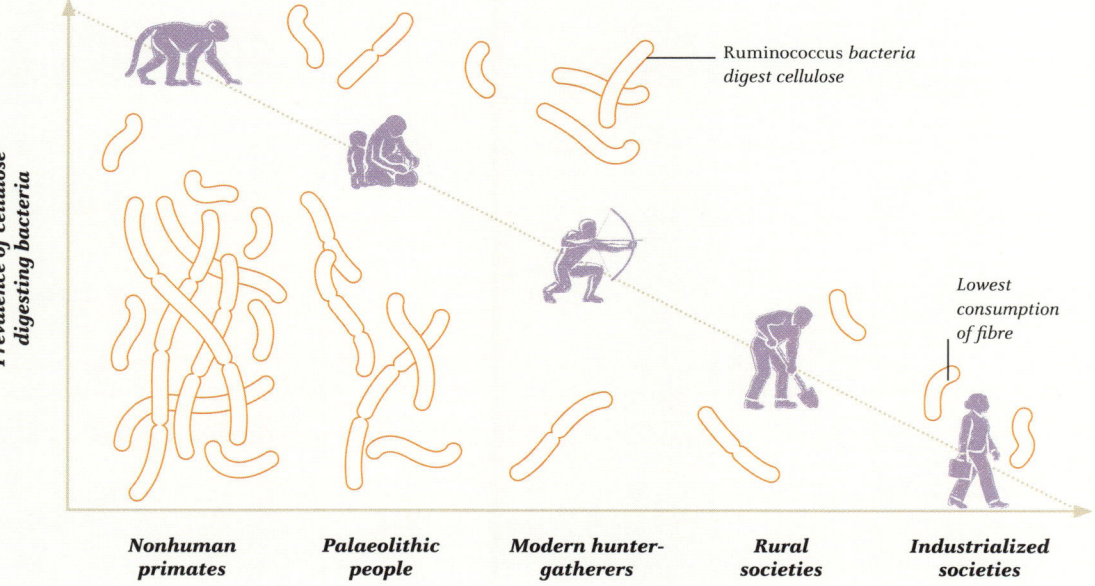

Ruminococcus *bacteria digest cellulose*

Lowest consumption of fibre

Prevalence of cellulose digesting bacteria

| **Nonhuman primates** | **Palaeolithic people** | **Modern hunter-gatherers** | **Rural societies** | **Industrialized societies** |

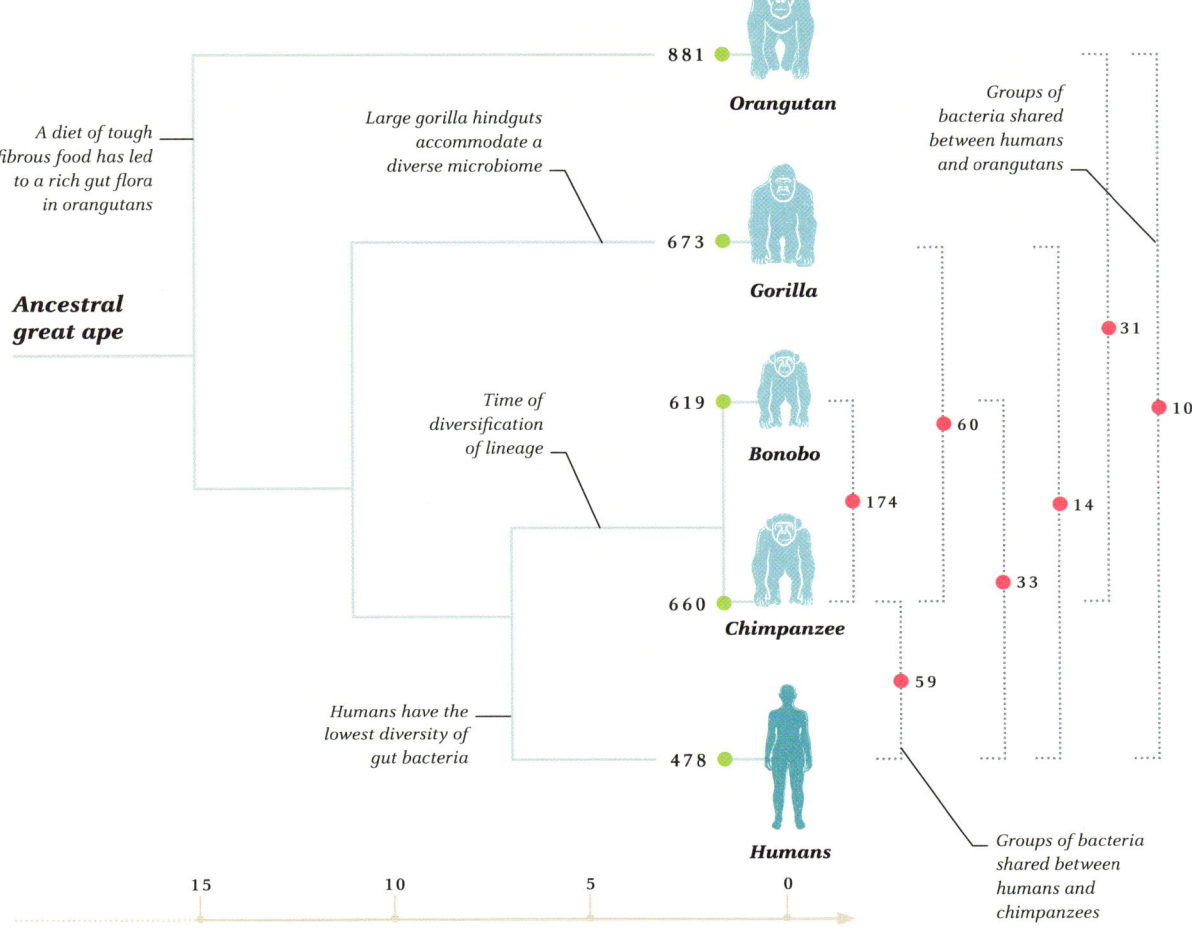

881 Orangutan

A diet of tough fibrous food has led to a rich gut flora in orangutans

Large gorilla hindguts accommodate a diverse microbiome

Groups of bacteria shared between humans and orangutans

Ancestral great ape

673 Gorilla

31

Time of diversification of lineage

619 Bonobo

60

10

174

14

660 Chimpanzee

33

Humans have the lowest diversity of gut bacteria

59

478 Humans

Groups of bacteria shared between humans and chimpanzees

15 10 5 0

Millions of years ago

EVOLVING WITH THE GUT

This chart shows diversification of hominids over the past 15 million years together with the parallel changes in gut flora. The numbers of taxonomic groups of bacteria present in the gut of each hominid species are shown, as are the numbers of groups in common between species.

● NUMBER OF BACTERIAL GROUPS PRESENT
● NUMBER OF BACTERIAL GROUPS SHARED

the transport of toxins from the gut to the blood, while propionate is involved in glucose metabolism and appetite regulation, and can prevent the formation of tumours in the colon.

Gut bacteria may generate "waste" products, some of which are valuable within the human body; these include vitamins B9 (folic acid) and K, which play critical roles in cell division and blood clotting, and fatty acids that strengthen the gut barrier. Products from some bacteria, such as lactic acid, can also suppress the growth of other pathogenic microbes, reducing infection.

Bacteria may be the simplest of life forms in terms of their structure, but their metabolism – their chemical reactions – can be complex and unique among organisms. They have a very short generation time, so can adapt and evolve to new environments very rapidly.

These two characteristics have made possible the swift transition of the human microbiome in its race to keep up with dietary shifts that have occurred in the last 10,000 years.

Beneficial or commensal bacteria can suppress the growth of pathogens

Food is key to human survival. Not only does it provide nutrition but it has a paramount impact on the lives of individuals and the structure of populations and societies. Some anthropologists have even asserted that developing the skills to prepare food helped to propel the evolution of our large, specialized brains. Most people in developed countries are largely insulated from the mechanisms of food production and its complex supply chains, but for our ancestors, these systems were a daily reality.

FOOD STORAGE AND SETTLEMENT

The hunter-gatherers of Europe's Mesolithic and the central African Middle and Upper Palaeolithic, were nomadic, and it is sometimes assumed that all people before 15,000 years ago lived similarly, in relatively small, mobile groups, moving several times per year in response to fluctuations in water and food availability. But archaeological investigations reveal that this is an oversimplified picture; some communities were able to store wild foods in large quantities, creating fixed

settlements long before the advent of full-time farming in the Neolithic. Acorns offer an example of how food storage affected human society: these nuts are available in autumn and can be dried and stored in huge numbers for later consumption. The Jōmon people of Japan (13,750–500 BCE) often lived in sedentary villages with huge storage pits for acorns, which

The food staples of the Jōmon people included plants, acorns, and other nuts.

Reservoirs have been found as far north as Hokkaido

HOKKAIDO

HONSHU

KYUSHU

SHIKOKU

More reservoirs are found in the east of Japan, suggesting that the eastern Jōmon were more sedentary

Distribution of Jōmon reservoirs

RYUKYU ISLANDS

CHESTNUT PROCESSING

Farming is not the only type of food production that encourages settled life. In the Jōmon period, many villages used wooden reservoirs to process chestnuts before drying and storing them for later consumption.

Food and society

All human societies, from hunter-gatherers to large nation-states, are organized around food production. Social bonding through the sharing of food is near-universal, and food culture may have even shaped brain evolution.

FISH–MAN SCULPTURE

Many elaborate sculptures have been discovered in the Iron Gates gorge, Serbia, on the river Danube. Although their functions are unclear, some point to a strong social and perhaps spiritual connection to the river – the main source of their food.

The texture of the figure recalls the scutes of a sturgeon, which could be fished from the Danube.

**Iron Gates gorge
sculpture, 6300–5900 BCE**

they processed using pestles and grinding stones. In the middle and late Jōmon periods, some villages built reservoirs for processing horse chestnuts, which need to be leached in water for two weeks before consumption to remove toxins.

Like acorns, marine resources also encourage sedentary living because they are difficult to deplete; catch one fish and another will come. It is no coincidence that many of the earliest permanent or semi-permanent settlements, such as the Iron Gates settlement on the Serbian Danube (active from the 11th to the 6th millennium BCE) were fishing communities.

THE FARMING REVOLUTION

The Neolithic marked the earliest transition from hunting, gathering, and fishing, to full-time farming. These developments were responsible for demographic changes on a massive scale. The increased food availability resulting from a shift to farming gave rise to higher birth rates and population sizes, and had a huge influence on settlement patterns and population movement. In southwest Asia, between 10,000 and 8000 BCE, Neolithic farming triggered early urbanization and the beginnings of state formation.

Farming reached Europe from western Asia along the Mediterranean coast and up the Danube river

The Falanan Agta (modern hunter-gatherers in the Philippines) share food between households. A household shares food unequally according to a hierarchy. Households ranked higher receive more food than those ranked lower in the hierarchy.

● PERCENTAGE OF DAYS IN WHICH FOOD IS SHARED WITH OTHER HOUSEHOLDS

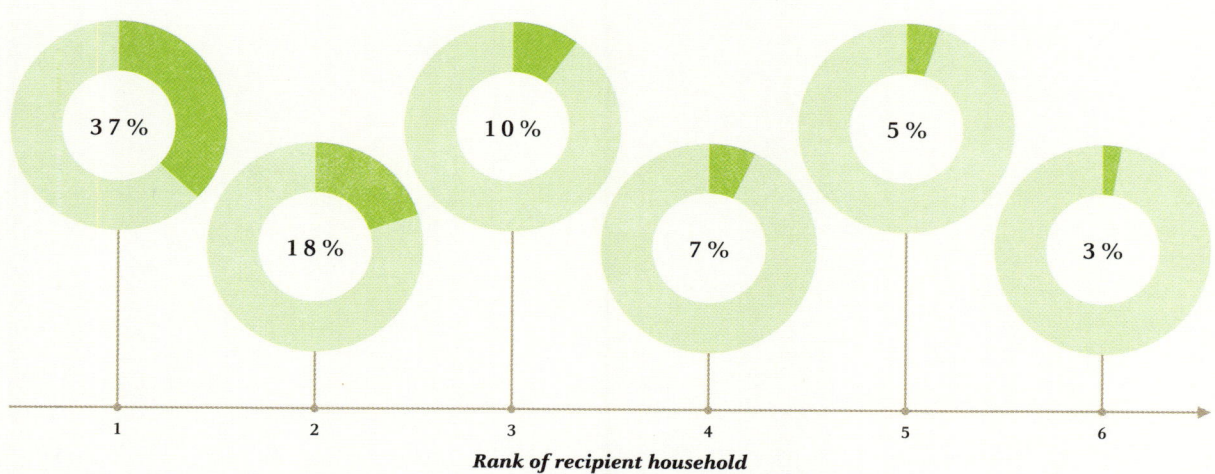

Rank of recipient household

Farming reached Europe via the shores of the Aegean around 6500 BCE and, propelled by population growth, spread throughout Europe over the next 2,000 years through a series of migrations. Wherever farming appeared, it caused large increases in the number and size of settlements, and in the complexity of their organization, also changing the material culture and genetic configuration of the continent.

FOOD SHARING AND SOCIETY

The term "commensality" derives from the Latin words com and mensa, meaning "with" and "dining table", so sharing a table

Food is fundamental to human sociality. Breaking bread together is the glue of all human societies, while pooling food allows us to insure against the risks of individual shortfall. Indeed, eating together is so ubiquitous that we even have a word for it: commensality. Many have argued that food sharing runs deep in our lineage, perhaps even preceding our genus *Homo*. Although we have many examples of ancient food-processing technology, the act of food sharing itself has left almost no archaeological trace. For this reason, researchers have often looked to modern hunter-gatherers to get answers. Among groups such as the Hadza, Ju/'hoansi, Agta, Batek, Mbendjele, and Mbuti, generous food-sharing is ubiquitous. These groups are often characterized by

"demand sharing"; when one person finds themselves with surplus, others will vociferously request a share. Among the Ju/'hoansi, relationships of mutual generosity even have an official name, being known as "hxaro ties". Anthropologists propose several explanations for such apparent generosity. Reciprocity – banking food in the bellies of neighbours and recalling the debt at a later date – is one popular theory. Sharing food to show-off your magnanimity or hunting prowess is another, as is the idea that sharing and gift-giving builds valuable social affiliations.

While such speculation is interesting, few studies have systematically addressed the motivations of hunter-gatherers directly. One such survey, led by Duncan Stibbard-Hawkes, asked Hadza foragers to explain their reasons: the answers were various and overlapping, from reciprocity to personal virtue to building friendships. Whether food-sharing has always been similarly generous is difficult to say, although it is such a fundamental aspect of contemporary human lives that it has almost certainly been important to our lineage for a very long time.

Breaking bread together is the glue of all human societies, while pooling food allows us to insure against the risks of individual shortfall.

Wild healing

Not everything taken by mouth is food. Humans and other animals also eat and drink for healing purposes. In historical times, remedies have been administered by shamans and medicine women and men, but they began as simple instinctive behaviour.

Early humans probably learned herbal medicine by trial and error – the same process by which natural selection has been shaping animal self-medication behaviour for millions of years. Across the animal kingdom, sick or injured animals seek out items that are not part of their usual diet and consume substances that have little or no nutritional value. The substances may even be avoided by healthy individuals and harmful in excess. Yet in small doses, they eliminate or prevent disease.

Wild animals eat or apply medicinal plants when they are sick or injured

The medicinal substance need not be ingested – it might be rubbed on the fur or skin and it may not always be plant in origin – it may be minerals or substances produced by animals. Red-fronted lemurs rub themselves with millipedes, which excrete the defensive substance benzoquinone. This is a known mosquito repellent, but because the lemurs rub it around their bottoms, scientists suspect it might repel itchy intestinal pinworms.

Ancient *Homo sapiens* seems to have taken insect-repelling measures too. In Sibudu Cave, South Africa, humans laid down bedding, the oldest of which dates to 77,000 years ago. It includes a top layer of Cape quince (*Cryptocarya woodii*). The aromatic leaves of this plant give off a camphor-like odour that would have repelled mosquitoes, fleas, and lice.

HERBAL REMEDIES

Self-medication is widespread in animals, and chimpanzees have shown the greatest repertoire. Since our closest living relatives treat themselves, we would expect our ancestors to do the same, but few traces of medicinal plants have survived from the Palaeolithic. One way they can be preserved is as charred remains, as in the case of *Ephedra* seed cones found in the Grotte des Pigeons, Taforalt, Morocco. At 15,000 years old, this is one of

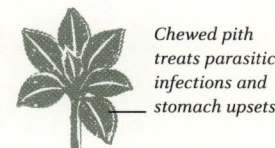

Chewed pith treats parasitic infections and stomach upsets

CHIMPANZEE
Bitterleaf
Vernonia amygdalina

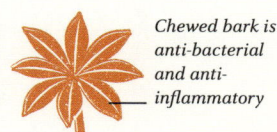

Chewed bark is anti-bacterial and anti-inflammatory

CHIMPANZEE
Dogbane
Alstonia boonei

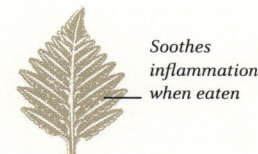

Soothes inflammation when eaten

CHIMPANZEE
Parasitic tri-vein fern
Christella parasitica

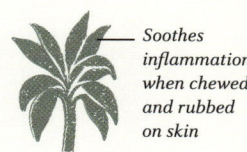

Soothes inflammation when chewed and rubbed on skin

ORANGUTAN
Dracaena cantleyi

APE MEDICINE
Great apes, like many other animals, treat themselves when feeling ill by picking out medicinal plants to eat instead of their usual food.

ANCIENT DRUG

The coniferous shrub *Ephedra* contains an active ingredient in its small, red seed cones that acts as a decongestant and bronchodilator, but can also narrow capillaries and reduce some types of bleeding.

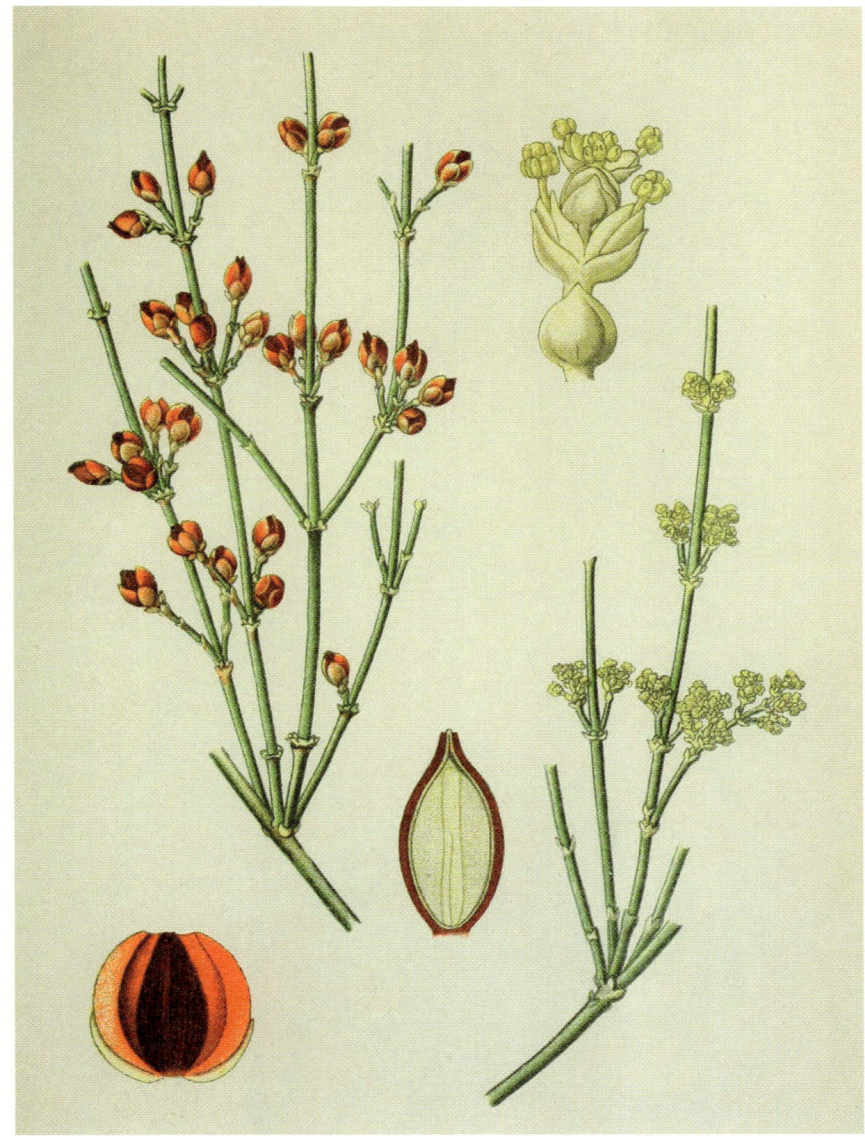

Africa's oldest known burials. *Ephedra* gave its name to the stimulant ephidrine – an alkaloid, like caffeine. Researchers think people may have taken it during a funerary ritual, or used it for its therapeutic effects. Writings from ancient China show *Ephedra* has been used in cold remedies for thousands of years, and it is still sold for that purpose in markets in parts of Africa and Asia.

We have no direct evidence that the Taforalt people took the stimulants. However, plant residues preserved in the plaque of ancient teeth are more telling. These have been detected in an older early human, a 49,000-year-old Neanderthal found in Spain. Researchers found chemicals from both yarrow and chamomile – plants that are of no use as food but with known bactericidal and

Remnants of herbal remedes can be found in fossil dental plaque

WILD HEALING

anti-inflammatory actions. They also found poplar DNA. Chewing on the bark or buds of this tree would yield salicylic acid, the active ingredient of aspirin. The scientists think this man was choosing painkillers to relieve aching in a visible dental abscess.

Some more recent remains feature evidence of the medicines themselves. The 5,000-year-old ice mummy, Ötzi (see p.100), lived with a litany of ailments and carried remedies to match. He packed dried sloes – helpful for alleviating symptoms of his suspected Lyme disease, and in his pouch was birch polypore fungus. This, and fragments of bracken found in his stomach, could both be attempts to kill off the parasitic worms that plagued him.

HEALTH CARE

Early humans did not only treat themselves – they cared for others. Humans are not alone in this either – caring has been observed in a wide range of animals, from ants to elephants. Animals usually care for close family members, but chimpanzees have been observed caring for unrelated injured individuals by licking, finger-pressing, and applying chewed medicinal plant material to their wounds. Evidence of human health care dates back 1.8 MYA to a toothless *Homo georgicus* individual (see p.245). Later, there are many instances of Neanderthals who lived with serious injuries for years. Several individuals buried 60,000 years ago at Shanidar, Iraq, for instance, exhibited damage in their skeletons. Shanidar I, in particular, had received a crushing blow to his face that likely damaged his eye, and had an atrophied, and possibly amputated, arm, hearing loss, multiple fractures, an injured leg and foot, and degenerative joint disease.

Health care given to close family members is widespread in animals

Chimpanzees have been seen treating the wounds of unrelated injured individuals.

ÖTZI'S AILMENTS
The natural ice mummy known as Ötzi, found in the Ötzal Alps, carried dried medicinal mushrooms and berries with him in his pouch.

Fronds of this fern are toxic, yet they were found in Ötzi's gut

Bracken

Sloes (blackthorn fruit) were found dried in Ötzi's pouch

Blackthorn

Birch polypore fungus was also carried in addition to his fire-lighting fungus

Birch polypore

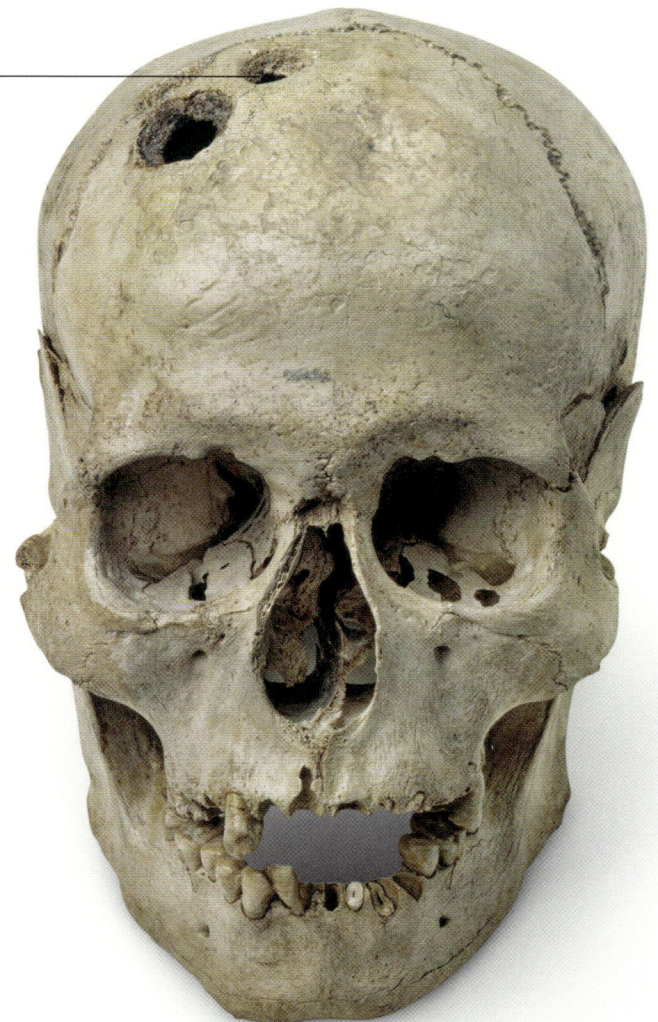

Rounded edges are signs that the bone has begun to heal

Trepanned skull

PREHISTORIC SURGERY
This Bronze Age skull was excavated from Jericho. It was drilled with holes when its owner was alive in around 2200–2000 BCE. We know this individual survived the procedure, since the bone shows signs of healing.

STONE BLADE
Obsidian, or volcanic glass, can be formed into such a fine blade that specialist surgeons use them to this day. This surgical knife comes from 19th-century Java, but similar tools may have been used in Neolithic surgery.

SURGICAL INTERVENTION

Other than feeding and protecting these sick and injured people, what medical care could others give them? One skeleton unearthed in Borneo had both lower leg bones – the tibia and fibula – ending in a clean cut, estimated to have been made 6–9 years before death and free from infection. At 31,000 years old, this is by far the earliest amputation known. More recently, a characteristic form of prehistoric surgery became mysteriously globally widespread – that of trepanning – the drilling or scraping of a hole through the skull, for reasons unknown. Many patients survived this treatment, with the earliest known healed trepanation made around 7300–6220 BCE in Ukraine.

From experiments, blades like this are known to be capable of trepanning

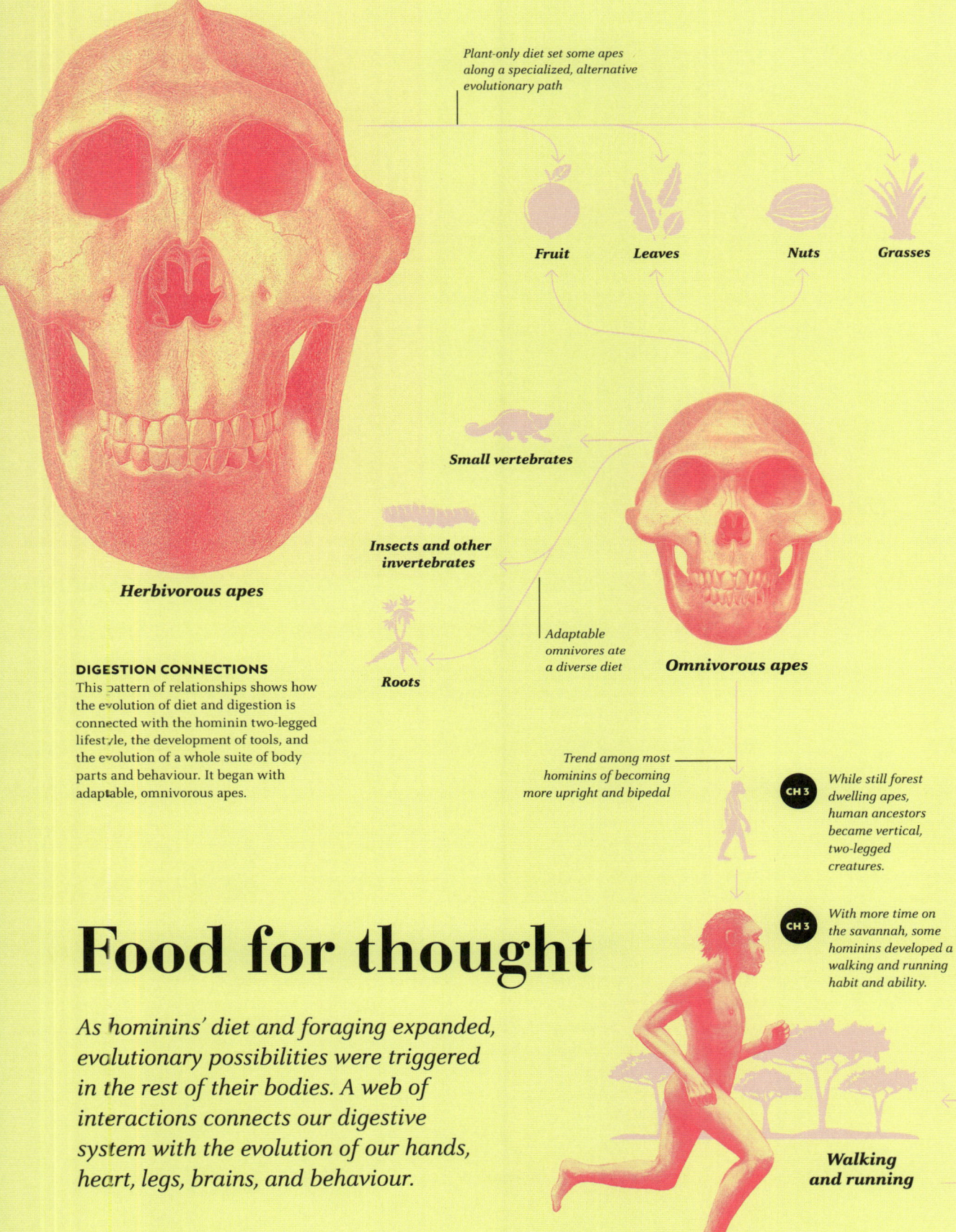

Plant-only diet set some apes along a specialized, alternative evolutionary path

Fruit

Leaves

Nuts

Grasses

Herbivorous apes

Small vertebrates

Insects and other invertebrates

Adaptable omnivores ate a diverse diet

Omnivorous apes

Roots

DIGESTION CONNECTIONS

This pattern of relationships shows how the evolution of diet and digestion is connected with the hominin two-legged lifestyle, the development of tools, and the evolution of a whole suite of body parts and behaviour. It began with adaptable, omnivorous apes.

Trend among most hominins of becoming more upright and bipedal

CH 3 *While still forest dwelling apes, human ancestors became vertical, two-legged creatures.*

CH 3 *With more time on the savannah, some hominins developed a walking and running habit and ability.*

Food for thought

As hominins' diet and foraging expanded, evolutionary possibilities were triggered in the rest of their bodies. A web of interactions connects our digestive system with the evolution of our hands, heart, legs, brains, and behaviour.

Walking and running

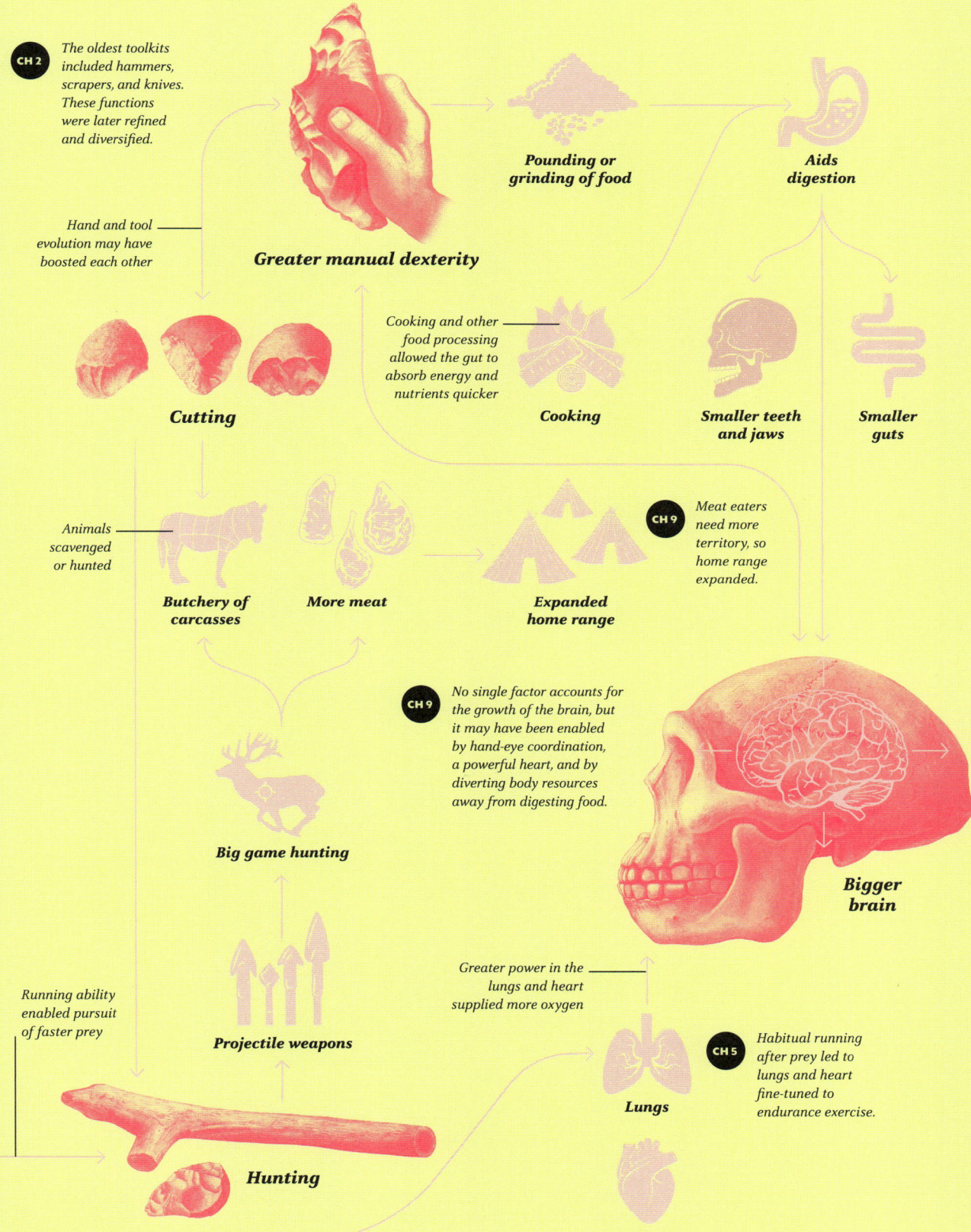

CH 2 The oldest toolkits included hammers, scrapers, and knives. These functions were later refined and diversified.

Pounding or grinding of food

Aids digestion

Hand and tool evolution may have boosted each other

Greater manual dexterity

Cooking and other food processing allowed the gut to absorb energy and nutrients quicker

Cutting

Cooking

Smaller teeth and jaws

Smaller guts

Animals scavenged or hunted

Butchery of carcasses

More meat

Expanded home range

CH 9 Meat eaters need more territory, so home range expanded.

CH 9 No single factor accounts for the growth of the brain, but it may have been enabled by hand-eye coordination, a powerful heart, and by diverting body resources away from digesting food.

Big game hunting

Bigger brain

Running ability enabled pursuit of faster prey

Greater power in the lungs and heart supplied more oxygen

Projectile weapons

Lungs

CH 5 Habitual running after prey led to lungs and heart fine-tuned to endurance exercise.

Hunting

Heart

179

FOOD FOR THOUGHT

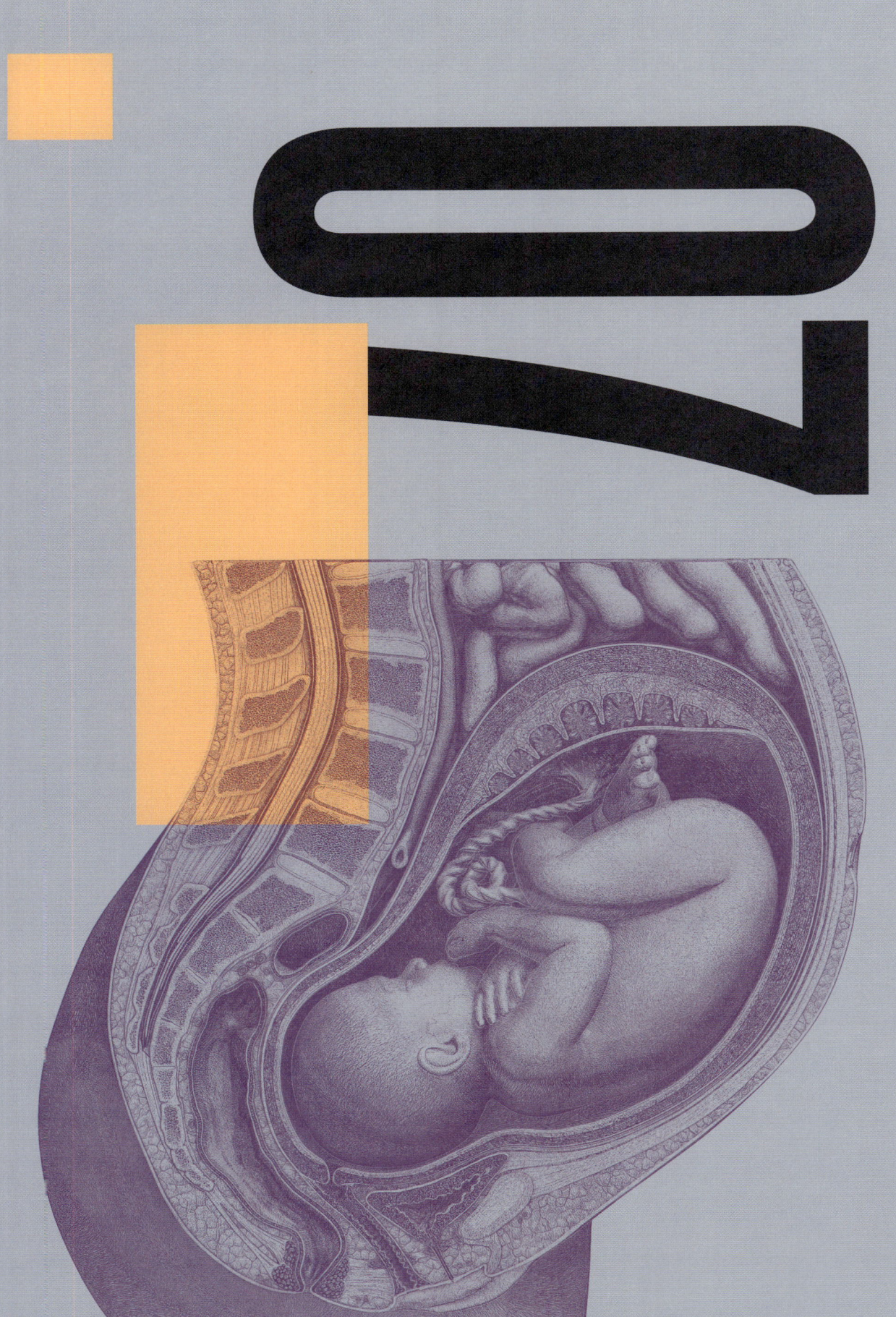

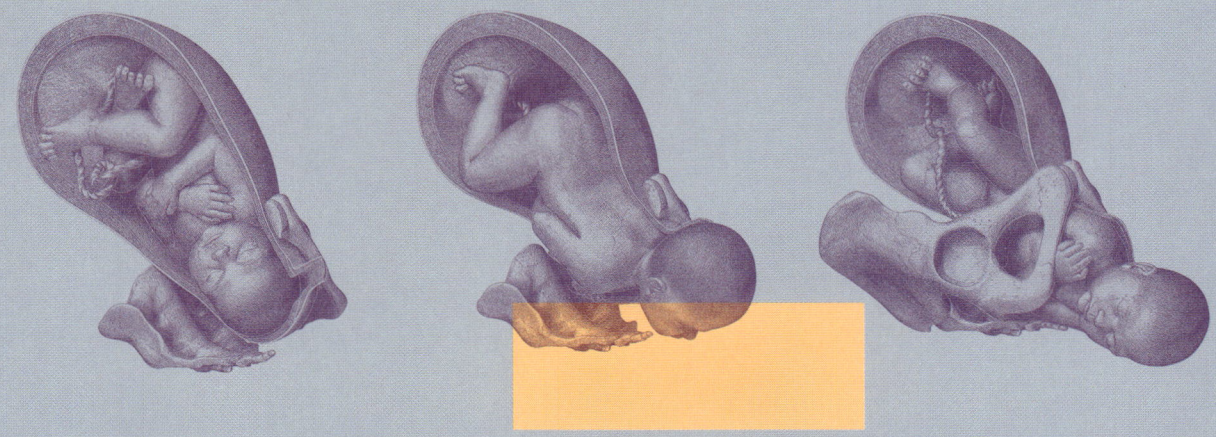

Reproduction

Human babies develop slowly and require nurture and teaching input from a supportive social group. This could be linked with the human tendency to pair-bond, and might even have extended our lifespan through the power of grandmotherly care.

MALES AND FEMALES

Two contrasting types of reproductive cells evolved: large female ones (eggs) contained resources for the offspring. Tiny male ones (sperm) contribute only genetic diversity.

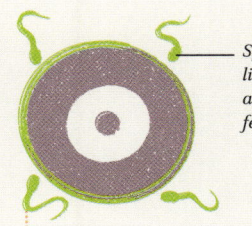

Sperm contain little else but DNA and compete to fertilize eggs

LIFE

Embryo is nurtured inside mother's body

4 BYA–220 MYA

New subunits join the dividing DNA chain, creating two daughter chains

PLACENTA

Some mammals evolved a structure that passes oxygen and nutrients from mother to unborn offspring. This innovation allowed young to develop more efficiently.

MAMMALS

DNA REPLICATORS

All life forms are based on self-replicating long-chain molecules of DNA. They contain coded instructions for making proteins and regulating processes that create the body of a living organism.

220–55 MYA

PRIMATES

Jurassic mammals may have oozed milk from modified sweat glands

55 MYA–PRESENT

DNA passed from bacterium to bacterium, mixing up their code

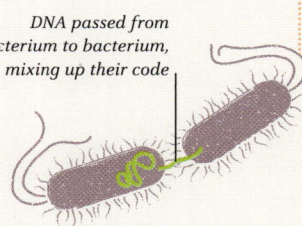

BREASTFEEDING

Early mammals extended their parental care by providing food to their young from modified sweat glands called mammary glands.

Embryo's blood vessels extend into the placenta

MIXING OF DNA

Some living organisms gained an advantage by mixing their DNA code into new instructions. Bacteria met to exchange their DNA. Eventually, life forms combined the DNA mixing process with their replication process – and in so doing, invented sexual reproduction.

INTIMATE PLACENTA

The placenta in humans and many of their primate relatives is of a very intimate kind, in which the embryo's blood vessels are bathed in pools of the mother's blood.

Carers may include grandparents and aunts

SHARING PARENTING

Social primates form long-term relationships with many members of their social group. Family bonds and friendships allow mothers to share the burden of childcare.

HOW REPRODUCTION CHANGES

Replication and the passing on of genes are the driving forces of life, and early in life's history, two alternative strategies evolved: female and male. Female mammals later evolved new ways of nurturing their young, while male mammals were often shaped by the competition for mates.

Two billion years of sex

For animals, the story of sex is a story of success, and humans are no exception. From breastfeeding to childcare, humans elaborate on ancient patterns in ways that have allowed us to number in the billion.

Some baboon societies feature large, dominant males that monopolize harems of females

SEXUAL DIFFERENCES

Primates are social animals and can pair up in a range of ways. In some species, the sexes differ in size and shape. In these species, the larger males often compete fiercely over mates.

PAIR BONDS

The recent ancestors of humans show a trend towards minimal sexual differences between males and females, which may be related to more parallel reproductive strategies.

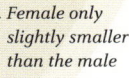

Female only slightly smaller than the male

Reproduction is one of the defining features of life, having been around since its origin, at least 3.8 billion years ago. However, the evolution of sexual reproduction between 2 and 1.5 billion years ago introduced an element of novelty that also supercharged a population's ability to adapt to environmental change. DNA was now recombined in new ways in every generation, increasing variation in species and giving some individuals a reproductive edge compared to their peers.

Sexual reproduction would diversify across animal groups, but one group, the mammals – emerging at the same time as the first dinosaurs – made innovations connected with intensive care of their offspring. Mothers would begin breastfeeding – nourishing their young by food excreted directly from glands in their skin – and eventually they would evolve "live births" – their young having already developed inside the mother's uterus instead of an external egg.

SOCIAL LIFE

One group of mammals, the primates, would evolve particularly complicated social systems. Although each mother and child was closely bonded, they also had lifelong relationships with other group members. Their social life even influenced their physical characteristics, since it affected the way individuals had to compete to determine who could mate with whom. These social bonds have also helped us survive and thrive globally.

Humans are social animals. As in any social creature, the size and structure of social groups, and the pattern of mating within them, are products of evolutionary forces. Those forces can be traced all the way back to the origin of the sexes – male and female.

The human pair bond

The sexes – males and females – occur in plants and animals and predate the origin of both, more than a billion years ago. The sexes are subject to opposing evolutionary forces, which affect their reproductive strategies and mating behaviour to this day – even in modern humans.

STARTING SMALL

Throughout the animal kingdom, females produce large, immobile sex cells called eggs, while males produce tiny, mobile sex cells called sperm. Sperm are much, much smaller than eggs, allowing males to create millions of them for every egg produced. As such, male reproductive success (which is all-important in terms of his evolution) is limited mainly by how many eggs he can fertilize, while female reproductive success is limited by the survival of her offspring. What's more, because fertilization occurs within the female reproductive system, females can generally guarantee they are the mother of their offspring, while males are often left wondering. These disparities have positive and negative consequences for both sexes and often lead to different priorities.

A lot of sexual behaviour can be linked to the disparity between female and male sex cells

MALES VERSUS FEMALES

While many animals have sexes that are virtually indistinguishable outside of their sex organs, the sexes may evolve further physical differences. In many egg-laying animals, females are larger than males, as larger females can lay more eggs. Males meanwhile may display bright colours or dramatic ornaments, as females prefer to mate with them. Males may also differ in having weapons, such as antlers or large canine teeth, that help them compete with other males for mates, food, or territory. Animals with notable sex differences are called sexually dimorphic.

Among primates, some species have nearly identical sexes, while others display some of the most extreme sexual differences seen in mammals. Sex differences can manifest in the size of canine teeth, or the size of the entire body, colouration in fur or skin, nose shape, call repertoire, coat length, and more. And because primates tend to be social animals, these differences are also linked with social behaviour.

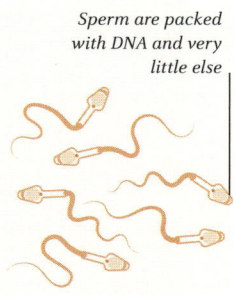

Sperm are packed with DNA and very little else

Sperm cells

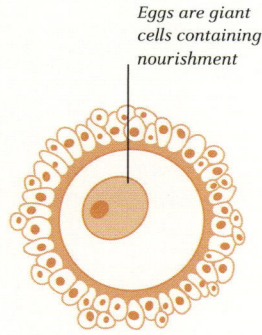

Eggs are giant cells containing nourishment

Egg cells

OPPOSITES ATTRACT

The evolution of eggs and sperm represent two opposite strategies of maximizing reproduction. Female eggs focus on nurturing the offspring. Male sperm concentrate on spreading their genes as widely as possible.

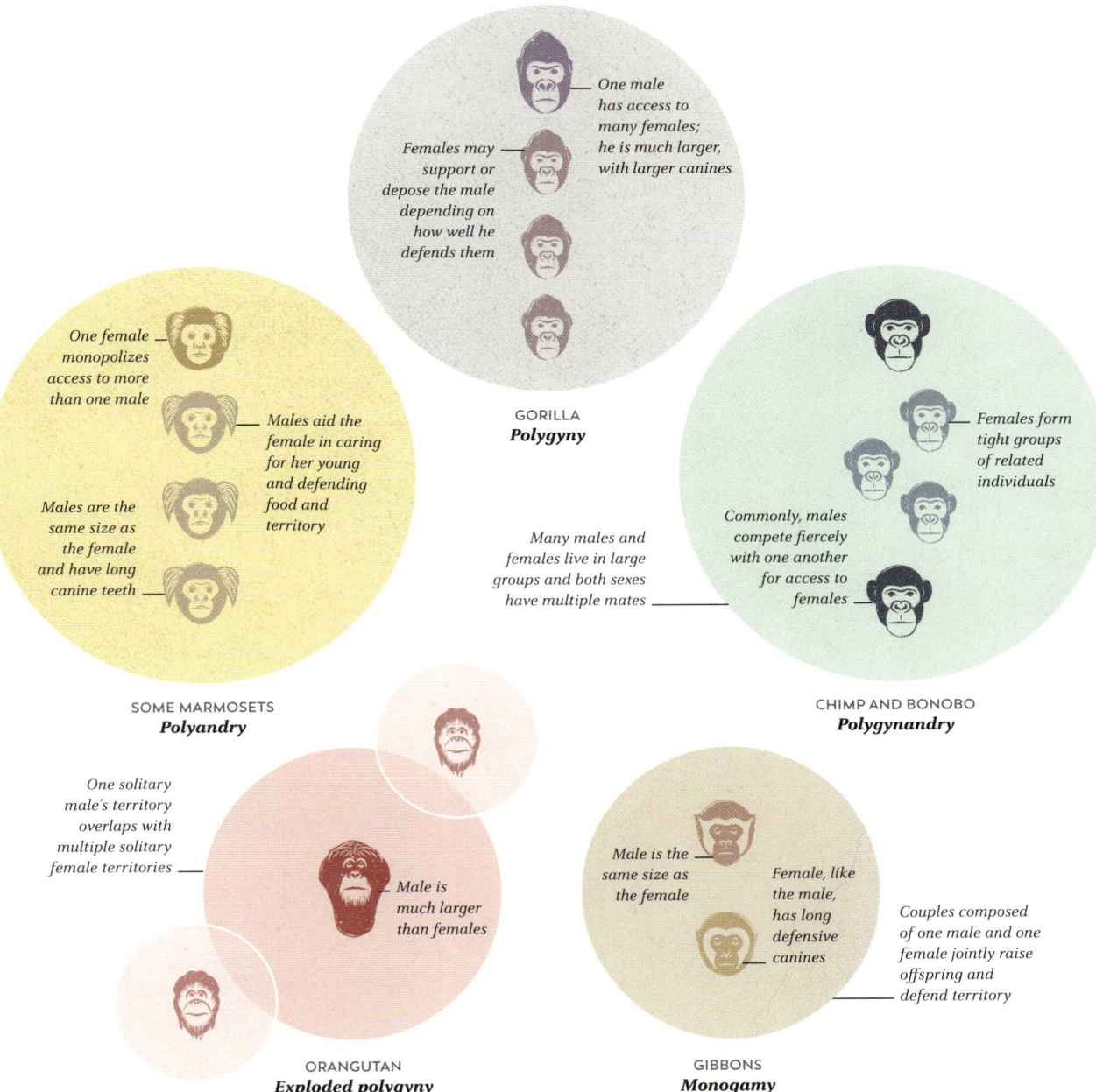

GORILLA
Polygyny

One male has access to many females; he is much larger, with larger canines

Females may support or depose the male depending on how well he defends them

SOME MARMOSETS
Polyandry

One female monopolizes access to more than one male

Males aid the female in caring for her young and defending food and territory

Males are the same size as the female and have long canine teeth

CHIMP AND BONOBO
Polygynandry

Females form tight groups of related individuals

Commonly, males compete fiercely with one another for access to females

Many males and females live in large groups and both sexes have multiple mates

ORANGUTAN
Exploded polygyny

One solitary male's territory overlaps with multiple solitary female territories

Male is much larger than females

GIBBONS
Monogamy

Male is the same size as the female

Female, like the male, has long defensive canines

Couples composed of one male and one female jointly raise offspring and defend territory

PRIMATE MATING SYSTEMS
Social primates show all the usual mating patterns of social animals: polygyny (one male and multiple females), polyandry (one female and multiple males), monogamy (one female and one male), and polygynandry (multiple males and multiple females).

While some primates are solitary, most live in groups due to the grouping patterns of females. Females tend to group to defend food, territory, and one another from predators and aggressive males. Depending on the species, females may be joined by one or multiple males, affecting how males and females pair up and reproduce. The pattern of mating is known as the species' "mating system".

Primates exhibit a wide diversity of mating systems, from strict monogamous pair bonding to free for alls! In fact, they show the full range of mating systems seen in social animals.

Which primate mates with whom constitutes the species' mating system

THE HUMAN PAIR BOND

PHYSICAL CLUES

Males and females within each mating system have very different strategies. Some mating systems and strategies can lead to physical differences, while others lead to males and females looking similar. Typically, high sexual dimorphism indicates females in these societies are picky and males are competing fiercely in fights for female attention. This is usual in a polygynous society, in which successful males control access to multiple females. It leads to a strong selection for long canines or large bodies in males. Not all differences point to male–male aggression, however. Contrasts in colour, calls, or ornaments such as beards are commonly due to courtship and female choice.

Low sexual dimorphism suggests either more peaceful males or similarly violent and well-armed females. These situations are more typical of monogamous societies. In species where females are promiscuous, males need not fight for attention or access, but their sex cells must be numerous and quick to beat other sperm to the egg. Bonobo males, for instance, have larger testes, proportionally, than other great apes and produce more sperm. Since physical characteristics are linked with mating patterns, we can estimate a primate's mating system without observing them – and this includes the extinct relatives of humans. We

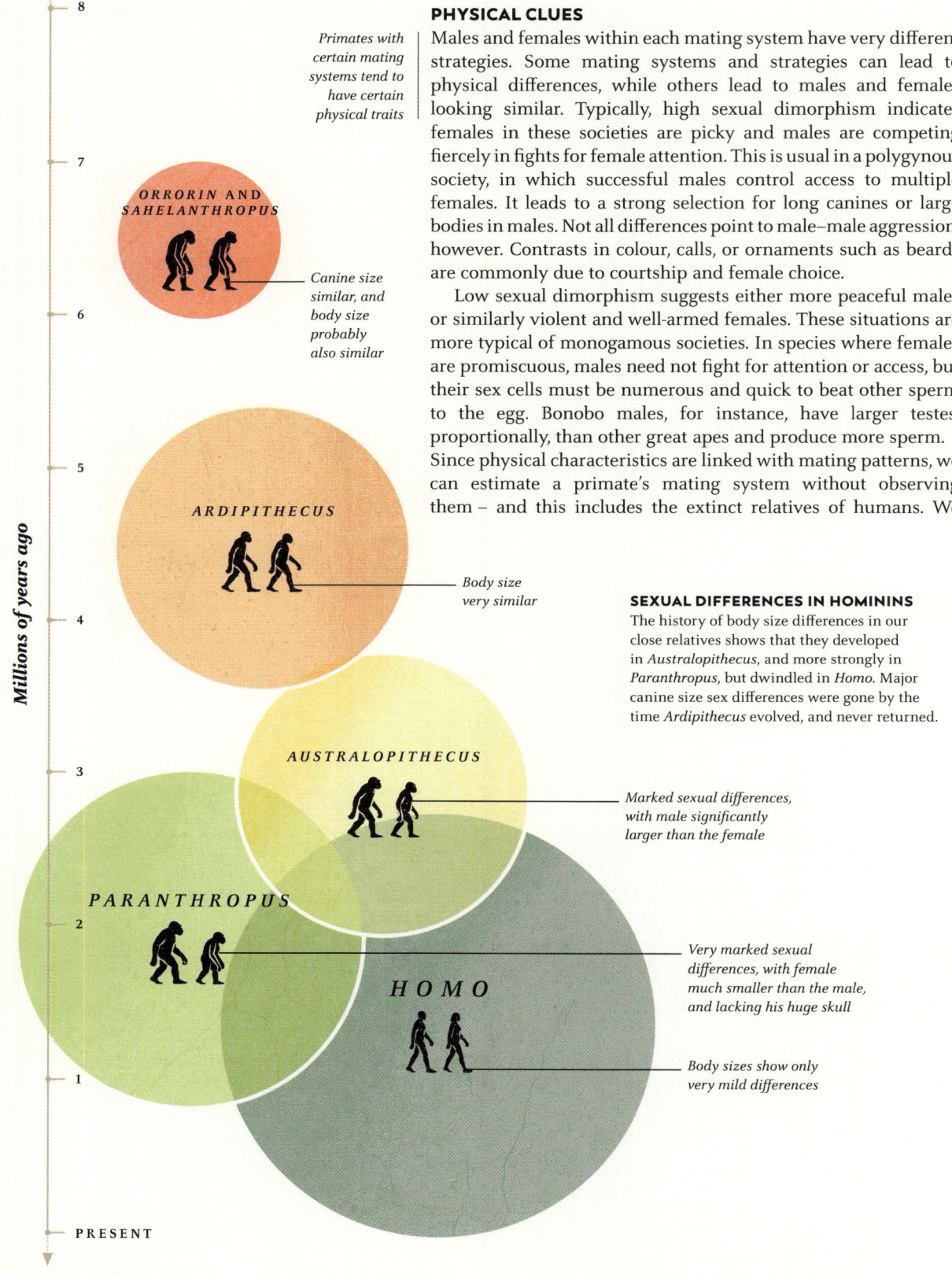

Primates with certain mating systems tend to have certain physical traits

ORRORIN AND SAHELANTHROPUS

Canine size similar, and body size probably also similar

ARDIPITHECUS

Body size very similar

AUSTRALOPITHECUS

Marked sexual differences, with male significantly larger than the female

PARANTHROPUS

Very marked sexual differences, with female much smaller than the male, and lacking his huge skull

HOMO

Body sizes show only very mild differences

Millions of years ago

8

7

6

5

4

3

2

1

PRESENT

SEXUAL DIFFERENCES IN HOMININS

The history of body size differences in our close relatives shows that they developed in *Australopithecus*, and more strongly in *Paranthropus*, but dwindled in *Homo*. Major canine size sex differences were gone by the time *Ardipithecus* evolved, and never returned.

Using physical traits, we can deduce how humans' extinct relatives may have mated

cannot observe their behaviour, nor do their testes size or sperm swimming speed preserve. However, we do have evidence of their comparative body size and canine tooth size. Canine dimorphism was lost in our earliest relatives, while body size differences seem to have hung around until the earliest members of our genus (*Homo*).

WHAT IS THE HUMAN MATING SYSTEM?

Today, humans exhibit low to no sexual dimorphism in stature and canine tooth size, with moderate sex differences in lean muscle mass. Male testes size and sperm swimming speeds are intermediate between a polygynous and a monogamous species. Taken together, our characteristics are most similar to pair-bonding species, with the traits suggesting polygyny or polygynandry perhaps pointing to a recent evolutionary history with such mating systems.

But we can also observe today's modern human behaviour – does it match our physical traits? Modern humans are exceptionally diverse in that virtually every mating system can be found in some form across the species. Our behaviour is flexible: social norms in human societies include

Humans are exceptionally diverse in that virtually every mating system can be found in some form.

monogamy, polygyny, and polyandry, and same-sex coupling. However, even within polygyny-friendly societies, there seems to be a tendency to pair-bond. A pair bond does not necessarily imply lifelong monogamy, since a range of human societies allow remarriage after the death of a spouse or divorce – but it is a long-term pair bond nonetheless.

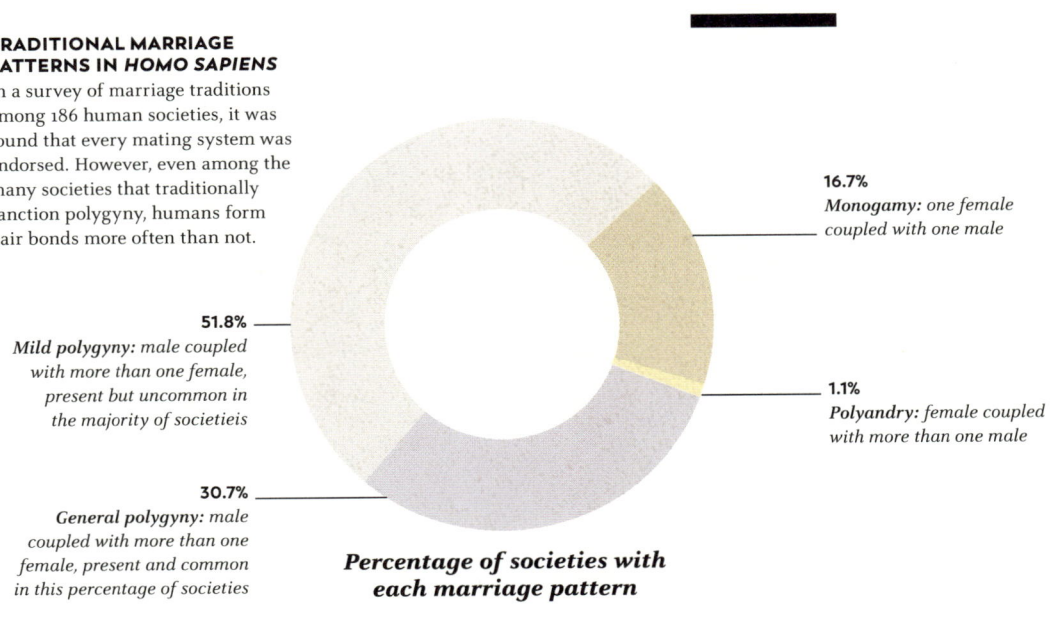

TRADITIONAL MARRIAGE PATTERNS IN *HOMO SAPIENS*

In a survey of marriage traditions among 186 human societies, it was found that every mating system was endorsed. However, even among the many societies that traditionally sanction polygyny, humans form pair bonds more often than not.

16.7%
Monogamy: one female coupled with one male

51.8%
Mild polygyny: male coupled with more than one female, present but uncommon in the majority of societieis

1.1%
Polyandry: female coupled with more than one male

30.7%
General polygyny: male coupled with more than one female, present and common in this percentage of societies

Percentage of societies with each marriage pattern

THE HUMAN PAIR BOND

A Jurassic mammal probably had a shell gland rather than a uterus

Ancestral Jurassic mammal

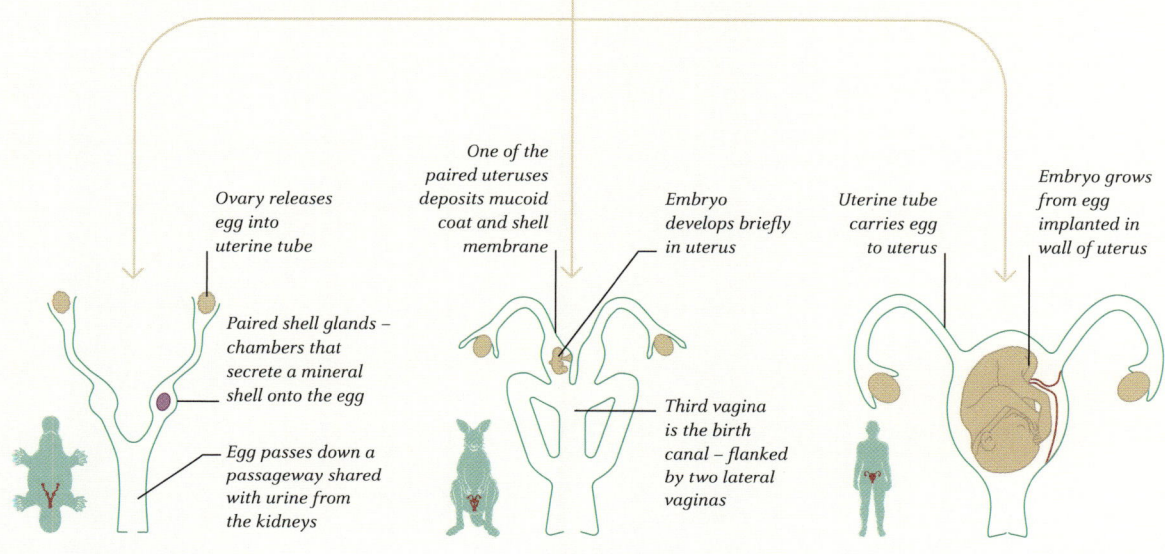

Ovary releases egg into uterine tube

Paired shell glands – chambers that secrete a mineral shell onto the egg

Egg passes down a passageway shared with urine from the kidneys

Platypus (monotreme)

One of the paired uteruses deposits mucoid coat and shell membrane

Embryo develops briefly in uterus

Third vagina is the birth canal – flanked by two lateral vaginas

Kangaroo (marsupial)

Uterine tube carries egg to uterus

Embryo grows from egg implanted in wall of uterus

Human (placental mammal)

Mothers of invention

Early female mammals were egg-laying, like their reptile ancestors. Placental mammals evolved later – and were very successful. The placenta is an incredible organ, which allows the mother to support her unusually demanding fetus in utero.

Humans, like most mammals, develop young within their bodies – the process known as pregnancy – before giving birth. Pregnancy involves the growth of a brand-new temporary organ, the placenta, that provides the developing young with oxygen and nutrients through an umbilical cord. How did such a system evolve?

EX-EGG-LAYERS
The ancestors of mammals did not give birth to live young, but laid eggs as most other animals do. In fact, some mammals – the monotremes, such as the platypus and echidnas – still lay eggs today. In their egg-laying system, the ovaries release large, yolk-filled eggs that are fertilized in the uterine tube, after which they travel to the shell gland – a chamber that secretes a leathery, protective layer on the egg's surface. The shell

EMBRYOS AND EGGS

Whether embryos develop in an egg or a uterus, they have the same basic structures in common, revealing the evolutionary continuity between egg layers and placental mammals.

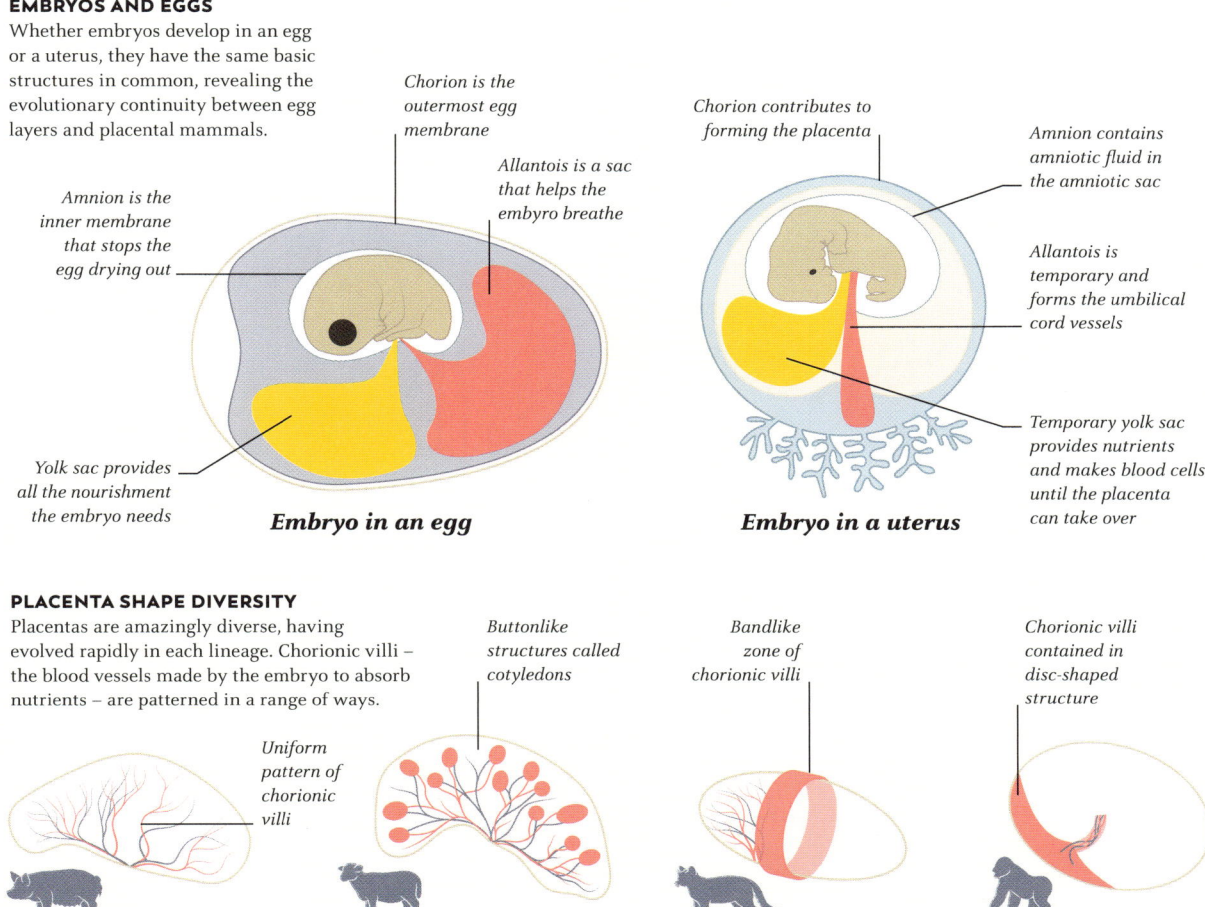

Chorion is the outermost egg membrane

Allantois is a sac that helps the embryo breathe

Amnion is the inner membrane that stops the egg drying out

Yolk sac provides all the nourishment the embryo needs

Embryo in an egg

Chorion contributes to forming the placenta

Amnion contains amniotic fluid in the amniotic sac

Allantois is temporary and forms the umbilical cord vessels

Temporary yolk sac provides nutrients and makes blood cells until the placenta can take over

Embryo in a uterus

PLACENTA SHAPE DIVERSITY

Placentas are amazingly diverse, having evolved rapidly in each lineage. Chorionic villi – the blood vessels made by the embryo to absorb nutrients – are patterned in a range of ways.

Buttonlike structures called cotyledons

Bandlike zone of chorionic villi

Chorionic villi contained in disc-shaped structure

Uniform pattern of chorionic villi

Diffuse

Cotyledonary

Zonary

Discoid

gland is also called a uterus, and it is indeed the evolutionary precursor of the uterus of other mammals. Marsupials retain some of the egg laying set-up, such as a shell-precursor called a mucoid coat and a shell membrane, but they give birth to tiny young, which develop further in a pouch. Placental mammals produce small, yolk-less eggs that are fertilized in the uterine tubes, as in monotremes and marsupials, and transported to the uterus. However, in placental mammals, the embryo implants in the uterus.

A PLETHORA OF PLACENTAS

Once a fertilized egg implants into the innermost lining of the uterus (the endometrium), the mother's body and the embryo begin forming the placenta. This embryonic stage is also when organs and the amniotic sac form. The placenta is one of the few examples in the animal kingdom in which an organ is made from two separate individuals. While one half is an extension of the mother's body, the other half is composed of the embryo's tissues and blood vessels, constructed and controlled by its genes.

The placenta in most primates is disc-shaped and haemochorial. Haemochorial placentas have a very limited cell barrier, which allows intimate contact between the mother's bloodstream and the chorion – the outermost membrane around the developing baby. This allows an efficient delivery of nutrients from mother to offspring, growing larger babies with larger brains faster than other placenta types. However, because of this close contact, the offspring are at risk of being rejected by the mother's immune system and the mother is at risk of dying during birth

The energy demands of a human baby have driven the evolution of an invasive and intimate placenta

MOTHERS OF INVENTION

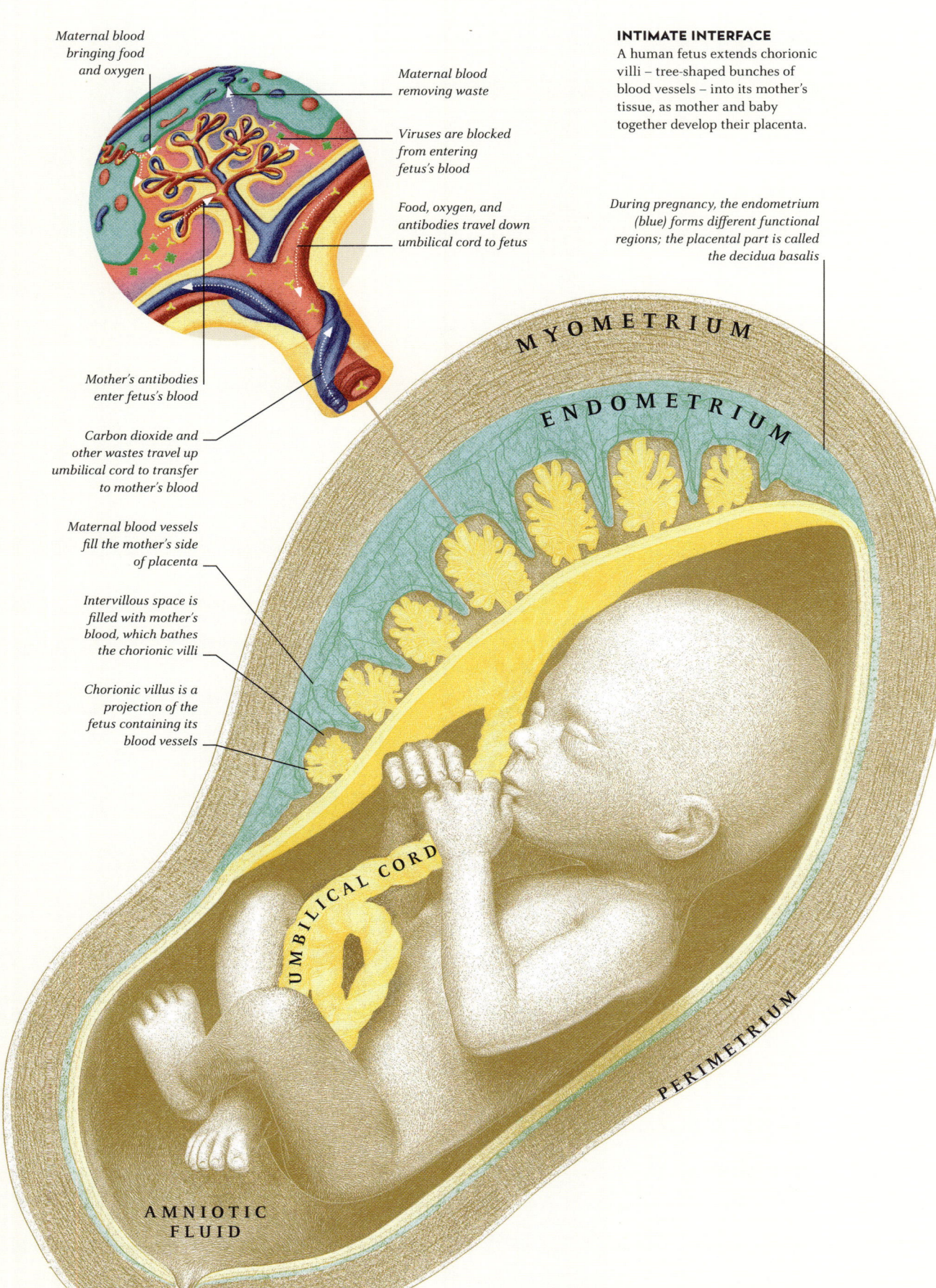

Maternal blood bringing food and oxygen

Maternal blood removing waste

Viruses are blocked from entering fetus's blood

Food, oxygen, and antibodies travel down umbilical cord to fetus

Mother's antibodies enter fetus's blood

Carbon dioxide and other wastes travel up umbilical cord to transfer to mother's blood

Maternal blood vessels fill the mother's side of placenta

Intervillous space is filled with mother's blood, which bathes the chorionic villi

Chorionic villus is a projection of the fetus containing its blood vessels

INTIMATE INTERFACE
A human fetus extends chorionic villi – tree-shaped bunches of blood vessels – into its mother's tissue, as mother and baby together develop their placenta.

During pregnancy, the endometrium (blue) forms different functional regions; the placental part is called the decidua basalis

MYOMETRIUM

ENDOMETRIUM

PERIMETRIUM

UMBILICAL CORD

AMNIOTIC FLUID

WHY IS OVULATION A SECRET?

Females might gain several potential evolutionary advantages by hiding their fertile period. One or more of these boosts could be the reason why humans evolved concealed ovulation.

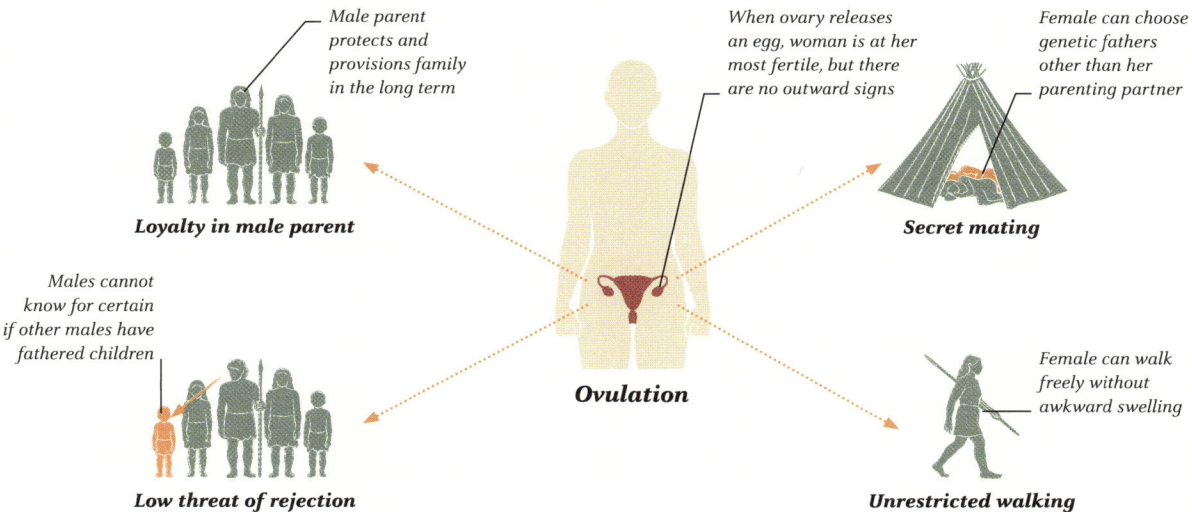

Male parent protects and provisions family in the long term

Loyalty in male parent

Males cannot know for certain if other males have fathered children

Low threat of rejection

When ovary releases an egg, woman is at her most fertile, but there are no outward signs

Ovulation

Female can choose genetic fathers other than her parenting partner

Secret mating

Female can walk freely without awkward swelling

Unrestricted walking

due to blood loss from placental detachment. The placenta must also maintain a delicate balance of resources: the baby must get enough to develop but not siphon too much to deplete the mother, killing them both.

SECRET FERTILITY

Humans do not advertise when they are fertile, and this could have numerous evolutionary advantages

Many primate females advertise when they are ovulating and receptive to mating with changes in coloration on their rump or chest. Humans conceal their ovulation – there are no external physical changes. Why this feature has evolved is not known with certainty, but there are several potential explanations. If a male does not know when babies are conceived, he is more likely to stick around after mating to exclude other males and guarantee fatherhood. This leads to dads protecting and provisioning their kids as well as bonding with their mother. When male primates are certain of their paternity, disaster can strike. Dominant males may threaten or even kill young that could compete with their own offspring. Concealed ovulation reduces this risk, due to the same uncertainty in male parentage that keeps male parenting partners loyal. There are still more advantages. A female can secretly mate outside her monogamous pairing, giving her choice in both her parenting partner and the father of her offspring (should they be different). Finally, many primates with obvious signs of ovulation have cumbersome genital swelling. Our ancestors may have lost these when they got in the way of walking upright!

WHY DO HUMANS BLEED?

The shedding of the uterine lining each month, known as menstruation in primates, is nearly unique to monkeys and apes, although it has evolved separately in the elephant shrew, the spiny mouse, and a handful of bats. The reason for expelling the lining (as opposed to absorbing it as most mammals do) is also a mystery. One idea is that it is due to the invasive nature of our placenta and high energy demands of our babies. By building up the lining of the uterus each month, the mother has a defensive "jump-start" in preventing the embryo from invading too deeply should she become pregnant – an expensive but worthwhile insurance policy.

MOTHERS OF INVENTION

The human birth dilemma

Humans face a unique struggle with childbirth due in part to the contrast between the large brains and wide shoulders of our offspring and our comparatively narrow hips and birth canals. Several theories attempt to explain why this situation evolved.

The pelvis is crucial in how an animal moves as well as how it reproduces. Young develop in the abdomen for both egg-layers and birth-givers alike. This means that the pelvis and birth canal must facilitate the passage of the egg or baby out of the mother and into the world no matter how large or small the baby or egg may be. For most mammals, labour is comparatively short and relatively uneventful.

Compared to the majority of other primates, the movement of the human fetus through the birth canal and bony pelvis is a tight squeeze. This is in part why labour in humans can last over a day, while birth for chimpanzees, whose newborn babies have small heads, are finished in a few hours. Other species of primate that give birth to relatively large-headed offspring have evolved different solutions to the problem, such as flexion or extension of the baby's neck or, in the case of squirrel monkeys, relaxation of the ligaments that hold the pelvic bones together, allowing the baby to pass through unimpeded.

Other primates either have smaller babies or have evolved other adaptations to large offspring

RISKY BUSINESS

Human birth, on the other hand, can be hazardous. The human fetus must undergo a series of different movements to manoeuvre through the birth canal, and getting stuck at any point risks the lives of both parties. Thanks to modern midwifery and obstetrics, the worldwide maternal mortality rate (maternal deaths during or within 42 days after pregnancy per 100,000 live births) was around 1.9 per cent in 2023.

RELATIVE SIZES AT BIRTH

Animals vary wildly in offspring size at birth or hatching. Larger babies are riskier for the mother to deliver (but may have a greater chance of survival afterwards), while smaller babies are easier to birth.

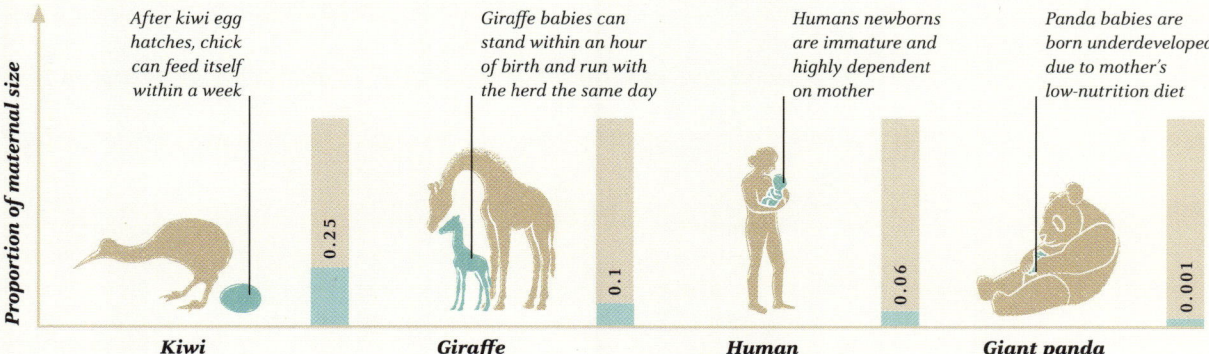

Proportion of maternal size

After kiwi egg hatches, chick can feed itself within a week

Kiwi — 0.25

Giraffe babies can stand within an hour of birth and run with the herd the same day

Giraffe — 0.1

Humans newborns are immature and highly dependent on mother

Human — 0.06

Panda babies are born underdeveloped due to mother's low-nutrition diet

Giant panda — 0.001

PRIMATE BIRTH ADAPTATIONS

Humans are not the only primates to give birth to large-headed offspring, but other species have evolved different solutions to the problem. For example, squirrel monkey mothers' pubic bones open up to allow their very large-headed babies to pass through.

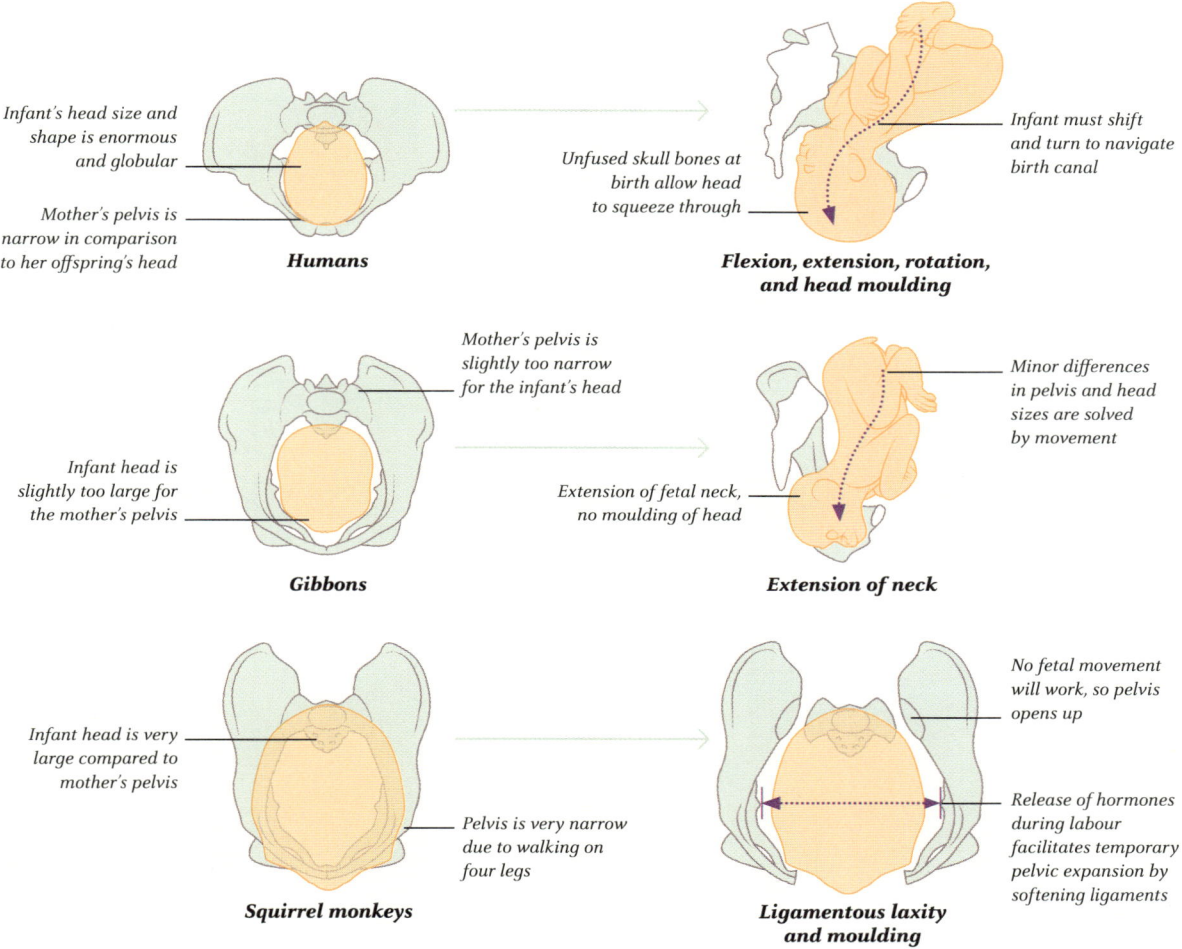

Infant's head size and shape is enormous and globular

Mother's pelvis is narrow in comparison to her offspring's head

Humans

Unfused skull bones at birth allow head to squeeze through

Infant must shift and turn to navigate birth canal

Flexion, extension, rotation, and head moulding

Mother's pelvis is slightly too narrow for the infant's head

Infant head is slightly too large for the mother's pelvis

Gibbons

Extension of fetal neck, no moulding of head

Minor differences in pelvis and head sizes are solved by movement

Extension of neck

Infant head is very large compared to mother's pelvis

Pelvis is very narrow due to walking on four legs

Squirrel monkeys

No fetal movement will work, so pelvis opens up

Release of hormones during labour facilitates temporary pelvic expansion by softening ligaments

Ligamentous laxity and moulding

Access to modern medicine reduces the dangers of human childbirth to some degree

In communities without access to modern medical facilities, such as most hunter-gatherer groups, maternal mortality can rise as high as 4 per cent. Current studies argue that 22 per cent of all maternal deaths are due to obstructed labour (inability of the fetal head or shoulders to fit through the maternal birth canal), but the incidence of non-lethal obstructed labour is far more common. It seems evolutionarily nonsensical for human birth to be so difficult and hazardous. The risk would be reduced were humans to evolve wider birth canals, babies

with smaller heads, or if birth occurred earlier, when young were less developed. Several theories have been proposed to explain why human birth is complicated compared to most other mammals, including our closest relatives.

WALKING UPRIGHT

Historically, it has been argued that our troubles with birth are due to the competing selective pressures of walking upright and having large-brained offspring. A narrow pelvis has been proposed to facilitate more efficient upright

193

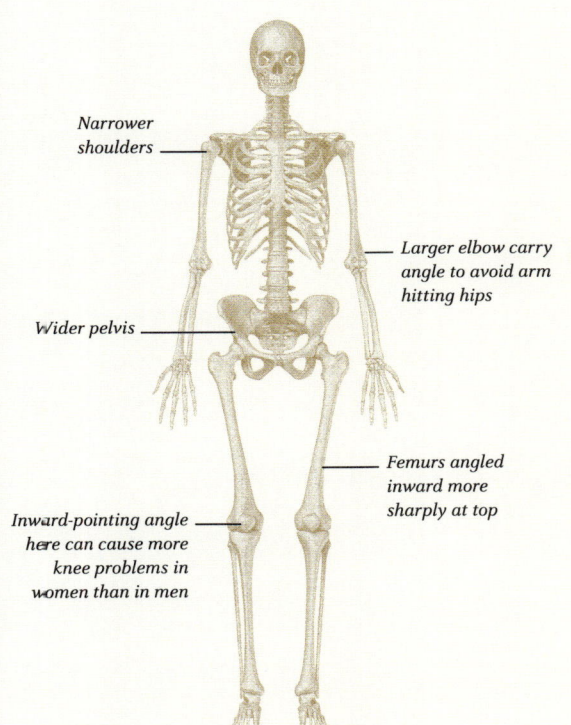

Narrower
shoulders

Larger elbow carry
angle to avoid arm
hitting hips

Wider pelvis

Femurs angled
inward more
sharply at top

Inward-pointing angle
here can cause more
knee problems in
women than in men

Female skeleton

Wider
shoulders

Smaller elbow
carry angle as arm
does not hit hips

Narrower,
taller pelvis

Femurs more
vertical

Male skeleton

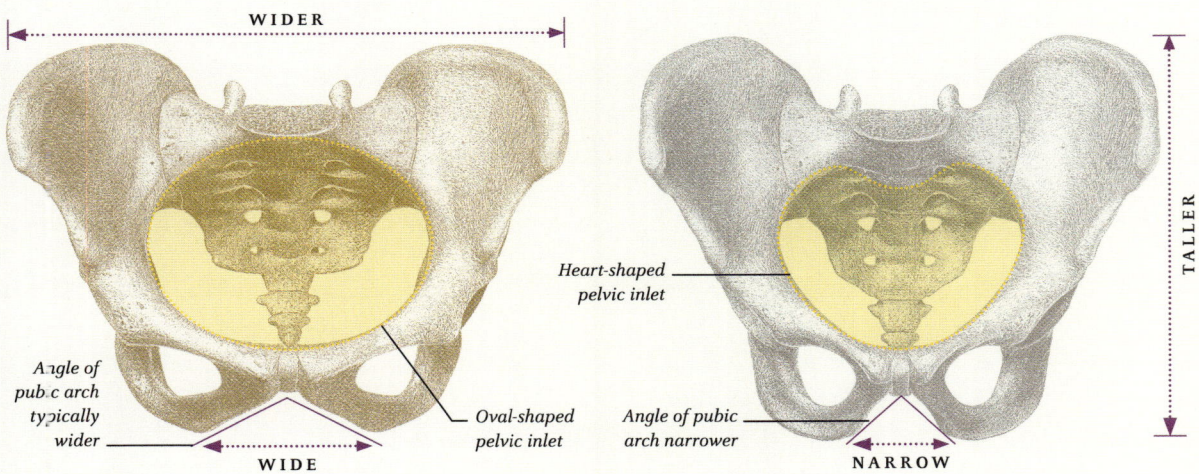

WIDER

Angle of
pubic arch
typically
wider

WIDE

Oval-shaped
pelvic inlet

TALLER

Heart-shaped
pelvic inlet

Angle of pubic
arch narrower

NARROW

FEMALE PELVIS
*The generalized female pelvis is
broader: from the pelvic inlet to the
pubic arch to allow for childbirth.*

MALE PELVIS
*The generalized male pelvis is
taller and narrower: from the
pelvic inlet to the pubic arch.*

SEXUAL DIMORPHISM

Humans are extremely diverse, but there are a few physical
traits that are typically different between males and
females. Some of these differences are due to hormonal
exposure during puberty or the requirements of birth.

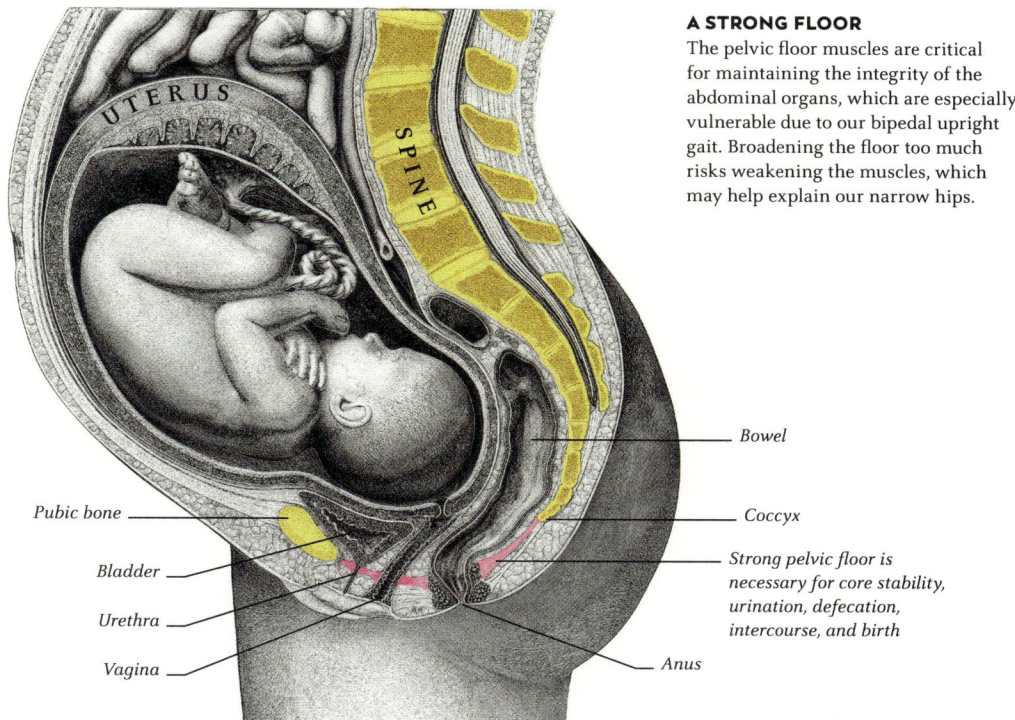

The pelvic floor muscles are critical for maintaining the integrity of the abdominal organs, which are especially vulnerable due to our bipedal upright gait. Broadening the floor too much risks weakening the muscles, which may help explain our narrow hips.

Bowel

Coccyx

Strong pelvic floor is necessary for core stability, urination, defecation, intercourse, and birth

Anus

Pubic bone

Bladder

Urethra

Vagina

UTERUS

SPINE

Abdominal organs

The original theory to explain the human birth dilemma simplified it to just two factors

walking and running, while a broader pelvis allows for bigger-brained and bigger-bodied babies to pass through unobstructed. Humans evolved to be both bipedal and big-brained, so this theory suggests that our pelvis (and birth canal) is "trapped" between two optimal shapes.

Known as the "Obstetric Dilemma", this theory supposedly explained the sexual dimorphism of the human pelvis: males, with narrower pelvises, were better adapted to upright walking, while females were held back from optimal locomotive efficiency by the constraints of childbirth. However, biomechanical studies show minimal differences in walking efficiency in males versus females. Indeed, a wider pelvis in females can even be associated with increased walking efficiency, particularly while carrying loads.

PELVIC FLOOR INTEGRITY

Another theory to explain the human birth dilemma suggests that a narrow pelvis is an essential prerequisite for upright movement on other health grounds. A pelvic opening that is too wide would reduce the strength of the pelvic

floor muscles, leading to an increased risk of organ prolapse and incontinence. Pelvic floor integrity may also explain why the human birth canal is so twisted compared to other apes: a uniform canal could cause strain or contort the mother's spine, or both. The best shape for an upright species' birth canal is thus the "sideways" oval that moves into a "forward" oval that human females have evolved, which optimizes health while requiring some fetal contortion.

This may also be the reason why humans have not evolved the mechanisms that other primates have to ameliorate the difficulties of birthing large offspring, such as the pelvic bones opening up during birth. Relaxin is the hormone that softens the ligaments to increase pelvic flexibility during birth. Too much relaxin in pregnant women can cause pelvic instability and immense pain, as well as an increased risk of organ prolapse and musculoskeletal injury. It may therefore be that our bipedal setup keeps our pelvis more rigid, and our birth canal constrained, but for a different reason than was originally proposed in the Obstetric Dilemma.

Habitual bipedalism evolved in our hominin ancestors, such as *Ardipithecus* and *Australopithecus*, some 5–4 million years ago, bringing with it constraints on the pelvis. Australopithecine mothers had similar pelvises in many ways to modern mothers, but their smaller brains may have made birth easier. Later increases in brain size seem to have led to the difficulties associated with human birth.

BORN BIG

Human babies are paradoxically born with large bodies, but neurologically underdeveloped brains. At birth, our brains are less than 30 per cent of their adult size, meaning most of our brain growth occurs outside of the womb during the first years of life. Human babies' skull bones are separated by thin sheets of fibrous tissue, and can overlap during childbirth, as the head squeezes though the birth canal – often leaving a newborn baby with a "squished" head shape, which then recovers.

Human babies are "altricial" (underdeveloped and highly dependent at birth), hence the need for the extra life stages in humans (see pp.202–07) . Giving birth to babies with underdeveloped brains has been suggested as an adaptation to our upright gait: if the birth canal cannot become wider, make the baby's brain, and thus head, smaller when passing through. Although this may be part of the story, altriciality in human offspring may have a different cause. During pregnancy, the mother's body spends unbelievable amounts of energy to develop and

Neurologically underdeveloped newborns may have evolved due to the narrow human pelvis

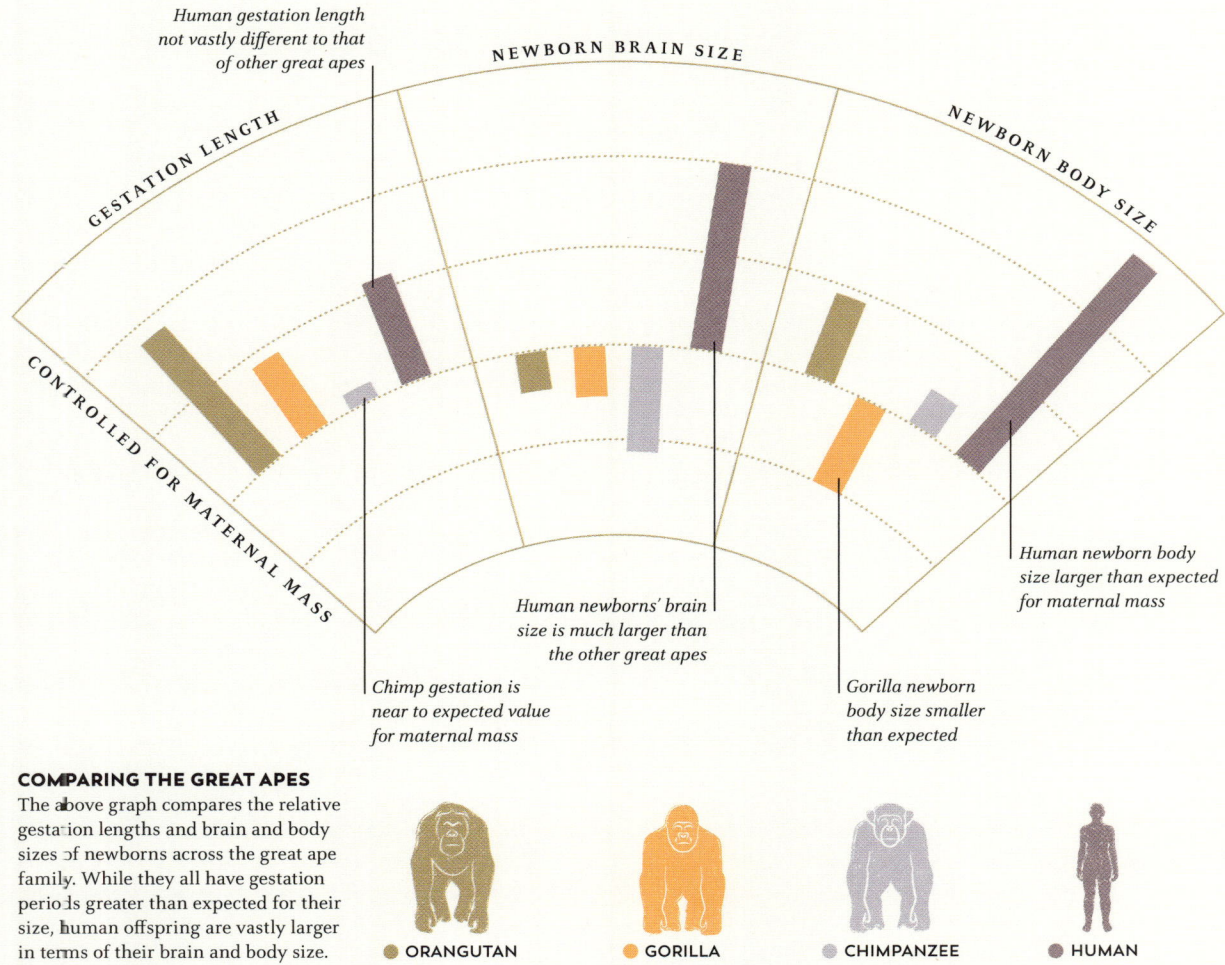

Human gestation length not vastly different to that of other great apes

GESTATION LENGTH

NEWBORN BRAIN SIZE

NEWBORN BODY SIZE

CONTROLLED FOR MATERNAL MASS

Human newborn body size larger than expected for maternal mass

Human newborns' brain size is much larger than the other great apes

Chimp gestation is near to expected value for maternal mass

Gorilla newborn body size smaller than expected

COMPARING THE GREAT APES

The above graph compares the relative gestation lengths and brain and body sizes of newborns across the great ape family. While they all have gestation periods greater than expected for their size, human offspring are vastly larger in terms of their brain and body size.

● ORANGUTAN

● GORILLA

● CHIMPANZEE

● HUMAN

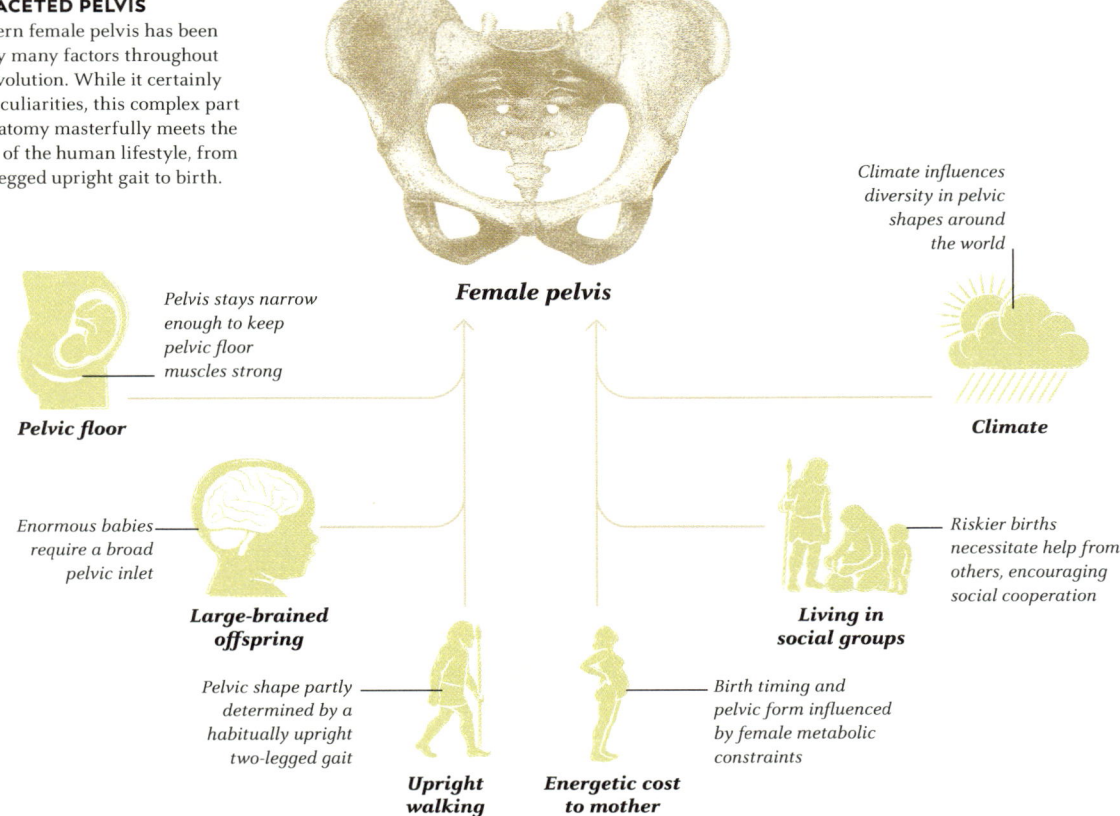

MULTIFACETED PELVIS
The modern female pelvis has been shaped by many factors throughout human evolution. While it certainly has its peculiarities, this complex part of our anatomy masterfully meets the demands of the human lifestyle, from our two-legged upright gait to birth.

Female pelvis

Climate influences diversity in pelvic shapes around the world

Pelvis stays narrow enough to keep pelvic floor muscles strong

Pelvic floor

Climate

Enormous babies require a broad pelvic inlet

Riskier births necessitate help from others, encouraging social cooperation

Large-brained offspring

Living in social groups

Pelvic shape partly determined by a habitually upright two-legged gait

Birth timing and pelvic form influenced by female metabolic constraints

Upright walking

Energetic cost to mother

Development of the fetus is energetically costly to the mother

nourish the growing baby inside her, including one of the most expensive organs: the brain. For a human baby to mature inside the mother to the corresponding neurological developmental milestones seen in chimpanzees at birth, human pregnancy would need to last 18–21 months. Instead, most of this neural and physical development takes place after birth, with energy coming from breast milk and food provisions during the early years of life.

A COSTLY BUSINESS

The "EGG" (energetics of gestation and growth) hypothesis proposes that the timing of birth, and the development of the baby at that time, is determined by the mother's ability to continue growing it inside her. Once a human baby becomes too "expensive" for the mother's body to upkeep and develop, her body produces hormones that start the run-up to birth. If correct, this theory adds yet another variable to the complex and multifactorial relationship

between human upright gait, anatomy, and reproduction. It must also be remembered that, as with the evolution of any living organism, our anatomy and physiology has been influenced by many factors and the human body has evolved in a mosaic, or piecemeal, way over time.

KEEPING COOL

Some scientists have suggested that climatic factors may also have influenced the dimensions of the human birth canal and pelvis. Animal populations that live in colder climates have stockier bodies than those in warmer climates, because thicker bodies help to conserve heat, and lankier bodies are more efficient at dispelling heat. Human populations follow this pattern to some extent, and therefore warmer climates may have kept early human pelvises from becoming too broad.

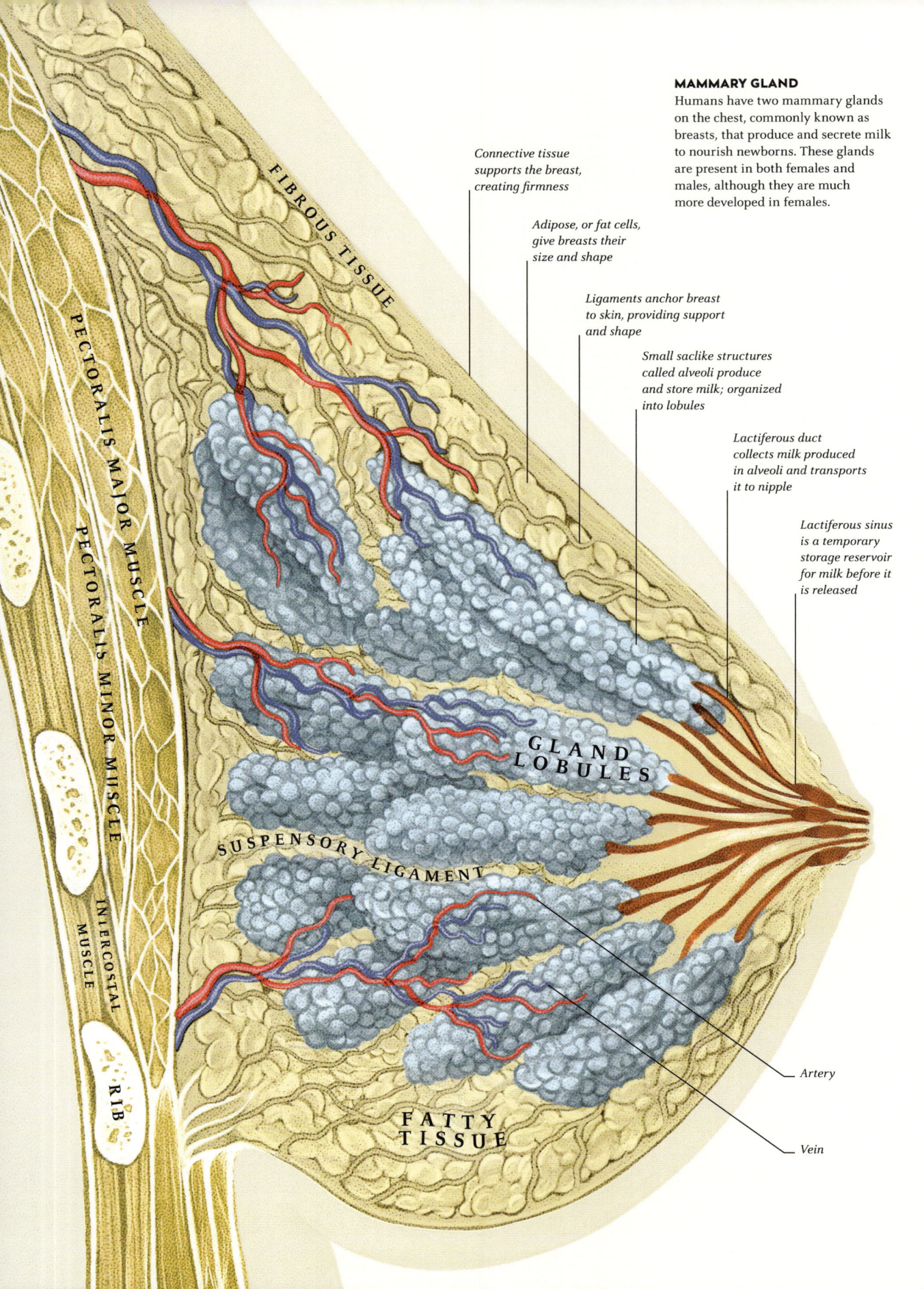

MAMMARY GLAND
Humans have two mammary glands on the chest, commonly known as breasts, that produce and secrete milk to nourish newborns. These glands are present in both females and males, although they are much more developed in females.

Connective tissue supports the breast, creating firmness

Adipose, or fat cells, give breasts their size and shape

Ligaments anchor breast to skin, providing support and shape

Small saclike structures called alveoli produce and store milk; organized into lobules

Lactiferous duct collects milk produced in alveoli and transports it to nipple

Lactiferous sinus is a temporary storage reservoir for milk before it is released

FIBROUS TISSUE

PECTORALIS MAJOR MUSCLE

PECTORALIS MINOR MUSCLE

INTERCOSTAL MUSCLE

RIB

GLAND LOBULES

SUSPENSORY LIGAMENT

FATTY TISSUE

Artery

Vein

Bonded at birth

Lactation is a unique evolutionary feature that continues to be a defining trait of all modern mammals. Through nourishment and connection, breastfeeding increases the survival chances of a mother's young.

Mammary glands are organs in mammals, including humans, that produce milk. A mother's milk nourishes her young, jumpstarts their immune system, and facilitates bonding. Milk is generally composed of fat, protein, sugar, vitamins, minerals, and water, keeping young fed and hydrated. Milk also contains antimicrobial and anti-inflammatory compounds, antibodies from the mother, hormones, and growth factors, all of which help develop an offspring's immunity.

FIRST MILK

The very first mammary glands are believed to have developed in ancient synapsids (mammal ancestors) around 310 million years ago (MYA), evolving from skin structures similar to apocrine sweat glands. In their earliest form, these glands likely served to coat newly laid eggs with antimicrobial mucus, which kept bacteria and fungus at bay while also preventing the eggs from drying out. By 200 MYA, therapsids, a more advanced group of mammal ancestors descended from synapsids, bore more refined versions of these glands, which were now producing nutrient-rich milk in place of the thinner secretions of their forebears.

The first forms of milk contained a wide range of nutritional substances, which allowed for more immature offspring at hatching. Eggs were incredibly vulnerable, and the earlier young hatched the sooner they became mobile, developed survival mechanisms, and avoided predators – all while continuing their growth on mother's milk.

SUCKLING

Some of the earliest mammals were similar to today's monotremes (a group of mammals that includes platypuses) – they laid eggs and lacked nipples. Instead, milk seeped from mammary patches on their underside, with young lapping it up as it was secreted.

Excluding monotremes, such as platypuses, all mammals secrete milk through teats or nipples

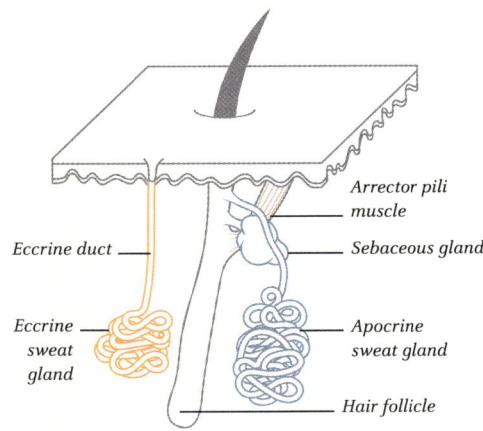

Arrector pili muscle

Eccrine duct

Sebaceous gland

Eccrine sweat gland

Apocrine sweat gland

Hair follicle

ANCIENT APOCRINE

Mammary glands evolved from apocrine glands (skin structures that are typically found in the skin, eyelid, ear, and breast). Apocrine glands generally secrete thick, oily fluid onto the skin. However, they are modified in the breast to secrete milk.

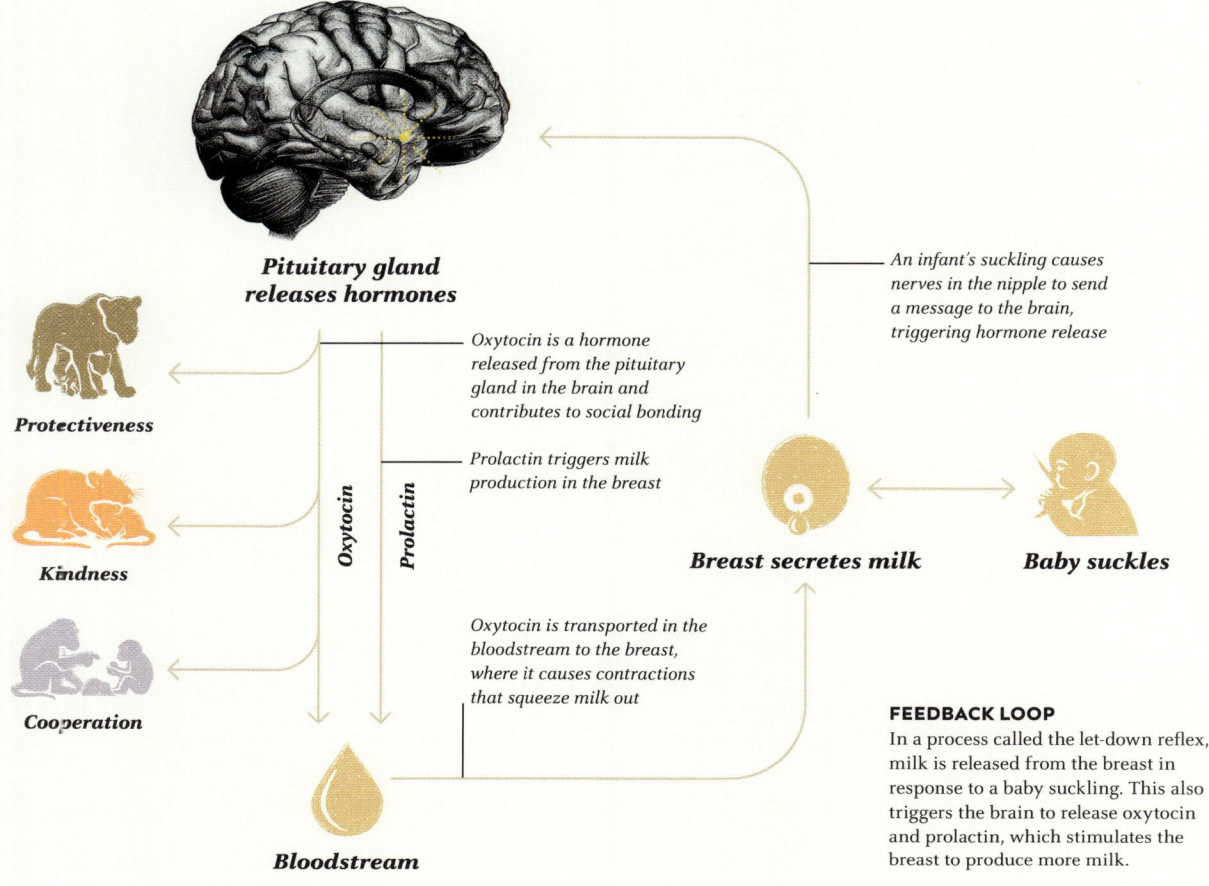

Pituitary gland releases hormones

Oxytocin is a hormone released from the pituitary gland in the brain and contributes to social bonding

Prolactin triggers milk production in the breast

Protectiveness

Kindness

Cooperation

Oxytocin

Prolactin

Oxytocin is transported in the bloodstream to the breast, where it causes contractions that squeeze milk out

Bloodstream

An infant's suckling causes nerves in the nipple to send a message to the brain, triggering hormone release

Breast secretes milk

Baby suckles

FEEDBACK LOOP
In a process called the let-down reflex, milk is released from the breast in response to a baby suckling. This also triggers the brain to release oxytocin and prolactin, which stimulates the breast to produce more milk.

Excluding this group, every other mammal secretes milk through teats or nipples. These protrusions of skin are connected to the milk glands, enabling direct flow and ensuring the milk is not wasted while nursing. Young initiate milk production when they suckle – the action causes the mother's brain to release the hormones prolactin, which triggers milk production, and oxytocin, which causes contractions that squeezes the milk towards the baby's mouth.

The roots of suckling can once again be found in the therapsids, with fossils from around 250 MYA displaying the beginnings of a secondary palate (made of bone and soft tissue) separating the nasal passages from the mouth. This evolved into a fully closed palate by around 247 MYA, as evidenced by fossils of cynodonts, a lineage that contains the ancestors of all mammals. A secondary palate allowed mammals to breathe while eating, and is a key mammalian adaptation that allows human newborns to breathe and nurse at the same time. The cheeks are used to form a vacuum seal that draws in milk.

THE FIRST BOND
In addition to stimulating mammary gland contraction, the hormone oxytocin also facilitates the oldest social bond in nature. In mammals, oxytocin is associated with affection, protection, trust, recognition, and a variety of other social bonds. However, it is also a crucial catalyst in

Oxytocin facilitates the oldest social bond in nature: between mother and offspring.

forging the connection between mother and offspring. Cultivating this bond was essential for the survival of young mammals. Mothers that formed stronger bonds with their infants were more protective and prioritized their nourishment, conferring a survival advantage onto their offspring over those with distant mothers. These young would have eventually passed this relationship onto their own children, cementing it as a fundamental part of mammalian existence.

COMPLETE NUTRITION

Human breastfeeding is not unique. Like other mammals, when we first start nursing we produce a thicker milk called colostrum. Yellowish in colour, colostrum is particularly nutrient-dense and loaded with antibodies and antioxidants that strengthens an infant's immune system. Once ingested, the colostrum coats a baby's intestines, introducing healthy gut bacteria and preventing harmful microbes from being absorbed. Colostrum also flows slowly from a mother's nipples, which helps offspring learn how to suck, swallow, and breathe during feeding.

Around 3 days after birth, the colostrum transitions into what is considered more typical breast milk – a thinner, creamy-white liquid that contains an array of vitamins, fat, and lactose. Transitional milk is high-calorie, which promotes

an offspring's growth, and continues to build up their gut strength by adding beneficial bacteria to the existing microbiome.

After around 15 days, mothers begin to produce mature milk, which is around 90 per cent water and key for a baby's hydration. The other 10 per cent is composed of carbohydrates, fats, and proteins, which are necessary for the continued growth and development of the infant.

MAMMAL MOTHERS

As a deep and fundamental part of mammal evolution, it is perhaps unsurprising that breastfeeding differs so little in humans compared to the rest of the world's mammals. Beyond nutrition, some studies have suggested that mother's milk may have a hand in shaping offspring's personalities and developing their social abilities. Researchers have suggested that high levels of cortisol, the stress hormone, in breast milk may result in negative temperaments in infants and risk-averse behaviour in adults. Observed in human children, this trend is also present in other mammals, such as rats and rhesus monkeys.

Like other apes, we hold our young in our arms with their faces angled towards our own while breastfeeding. This provides offspring a critical opportunity to learn how to read and interpret facial expressions, crucial for social cohesion and, in the wild, threat assessment.

In many mammal populations, breastfeeding is shared across multiple lactating females, nursing another mother's offspring if she is hunting or gathering. In the case of other primates, such as the golden snub-nosed monkey, over 87 per cent of infants are nursed by females that are not their mother.

Breastfeeding is a fundamental part of mammal evolution. Human nursing patterns are similar to other mammals'

Transitional milk is nutritious, calorie-dense, and stimulates an infant's growth

DAYS 3~15

DAYS 1~3

DAYS 15+

The first stage of breast milk, colostrum, typically lasts up to 3 days after birth

The final stage of breast milk can last up to 2 years or beyond

COMPARING MILK

The composition of breast milk changes over time. The three main types of milk are colostrum, transitional milk, and mature milk. Even during a feed, the texture and contents of breast milk will change. Foremilk is thin, hydrating, with a low fat content. Hindmilk is rich, fatty, and crucial for growth.

Every animal life is a complicated tale. Although continuous, lives involve a constant process of change – and we can identify chapters marked by certain physical and behavioural markers.

Life stories

DIFFERENT TALES
One of the biggest differences between us and our closest relatives is the time it takes to mature. So, while we recognize just one phase between infancy and adulthood in other apes, in humans we can divide that time into three chapters.

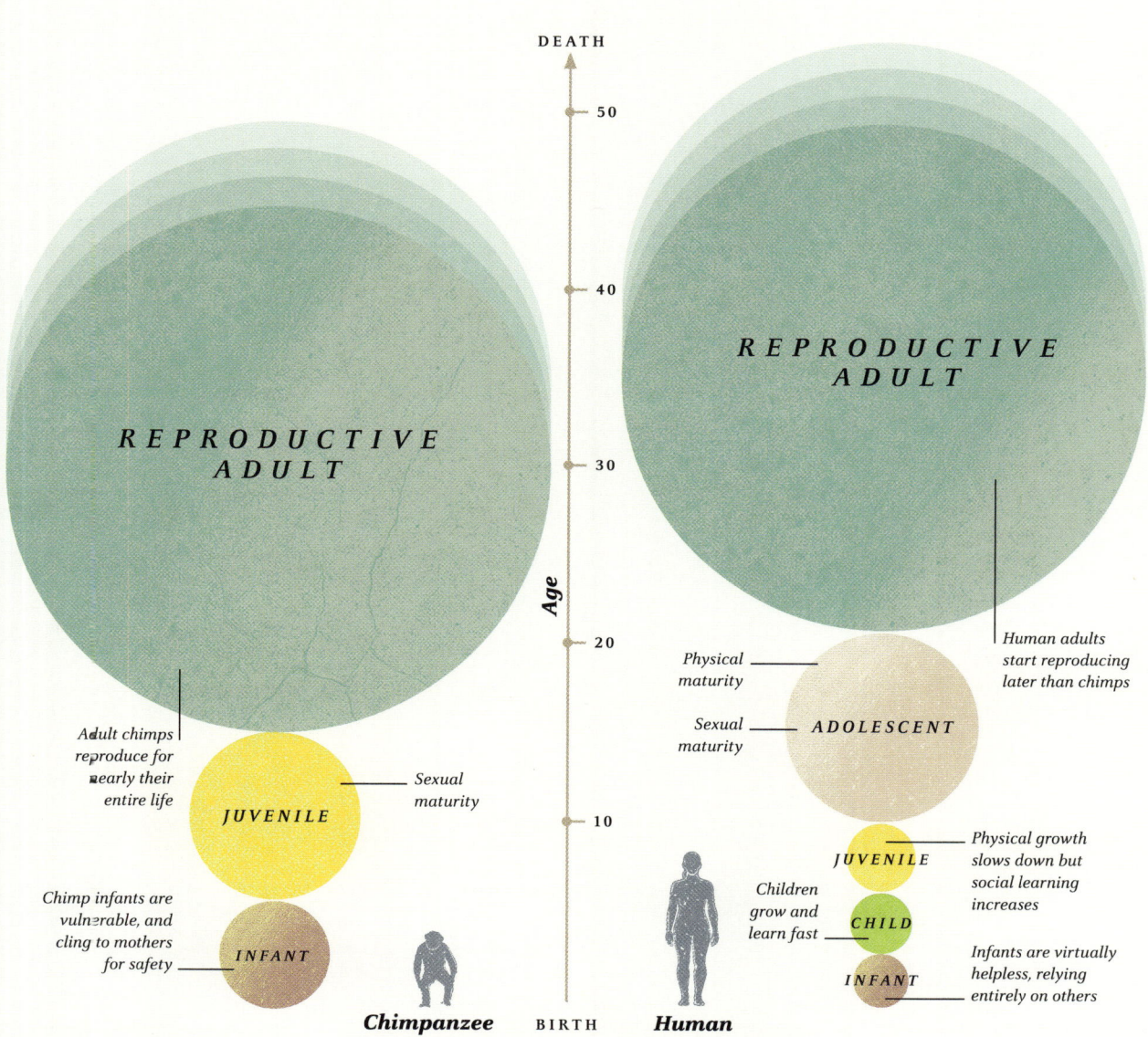

DEATH

50

40

30

Age

20

10

BIRTH

REPRODUCTIVE ADULT

Adult chimps reproduce for nearly their entire life

JUVENILE

Sexual maturity

Chimp infants are vulnerable, and cling to mothers for safety

INFANT

Chimpanzee

REPRODUCTIVE ADULT

Human adults start reproducing later than chimps

Physical maturity

Sexual maturity

ADOLESCENT

Physical growth slows down but social learning increases

JUVENILE

Children grow and learn fast

CHILD

Infants are virtually helpless, relying entirely on others

INFANT

Human

Humans are not unique in growing up slowly, but we do take this trend to the extreme in comparison to much of the animal kingdom. It takes time to grow large brains, and learning how to use them requires an extended period of development outside the womb. Over time, as human brains became larger, this period of early learning elongated, and was even supplemented with additional developmental stages.

LIFE'S A STAGE

An animal's "life history" is the pattern of major events surrounding its survival and reproduction, usually marked by physical and behavioural milestones. However, animals can differ wildly in their life histories, from the length of stages to the number of discrete stages we may be able to recognize. Infancy is universal, and in mammals it is defined as the stage that begins with birth and ends at weaning. The juvenile stage in most mammals spans from weaning to sexual maturity, and represents the time when young and maturing animals learn to survive and secure food on their own, but are not yet ready to reproduce. Adulthood is the final stage, when animals have reached physical maturity and are ready to reproduce. Humans experience all these stages, but take roughly twice as long to reach reproductive adulthood compared to other apes. The human juvenile period is so long and varied that we tend to divide it into three stages, including two strongly related to learning – childhood and adolescence. We also have a third, post-reproductive stage of adulthood (see p.210).

Human life history includes three additional stages: childhood, adolescence, and an adult stage post reproduction

WATCH AND LEARN

Many socially complex, tool-using mammals take some time to reach maturity, including monkeys and apes. Some hypotheses suggest that this is because brains capable of navigating complicated social relationships, or making tools, take time to grow.

In infancy, the first teacher for many young mammals is their mother. Chimpanzee young have been seen to observe their mother's behaviour while termite fishing. As both male and female offspring grow, they also learn from other members of their group. Capuchin monkey juveniles preferentially observe the most successful nut-crackers in their group, learning from the best. Learning is lifelong in all social species, as the ability to pick up new skills never stops being valuable. These skills may take the

Infant observes mother making a tool to catch termites

Mother and infant

Juvenile observes adult cracking nuts

Adult and juvenile

Adult learns from another adult how to use a tool to forage

Two adults

LIFELONG LEARNING
Primates observe and learn skills from other group members for the duration of their lives. In infancy, they may rely primarily on their mother to show them basic skills. As they mature, they may turn to close relatives, or even unrelated group members, to learn new skills.

form of social interactions – forming alliances, bartering for food, mitigating conflict, or technical innovation, such as designing a new tool. But the most important stage of learning is early in life, and humans draw this stage out more than any other ape – compared to chimpanzees, who take around 8 years to reach full cognitive development, humans require around 20 years (or more) for all of our brain to come fully online.

KIDDING AROUND

Childhood is universal across human cultures, and is the stage when humans are weaned but still dependent on older members of the group for food, protection, and guidance. Compared to

LEARNING TO COMMUNICATE

Language processing is a prime example of lengthy brain development. These scans made in 2022 by Olumide Olulade and colleagues show brains that are busy processing language. In adulthood, this task is typically confined to the left hemisphere – but it takes up to 19 years to develop this mature pattern of brain activity.

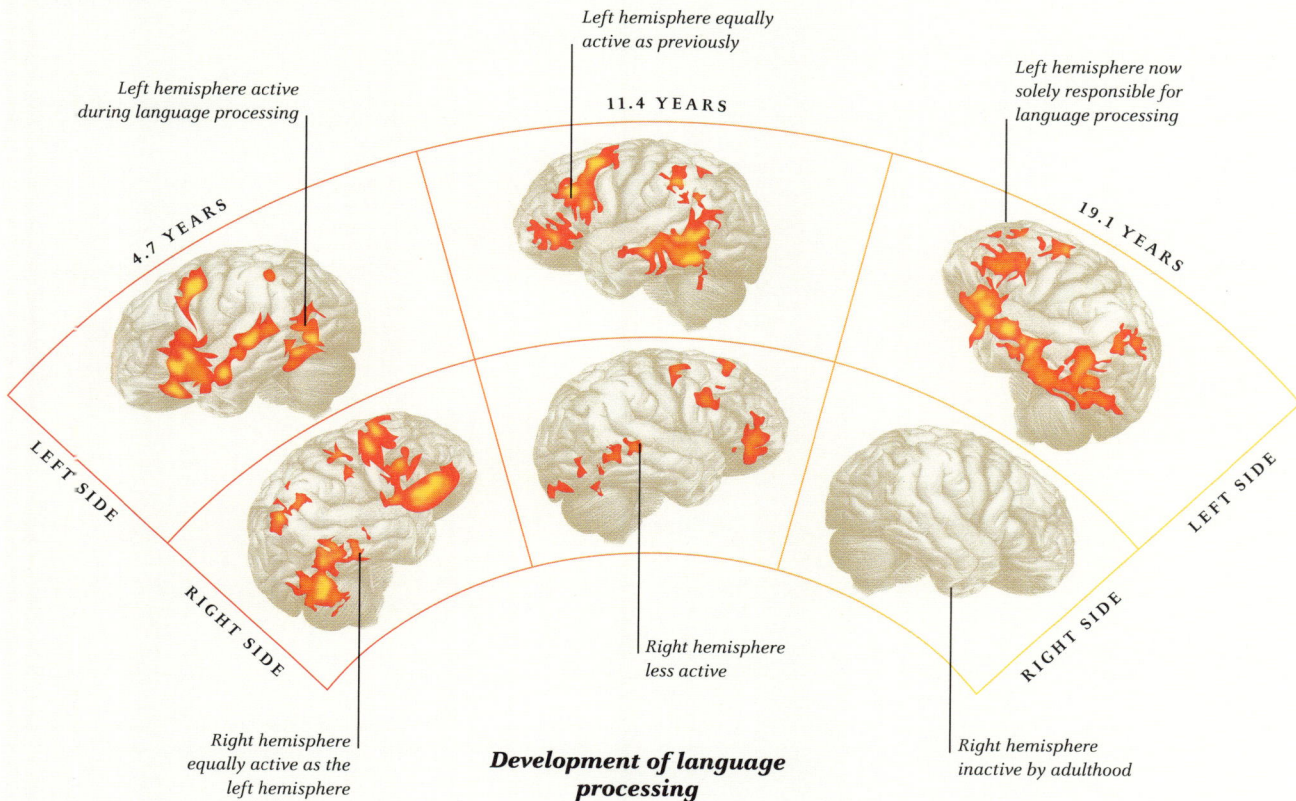

Left hemisphere equally active as previously

11.4 YEARS

Left hemisphere now solely responsible for language processing

Left hemisphere active during language processing

4.7 YEARS

19.1 YEARS

LEFT SIDE

RIGHT SIDE

LEFT SIDE

RIGHT SIDE

Right hemisphere less active

Right hemisphere inactive by adulthood

Right hemisphere equally active as the left hemisphere

Development of language processing

chimpanzees and other apes, modern humans tend to wean their young early, long before they are capable of feeding themselves, at 18 months to 2 years. In contrast, chimpanzee infants continue to ingest their mother's milk, alongside increasing quantities of other foods, up until around 5 years of age.

This early weaning by humans allows shorter intervals between births because the mother can breastfeed another baby while feeding the previous child on foraged food. In evolutionary terms, it has been suggested that this model would only have been possible once humans began to live socially, in more complex societies where foraging was communal and resources shared. This increase in reproductive fitness may have led to women having children in quicker succession, supporting population expansion.

RAPID BRAIN GROWTH

The majority of human brain growth and neural development occurs between birth and 5 years of age, mainly due to our brain's immaturity at birth (see pp.192–97). During these early years, one million neural connections form every second. Childhood is a time when young children learn to walk, talk, navigate their first social relationships, and begin to understand cultural and emotional boundaries, all aided by their rapidly growing and flexible brains.

Human childhood is the time of greatest learning due to our rapidly developing brain

PREPARING FOR PARENTHOOD

During adolescence, both sexes experience physical changes. Primary characteristics are development of the sex organs. Secondary characteristics include growth of body and pubic hair. Males experience changes in their voice and growing facial hair, while females develop breasts and begin menstruation.

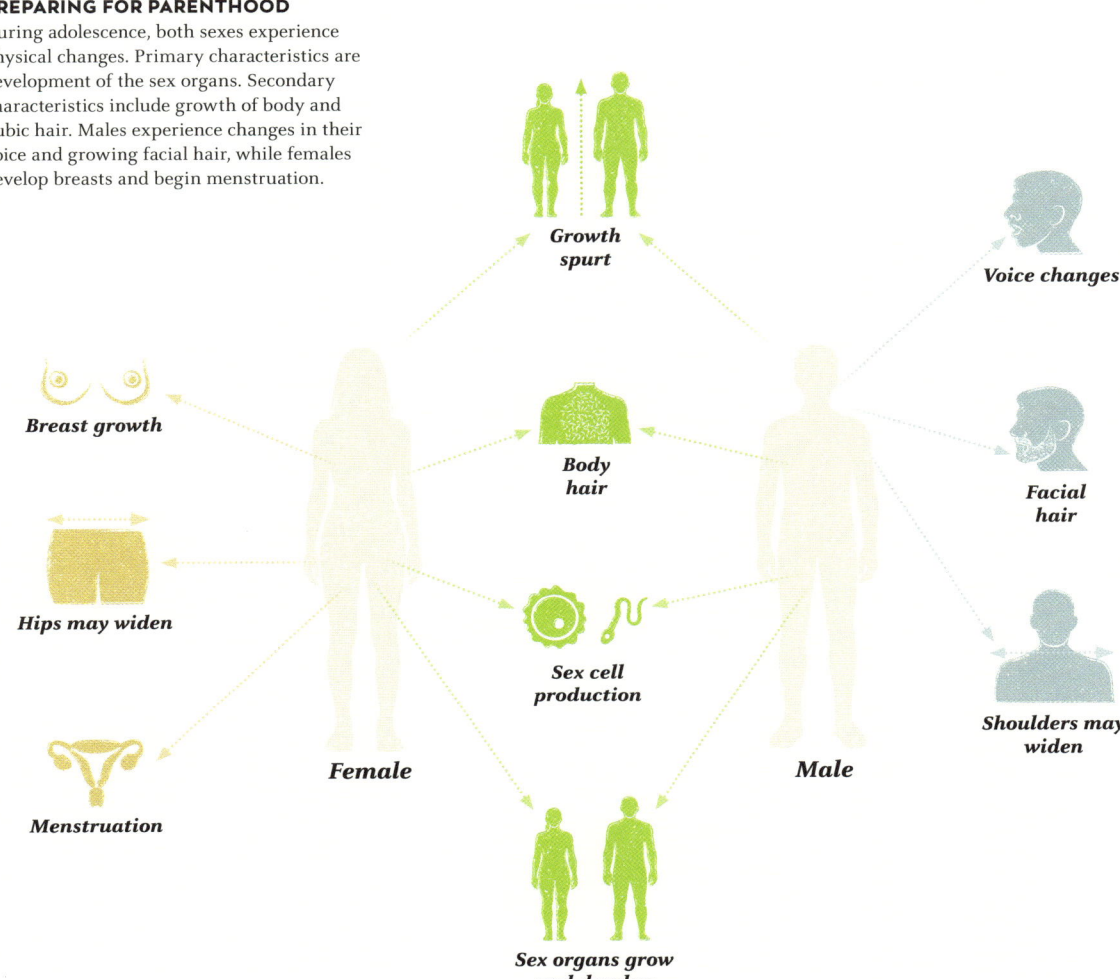

Growth spurt

Voice changes

Breast growth

Body hair

Facial hair

Hips may widen

Sex cell production

Shoulders may widen

Menstruation

Female

Male

Sex organs grow and develop

Other great apes have a juvenile stage when they learn social and survival skills

Humans are not distinct in having a pivotal learning period, however we are notable in how long we learn and how much we learn – a direct result of our larger brain size. Chimps and orangutans spend a significant portion of their juvenility learning social and survival skills from their mothers and other group-mates.

Humans also continue learning into the juvenile stage, which begins at about 7 years of age. In modern hunter-gatherer communities, youngsters at this stage develop their foraging abilities alongside practiced adults, rapidly learning how to collect difficult foods, such as tubers and small game. Therefore, juvenility in humans is not so dissimilar to our ape cousins,

in that it is a life stage when individuals learn necessary social and survival skills in order to thrive more independently.

BECOMING AN ADULT

When juvenility ends, humans enter another life history stage that some experts have suggested is unique to us – adolescence. Some primatologists argue that chimpanzees also have an adolescence, because chimps reach sexual maturity at around 11 years of age but do not reproduce until they are about 15 years old. Like childhood, adolescence is a time marked by intensive learning. However, unlike childhood, much of it is tied to the social and reproductive

The most important stage of learning is early in life, and humans draw this stage out more slowly than any other ape.

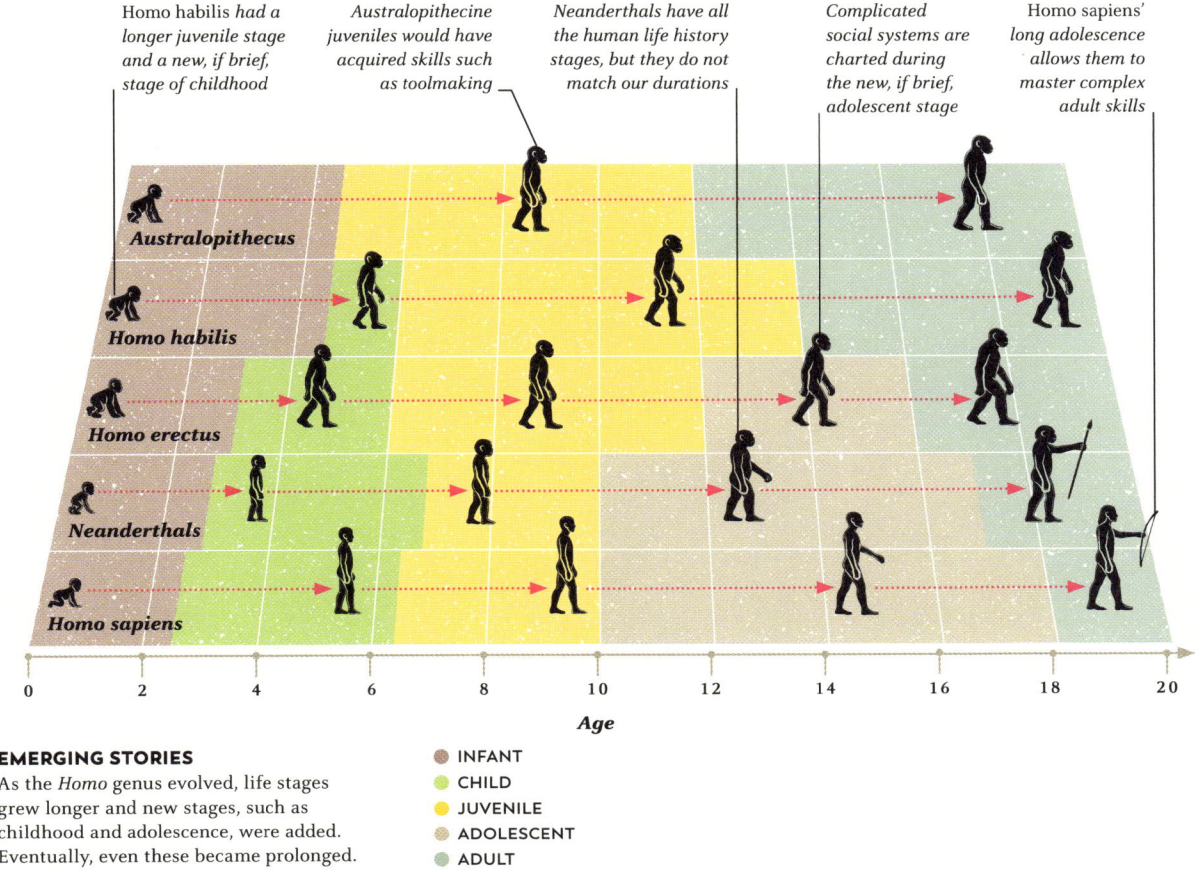

Homo habilis *had a longer juvenile stage and a new, if brief, stage of childhood*

Australopithecine juveniles would have acquired skills such as toolmaking

Neanderthals have all the human life history stages, but they do not match our durations

Complicated social systems are charted during the new, if brief, adolescent stage

Homo sapiens' *long adolescence allows them to master complex adult skills*

Australopithecus

Homo habilis

Homo erectus

Neanderthals

Homo sapiens

0   2   4   6   8   10   12   14   16   18   20

Age

EMERGING STORIES

As the *Homo* genus evolved, life stages grew longer and new stages, such as childhood and adolescence, were added. Eventually, even these became prolonged.

- INFANT
- CHILD
- JUVENILE
- ADOLESCENT
- ADULT

Adolescents may be physically capable of reproducing before adulthood, but they rarely do

behaviours of approaching adulthood. Human males and females may become physically capable of reproduction in their mid-teens, but it is rarely the age that either has their first child, often due to cultural norms. Instead, most human societies see adolescence as a time to prepare for adult relationships and activities, including child-rearing. In industrial and hunter-gatherer societies, individuals typically wait until what is considered adulthood (around 20 years of age) to have their first child. As well as the physical changes that occur during adolescence (see p.205), it is a time spent learning valuable cultural skills. This life stage offers youths an additional period of social and technical learning before diving into adulthood, and parenthood.

STORIES THROUGH TIME

Because the human life story is so unique, palaeobiologists have tried to determine at what point it evolved. By studying the physical

characteristics of each stage, including dental evidence and brain size, researchers have been able to estimate when our distinctly human life pattern began to emerge.

Analysis of when adult teeth erupted in australopithecines, as well as their brain size at various ages, suggests that they grew at a similar pace to chimpanzees, whereas early members of our genus, *Homo*, appear to have had an early life pattern similar to modern humans, while emulating chimp patterns later on. This implies that, in humans, childhood as a life stage may have begun to evolve before adolescence. Even some of our closest relatives, the Neanderthals, seem to have grown up faster than us, with their adolescence being shorter than ours. This implies that the life history of *Homo sapiens*, while not exceptional, may be uniquely long.

Families and social units

To understand how early human societies may have survived, developed, and grown to become the huge metasocieties that we live in today, it is useful to look at the structure of modern hunter-gatherer societies as, for 95 per cent of human evolution, that was the predominant lifestyle.

Today human society is organized around enormous groups: countries are composed of states, counties, or provinces, each with many cities and towns spread across hundreds of kilometres. The scale of these settlements is a recent invention, on an evolutionary timescale.

It seems likely that humans first evolved in small groups of no more than 150 people, in which everyone was familiar with one another and each member played an important role in the community. Large social networks are common in primates, and the basic pattern seen in modern hunter-gatherer societies is no different: adults of all ages, and dependent offspring, live together as they forage, bond, seek shelter, and defend themselves from danger. The trends that make human groups stand out are not entirely unique, but the degree of group cohesion is extreme.

SHARING THE LOAD

Modern nonhuman ape societies are small-scale and often hierarchical, dominated by one or several males or females. Many industrialized societies are also hierarchical, but for different reasons: larger population sizes, more complex societal structures, and the need for resource control. However, of all contemporary societies, hunter-gatherer communities are probably the most instructive to study, since for most of our evolutionary lifetime (95 per cent), hunting and gathering is how humans survived.

We spent most of our recent evolutionary history in small hunter-gatherer societies

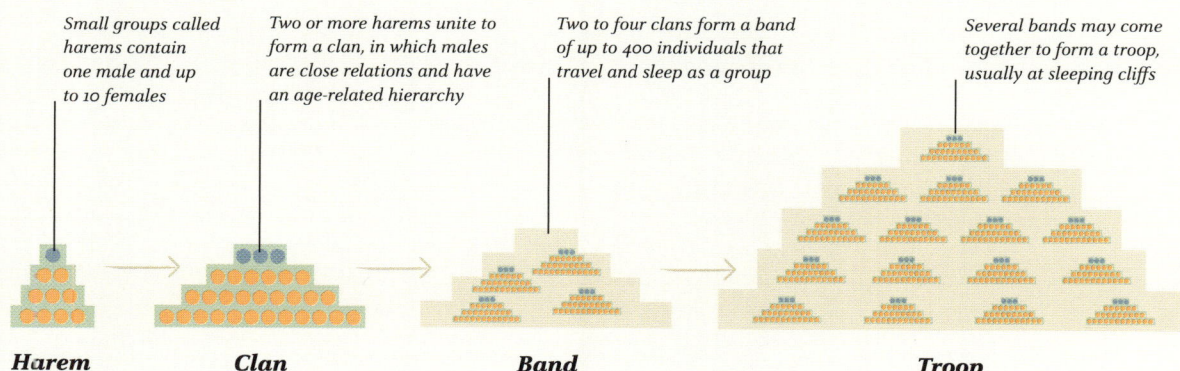

Small groups called harems contain one male and up to 10 females

Two or more harems unite to form a clan, in which males are close relations and have an age-related hierarchy

Two to four clans form a band of up to 400 individuals that travel and sleep as a group

Several bands may come together to form a troop, usually at sleeping cliffs

Harem   **Clan**   **Band**   **Troop**

EVOLUTION OF SOCIAL ORGANIZATION

Hamadryas baboons live in multilevel societies, with four distinct layers: harem, clan, band, and troop. The society is highly patriarchal, with a single male controlling a harem, and several closely related male patriarchs leading a clan.

 ● MALE BABOON    ● FEMALE BABOON

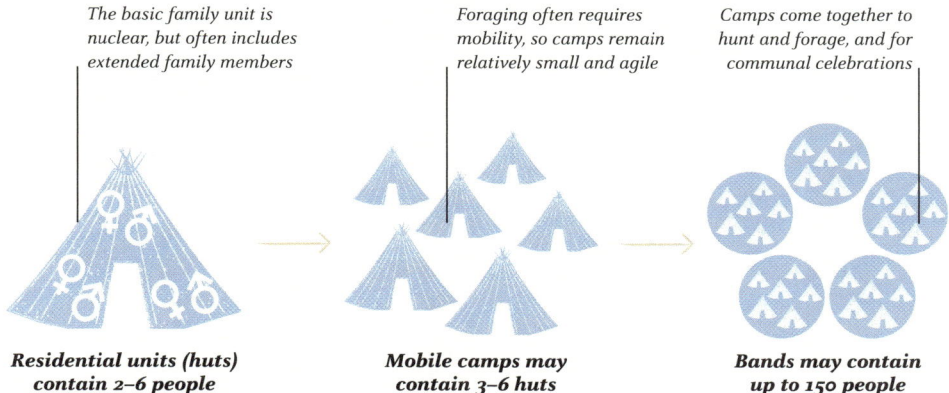

The basic family unit is nuclear, but often includes extended family members

Foraging often requires mobility, so camps remain relatively small and agile

Camps come together to hunt and forage, and for communal celebrations

Residential units (huts) contain 2–6 people

Mobile camps may contain 3–6 huts

Bands may contain up to 150 people

MODERN HUNTER-GATHERERS

Archaeology reveals some past hunter-gatherer societies may have been hierarchical, but most modern hunter-gatherer societies are egalitarian. They have a social structure based on extended family units within camps, high between-camp mobility, and strong multi-camp cohesion.

Hunter-gatherer childcare involves a high percentage of care by people other than the parents

One of the key differences highlighted by studies of today's hunter-gatherer societies is the way children are cared for compared with industrialized societies. Many hunter-gatherer communities show widespread alloparenting – the care of offspring by individuals who are not the biological parents. Alloparents, such as older siblings, aunts, cousins, grandparents, and other unrelated members of the community, can provide as much as half of an infant or toddler's care in the first years of life. Several recent ethnographic studies have concluded that the reason for this is two-fold. Firstly, resources are pooled and shared among everyone. Individual calorie contributions differ by the day and season, revealing the importance of collaboration among group members. Sometimes hunters bring in big kills and gatherers haul back baskets of fruits or tubers, but on other days they are less successful. No matter who succeeds, the group is fed due to strong societal cohesion.

Secondly, the calorific contribution of mothers drops at precisely the time when they need extra calories to feed their newborns.

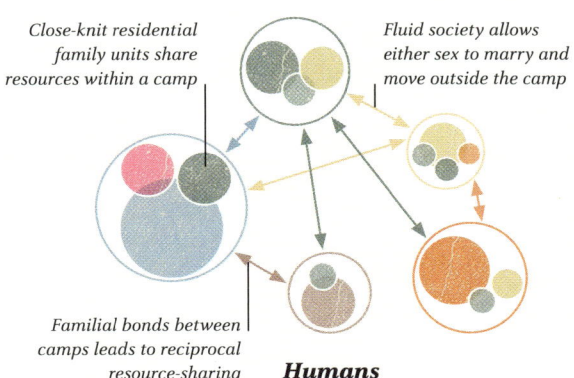

Close-knit residential family units share resources within a camp

Fluid society allows either sex to marry and move outside the camp

Familial bonds between camps leads to reciprocal resource-sharing

Humans

Only one sex leaves the group to find a mate, leading to small, isolated groups

Lack of cooperation between family units or harems

Other primates

BUILDING FAMILY TIES

Many human social structures features two-way marriage exchange of both sexes between groups, leading to a wider network of bonds and social cohesion. In other primate societies, only one sex leaves the group to find a mate.

- ○ RESIDENTIAL FAMILY UNIT
- ○ CAMPS OF MULTIPLE FAMILY UNITS
- ○ COOPERATION BETWEEN FAMILY UNITS
- ⟷ INTERMARRIAGE AND RESOURCE EXCHANGE
- ⋯▸ ONE SEX LEAVES TO FIND A MATE

FAMILIES AND SOCIAL UNITS

GRANDMOTHER HYPHOTHESIS

Caring for grandchildren is instinctive, because it favours the grandmother's genes carried by her progeny. She switches care from children to grandchildren, even though they are less closely related, genetically. She may also pass on genes for a long life.

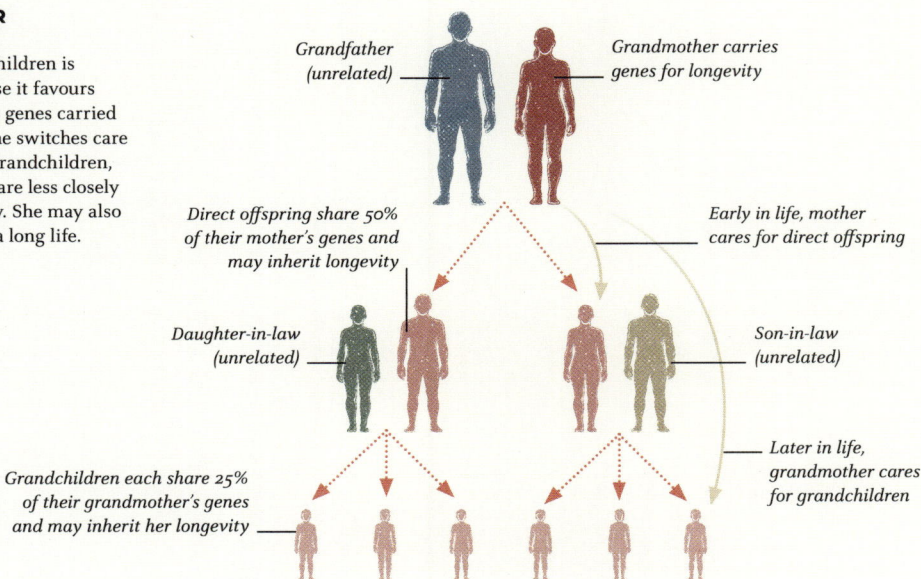

Grandfather (unrelated)

Grandmother carries genes for longevity

Direct offspring share 50% of their mother's genes and may inherit longevity

Early in life, mother cares for direct offspring

Daughter-in-law (unrelated)

Son-in-law (unrelated)

Later in life, grandmother cares for grandchildren

Grandchildren each share 25% of their grandmother's genes and may inherit her longevity

Couples with children receive free cooperative childcare at a time of high calorific need

A mother can take her breastfeeding baby with her while foraging, but her other young children need to stay behind in camp, and this is where the wider community steps in, to supplement her childrearing, as well as her calories.

LISTEN TO YOUR ELDERS

Older generations also play a vital role, acting as teachers, mediators, and knowledge repositories, as well as contributing to foraging and childcare. Women who have experienced the menopause, and no longer have a chance to reproduce, continue being productive and valued members of the group.

A post-reproductive stage of life is rare among mammals – it is seen in some toothed whales, for instance – but in humans, it is an important one. Grandmothers remove some of the parental burden from mothers and fathers. Studies have shown that children with living grandmothers survive and thrive at a greater rate than those without. The evolutionary benefits are clear: grandmothers increase the reproductive fitness of their children by helping to raise their grandchildren, who also carry their DNA. This is a safer way for older females to promote their genetic legacy than continuing to bear children into later life, which becomes increasingly risky. What's more, they do not bestow only care on their grandchildren. They also pass on genes for

longevity and a post-reproductive life. Some of their granddaughters will grow old themselves and become active grandmothers. Even their grandsons may inherit the increased lifespan.

If this evolutionary story – known as the "grandmother hypothesis" – is correct, it has further benefits. Elders of both sexes aid the whole group, remembering solutions to problems from decades ago, such as critical foraging sites used during rare droughts or floods. This promotes behaviours supporting and respecting elders in the community. In turn, this eases the evolutionary selection pressure (so this reasoning goes) for women to remain reproductively active, and so evolved our post-menopausal stage of human life.

Grandmothers give a survival edge to their children and grandchildren, which could explain the evolution of our post-menopausal lifespan.

Interbreeding

In the not-too-distant past, our ancestors lived alongside other hominin species, and these encounters led to exchanges of not just resources and technology, but genes. Palaeogenetics is beginning to reveal to what extent these different species interbred with each other and us.

Many hominin species lived at the same time during the last 2 million years, giving these populations plenty of opportunities to encounter and mate with one another. Some biologists argue that if two species can produce healthy, fertile offspring, perhaps they should not be considered different species at all. Biologists define a species as a reproductively isolated population – an insulated gene pool that evolves in its own unique direction, without mixing its genes with others. Over geological time, however, this definition becomes hazy. Populations that were once a single species become isolated and begin diverging, sometimes evolving at different rates. The fossils we find could be just snapshots of any point in this process.

Some of today's animal species, such as carrion crows and hooded crows, sometimes interbreed in the wild, while others can be induced to interbreed in rare or unnatural situations, such as in zoos. This is because reproductive isolation can come about by many means, and the isolation may be incomplete. If two species have recently split, they may be unable to mate due to acquired

Many diverse hominins lived on Earth during the Pleistocene epoch and would have met each other

BEFORE CONCEPTION
Species may be reproductively isolated due to some kind of barrier that prevents them from mating.

One frog mates in spring, while the other mates in summer

TEMPORAL
The two species mate at different times of year.

Lions live on the African savannah, and tigers live in Asian forests

ECOLOGICAL
They do not meet due to living in different habitats.

Birds in particular only respond to certain mating calls

BEHAVIOURAL
The two species have different courtship behaviours.

Physical incompatibility between some domestic dogs prevents mating

MECHANICAL
Mating between the two species is morphologically impossible.

AFTER CONCEPTION
This form of reproductive isolation is caused by an incompatibility in the genes of the two species, meaning that their hybrids either cannot reproduce or die.

Hybrid tadpole dies before becoming a frog

HYBRID INVIABILITY
Two species mate and produce offspring, but the offspring die before maturity.

Mules – hybrids between a horse and a donkey – are sterile

HYBRID INFERTILITY
The hybrid offspring live until maturity but are unable to reproduce.

Offspring of hybrid copepods are less likely to survive

HYBRID BREAKDOWN
The second generation after hybridization die or are unable to reproduce.

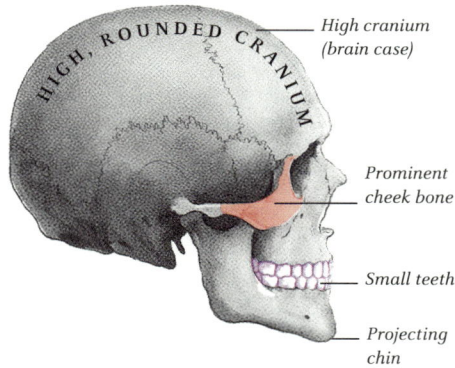

High cranium
(brain case)

HIGH, ROUNDED CRANIUM

Prominent
cheek bone

Small teeth

Projecting
chin

Homo sapiens

*Human skulls are generally more lightly built than
Neanderthal skulls, but their prominent cheek bones
and small teeth are two key indicators.*

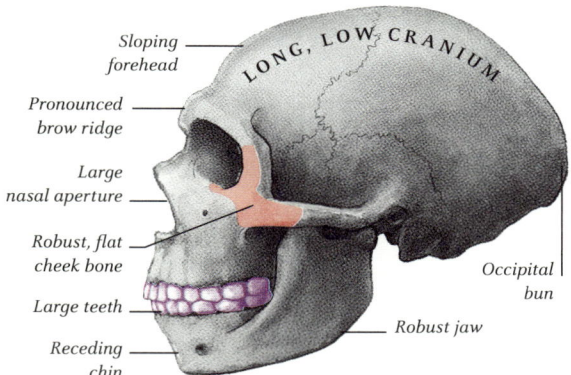

Sloping
forehead

LONG, LOW CRANIUM

Pronounced
brow ridge

Large
nasal aperture

Robust, flat
cheek bone

Large teeth

Receding
chin

Occipital
bun

Robust jaw

Homo neanderthalensis

Neanderthals had much larger teeth than Homo
sapiens, *plus a strong, square jaw – probably due
to the difference in their diets.*

physical differences, they may not want to mate due to behavioural changes, or mating and fertilization may successfully occur, but evolved genetic differences prevent the fertilized egg from developing. These barriers before or after fertilization are a sign that enough time and change has taken place for most to agree that reproductive isolation is now in progress. However, since it is a process, many animals we already consider separate species can still sometimes produce viable young. This was clearly the case with many hominins as well, as we have recently discovered.

ANCIENT DNA

Recent developments in DNA sequencing have allowed researchers to create hominin family trees

Modern techniques have allowed scientists to recover genetic material from well-preserved fossils, making it possible to sequence the genomes of our closest relatives. Proteins, mtDNA (DNA in cell structures called mitochondria), and nuclear DNA (in a cell nucleus) have allowed researchers to create family trees that reveal much about how and when our ancestors interacted with earlier human species such as Neanderthals. One of the most surprising discoveries from ancient DNA was not just this widespread evidence of interbreeding, but that many of us still carry genes from those other species today.

HUMANS, NEANDERTHALS, AND DENISOVANS

Interbreeding between humans and Neanderthals took place many times across Europe and Asia. Today, up to 2 per cent of most people's genomes are inherited from Neanderthals, with higher percentages in modern European and east Asian populations. Likewise, up to 6 per cent of some ancient Neanderthals' genomes are human. Genetic exchange is thought to have taken place periodically across hundreds of thousands of years. Neanderthal and human lineages diverged some time 800,000–500,000 years ago (YA) – with the most recent liaisons being a little before

COMPARATIVE FEATURES

When hominin skull bones are found, the species needs to be identified. For Neanderthals there are several tell-tale signs that palaeontologists use to determine their lineage – from the cheek bone angle to teeth size.

ONLY HUMAN MOTHERS

The genetic record preserves no trace of Neanderthal mothers rearing hybrid children, so some kind of reproductive barrier, either before or after conception, must have prevented successful matings with human males. Only human mothers seem to have successfully reared hybrid offspring.

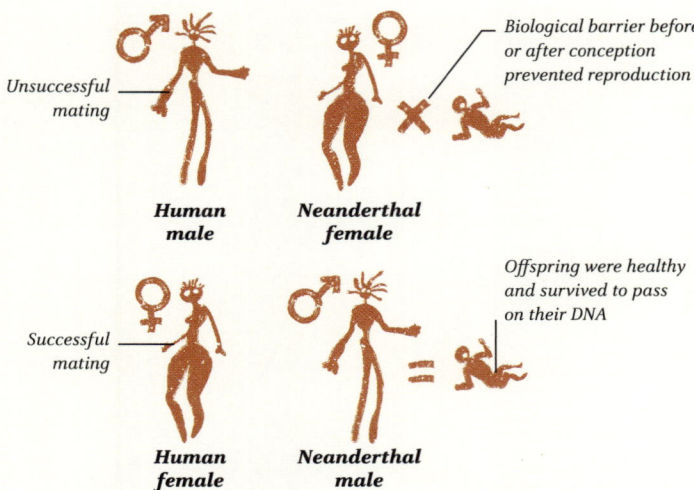

Unsuccessful mating

Human male

Neanderthal female

Biological barrier before or after conception prevented reproduction

Successful mating

Human female

Neanderthal male

Offspring were healthy and survived to pass on their DNA

40,000 YA. However, some evidence suggests that our two species were on the way to being reproductively isolated. Mitochondrial DNA (mtDNA) is only passed down through the maternal line, and no known modern humans have Neanderthal mtDNA in their mitochondria. This suggests that only some human–Neanderthal pairings produced healthy and fertile offspring.

Other early humans lived on Earth during this time, and many of them were interbreeding, too. Another species whose DNA we still carry is an enigmatic group called the Denisovans (*Homo longi*); about 5 per cent of modern Pacific Islanders' genome is Denisovan. Neanderthals also bred with Denisovans, as fossils

About 5 per cent of Pacific Islanders' genome can be traced to Denisovans.

belonging to a first-generation hybrid have been discovered in the Altai mountains in Siberia. Genes from "ghost" hominins (species that are as yet unknown from the fossil record) have also been discovered in humans, Neanderthals, and Denisovans.

THE NEANDERTHAL IMPACT

Some human diseases can be traced to Neanderthal gene variants

Recent discoveries suggest that Neanderthal genes are linked to several medical conditions in modern humans. Genes from Neanderthals seem to increase the risk of autoimmune diseases, such as lupus, Crohn's disease, and IBS (irritable bowl syndrome), and a Neanderthal mutation in a gene called SLC16A11 is also associated with a greater risk of Type 2 diabetes. Neanderthal gene variants also seem linked to depression, fertility issues, hair loss, and sun sensitivity. All these genes were possibly benign or even beneficial to Neanderthals, adding weight to the argument that the two species were already evolving reproductive isolation.

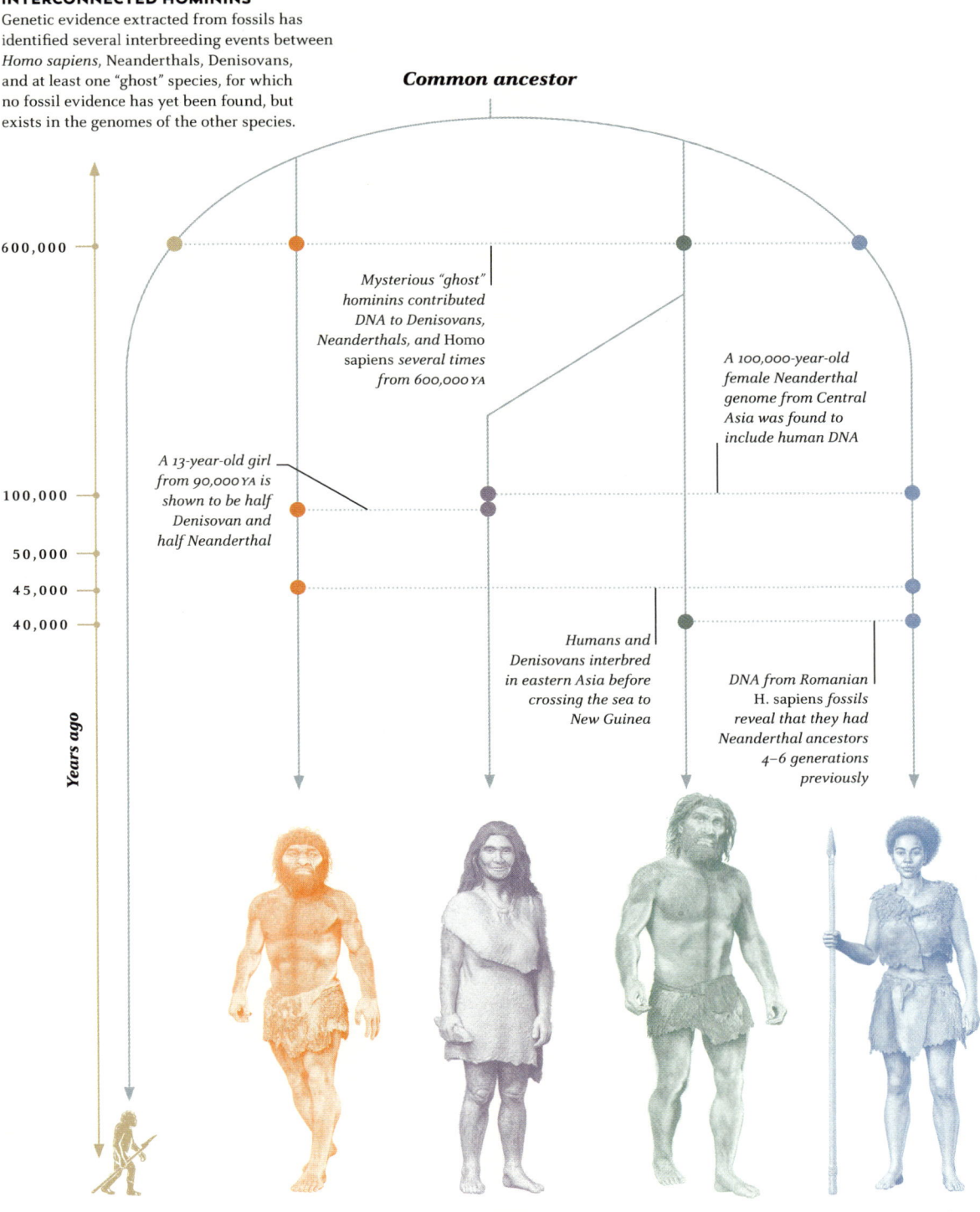

INTERCONNECTED HOMININS

Genetic evidence extracted from fossils has identified several interbreeding events between *Homo sapiens*, Neanderthals, Denisovans, and at least one "ghost" species, for which no fossil evidence has yet been found, but exists in the genomes of the other species.

Common ancestor

Mysterious "ghost" hominins contributed DNA to Denisovans, Neanderthals, and Homo sapiens *several times from 600,000 YA*

A 100,000-year-old female Neanderthal genome from Central Asia was found to include human DNA

A 13-year-old girl from 90,000 YA is shown to be half Denisovan and half Neanderthal

Humans and Denisovans interbred in eastern Asia before crossing the sea to New Guinea

DNA from Romanian H. sapiens *fossils reveal that they had Neanderthal ancestors 4–6 generations previously*

600,000

100,000

50,000

45,000

40,000

Years ago

"GHOST" HOMININS
These are as yet unidentified, as we have only found traces of them in other species.

DENISOVANS
Fossils have been found in Russia, China, and Laos, and their genes exist in Pacific Islanders.

EASTERN NEANDERTHALS
Mostly lived in Central Asia, especially Siberia and east Asia.

WESTERN NEANDERTHALS
Lived in Europe; low genetic variation suggests a small population.

HUMANS
After leaving Africa, H. sapiens migrated north and eastwards, meeting other hominins.

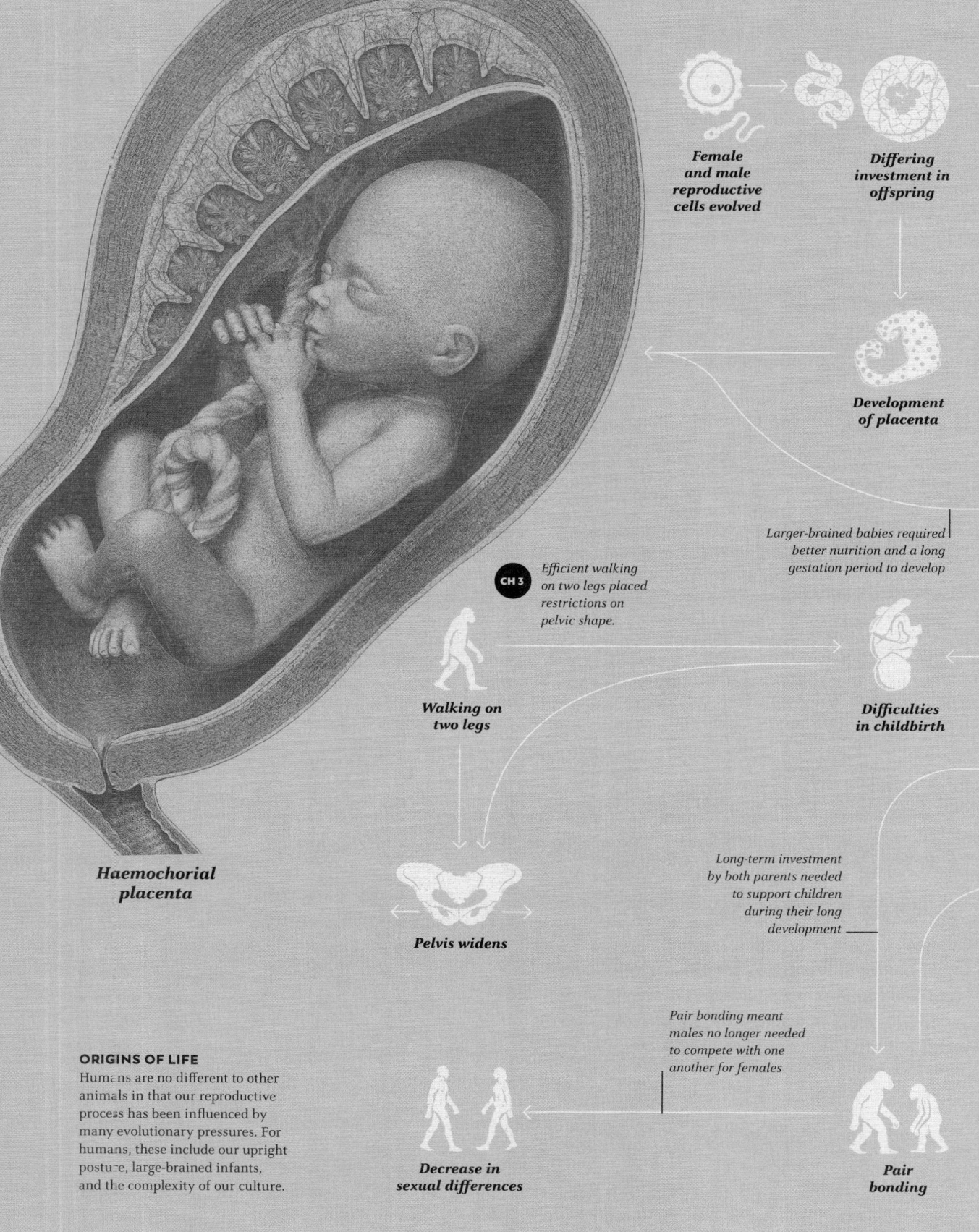

Haemochorial placenta

Female and male reproductive cells evolved

Differing investment in offspring

Development of placenta

Larger-brained babies required better nutrition and a long gestation period to develop

CH 3 *Efficient walking on two legs placed restrictions on pelvic shape.*

Walking on two legs

Difficulties in childbirth

Long-term investment by both parents needed to support children during their long development

Pelvis widens

Pair bonding meant males no longer needed to compete with one another for females

ORIGINS OF LIFE
Humans are no different to other animals in that our reproductive process has been influenced by many evolutionary pressures. For humans, these include our upright posture, large-brained infants, and the complexity of our culture.

Decrease in sexual differences

Pair bonding

Mother and baby bonding

Close contact and breastfeeding taught offspring facial expressions, which are key in wider social contexts

Complex social interaction

Producing the next generation

Our reproductive system has been shaped by many factors. Its evolution is intertwined with the way human brains have grown over time, how our pelvises have changed shape to support upright walking and childbearing, and the fact that bringing up large-brained children involves many members of a complex community.

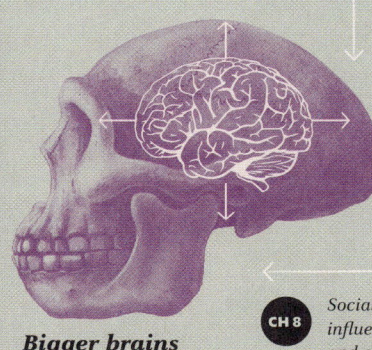

Bigger brains

CH 8 *Social and cultural learning influenced the development and evolution of the brain.*

Cultural exchange

Cooperation

Growing up slowly

Parenting by helpers

Post-reproductive care by grandmothers may have helped the evolution of longevity and a post-menopausal stage in life

Menopause

Longer lifespan

Increasing sociality

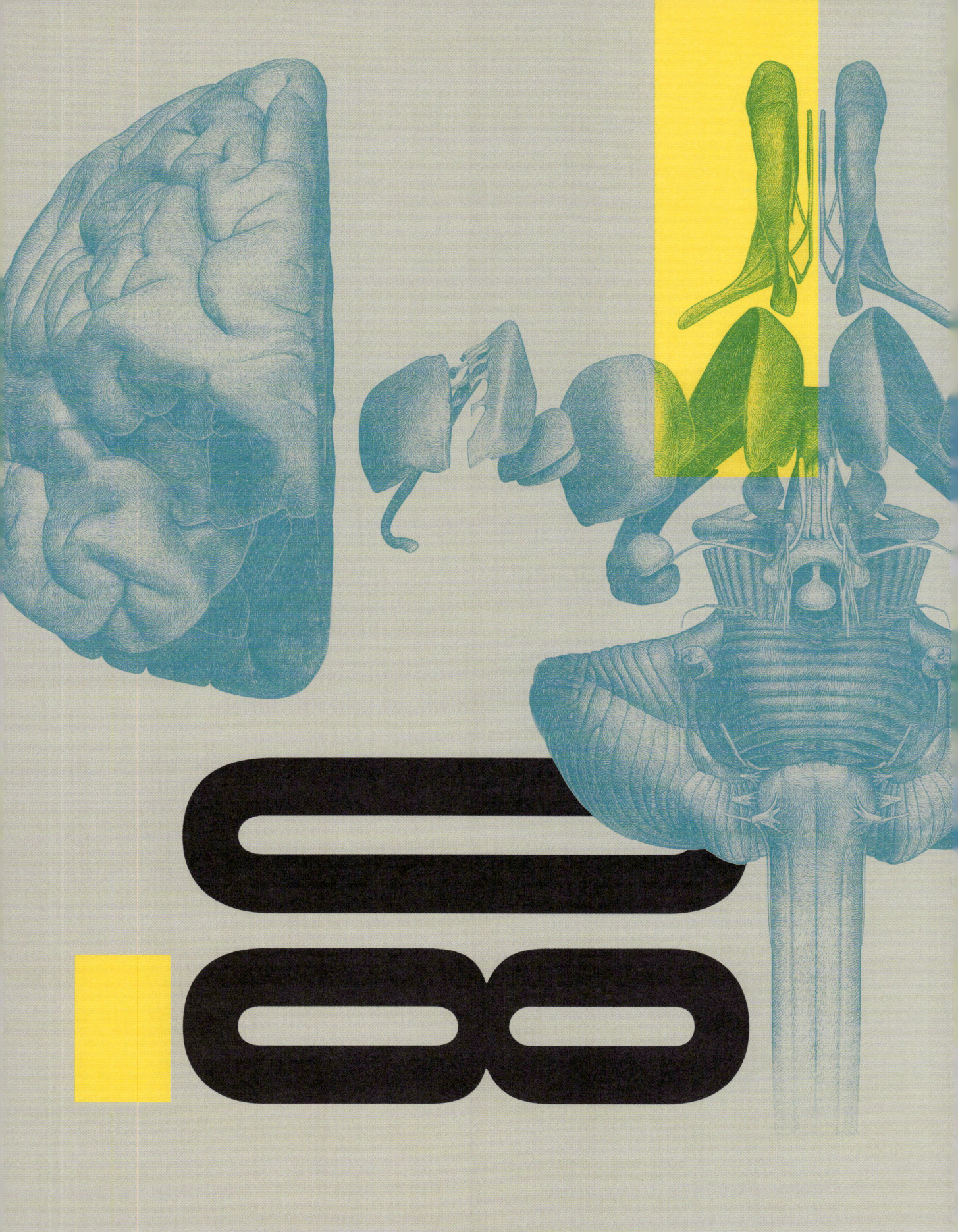

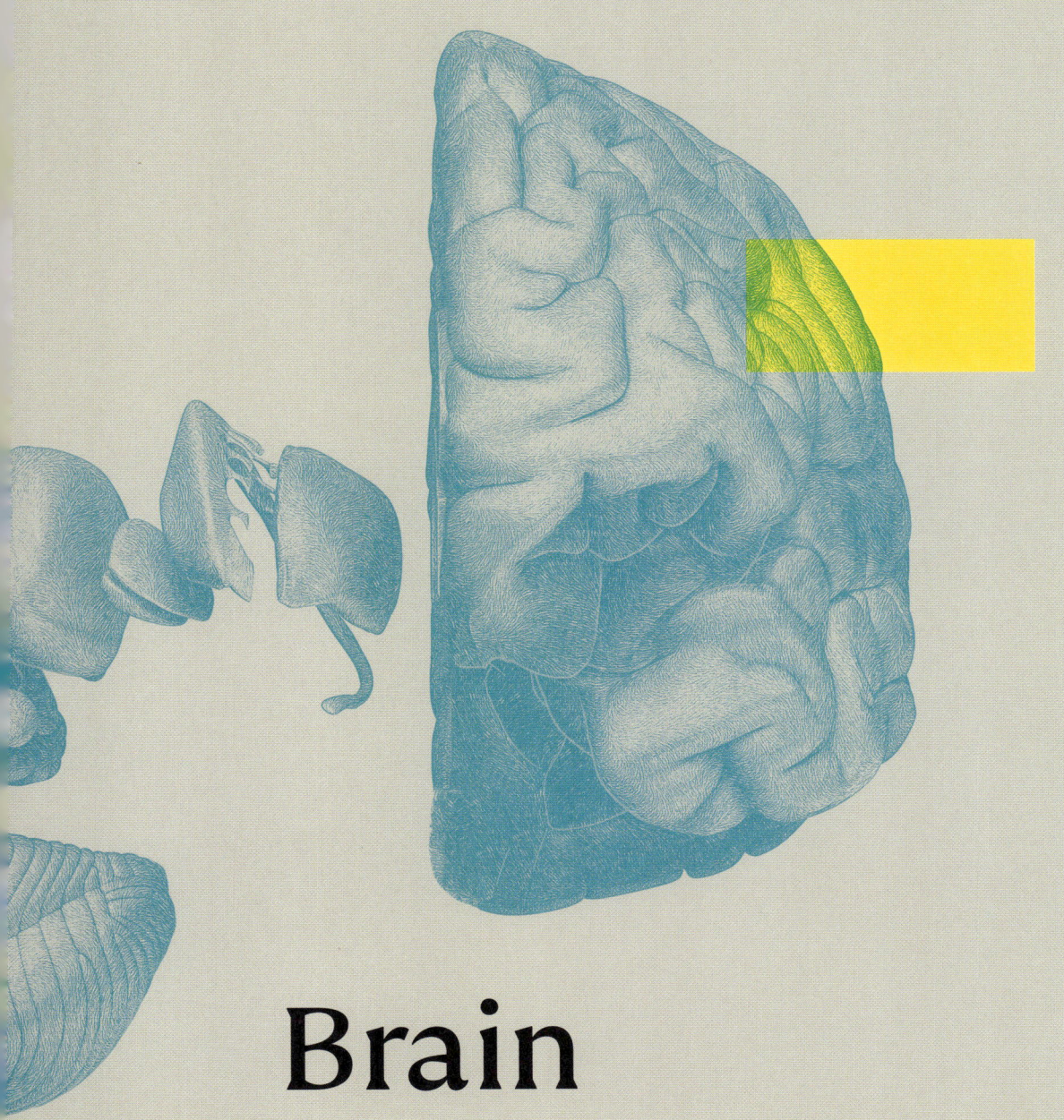

Brain

Big brains helped early humans solve problems and find new ways to live in new habitats. As they developed, humans passed on skills and knowledge to other receptive minds, leading to progressive cultural change, language, art, and global expansion.

The evolution of the brain spans more than 500 million years, starting with the origin of neurons (nerve cells) in simple organisms such as jellyfish and sea anemones. These specialized cells revolutionized animal life by transmitting electrical impulses, enabling fast information transfer in early multicellular creatures.

At first, neurons were arranged in a network without centralized organization. As animals evolved directional movement, with distinct front and rear ends, neurons began clustering at the front end. This concentration improved communication efficiency between neurons, eventually forming what we recognize as brains: complex neural clusters with sophisticated arrangements that enhance connectivity.

The rise of vertebrates marked a profound increase in brain complexity. Fish, amphibians, and reptiles had brains with distinct regions specialized in functions, such as vision, smell, and memory. Brain size increased alongside body size, as larger organisms required more processing power. However, some vertebrate brains expanded far beyond what body size alone would predict: mammals and birds evolved proportionally larger and more complex brains than similarly-sized reptiles.

This trend led to large-brained mammals, notably great apes and early hominins, eventually giving rise to *Homo sapiens*, whose exceptionally large brains support advanced problem-solving, the cultural accumulation of knowledge, and the self-awareness to reflect on their own evolutionary origins.

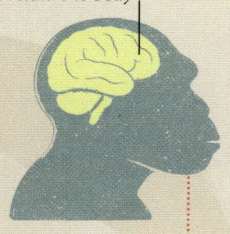

Enlarged prefrontal cortex and increased overall brain size relative to body

APES AND ANCESTORS
Between 15–2 million years ago, great apes and early hominins evolved larger prefrontal cortices. Tool use, empathy, and cultural transmission emerged, hinting at human cognition.

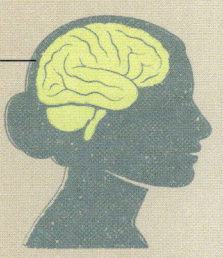

Very large brain with highly folded neocortex and specialized language centres

15 MYA–PRESENT HOMINIDS

MODERN HUMANS
From around 300,000 years ago, Homo sapiens exhibited unprecedented brain complexity, with highly folded neocortices and language centres enabling symbolic thought, abstract reasoning, and culture.

Vertebrate brains grew larger and more complex, with specialized regions

Large brain-to-body ratio with dense arrangement of neurons

AVIAN INTELLIGENCE
The evolution of bird brains shows that bigger isn't always better. Crows and parrots evolved dense, complex neural structures that enable remarkable intelligence.

From nerve cells to comprehension

Over millions of years, brains evolved from scattered nerve cells in simple organisms to increasingly centralized and structured networks in vertebrates, culminating in the highly complex organ powering our unique cognitive abilities.

THE EVOLUTIONARY JOURNEY
Brain evolution progressed from simple nerve nets in jellyfish through clustered neurons in flatworms, then increasingly complex structures in fish, amphibians, reptiles, and birds, and reaching exceptional size and sophistication in mammals, great apes, and humans.

Enlarged neocortex and increased brain-to-body size ratio

SOCIAL MINDS
From 65–15 million years ago, mammals including dolphins, elephants, and primates evolved large brains with expanded neocortices, enabling social learning, memory, communication, and playful behaviour.

Expanded cerebral cortex

BRAIN EXPANSION
Between 300–100 million years ago, reptiles and early birds and mammals showed significant brain growth. The expanding cerebral cortex improved instincts, navigation, and coordination, enabling complex environmental interactions.

INVERTEBRATE INTELLIGENCE
The ancestors of octopuses had simple brains with two lobes that coordinated swimming, tentacle movements, and hunting behaviour.

Simple bilobed brain with early neural clustering that coordinated movement

375 – 15 M Y A

LAND VERTEBRATES

520 – 375 M Y A

VERTEBRATES

BEGINNINGS OF BRAINS
Between 550–500 million years ago, flatworms developed a front-back body plan and began clustering neurons at the head. Their primitive brain allowed directional movement and information processing.

Neurons at front end forming ganglia (small neural clusters)

Three-part structure with specialized regions for sensory processing and motor control

800 – 520 M Y A

ANIMALS

Long axon transmits electrical impulses

FIRST VERTEBRATE BRAINS
Around 500–300 million years ago, early fish developed segmented brains consisting of forebrain, midbrain, and hindbrain.

ORIGINS OF NEURONS
Around 600–550 million years ago, animals including jellyfish evolved the first neurons, forming simple nerve nets throughout their bodies. These enabled basic reflexes and coordination, marking the earliest nervous systems in animals.

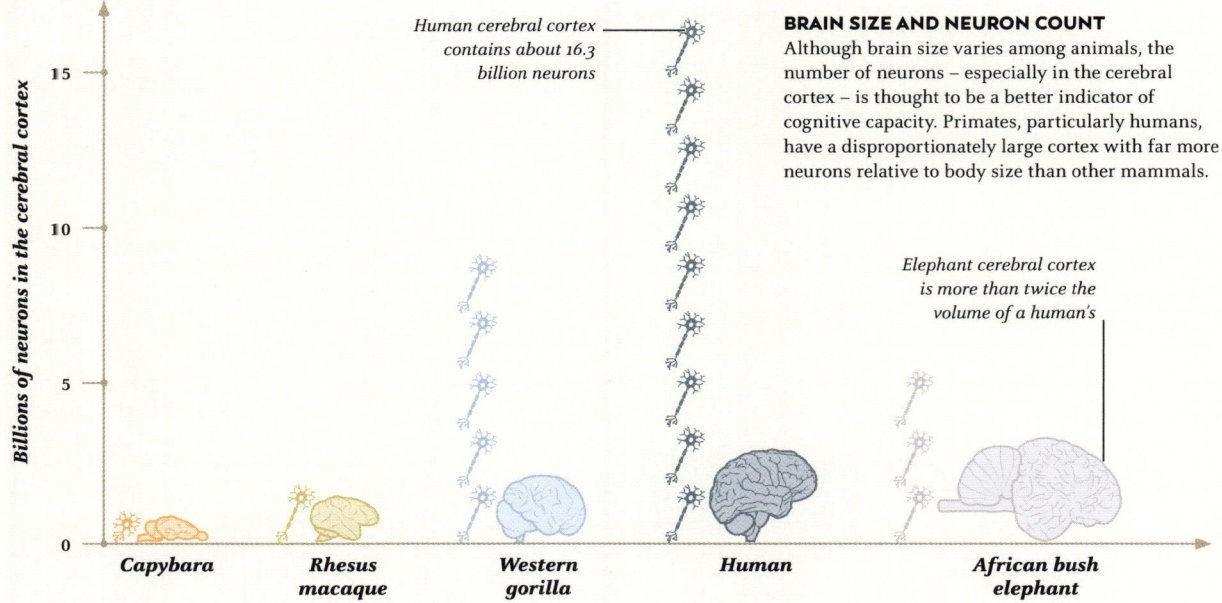

BRAIN SIZE AND NEURON COUNT
Although brain size varies among animals, the number of neurons – especially in the cerebral cortex – is thought to be a better indicator of cognitive capacity. Primates, particularly humans, have a disproportionately large cortex with far more neurons relative to body size than other mammals.

Human cerebral cortex contains about 16.3 billion neurons

Elephant cerebral cortex is more than twice the volume of a human's

Billions of neurons in the cerebral cortex

15
10
5
0

Capybara   **Rhesus macaque**   **Western gorilla**   **Human**   **African bush elephant**

Growing brains

The evolution of large brains is one of nature's most remarkable experiments, shaping intelligence, behaviour, and culture. Why some species evolved such costly organs, and how, reveals much about what it means to be human.

Across the animal kingdom, larger brains tend to contain more neurons (nerve cells) and support more complex information processing, planning, and sophisticated problem solving. In this sense, brain expansion can be seen as evolution's way of increasing computational power. The parallel with artificial intelligence is striking: in both cases, scaling up networks unlocks new capabilities.

During mammalian evolution, some lineages gained disproportionately large brains relative to body size, a trend most pronounced in primates. In the hominin lineage, early australopithecines had brains similar in size to modern chimpanzees, yet within a few million years, their descendants had evolved brains

more than three times as large. However, this expansion was not universal. *Homo floresiensis* and *Homo naledi* retained small, chimpanzee-sized brains, showing that enlargement was neither inevitable nor uniform.

RELATIVE BRAIN SIZE
A useful measure of relative brain size is the encephalization quotient (EQ), which compares an animal's actual brain mass to what would be expected for an animal of its body size. An EQ of 1 is average; higher values indicate a disproportionately large brain. Humans have an EQ of about 7, meaning our brains are roughly seven times larger than what is typically expected for a mammal of our size (compared

Human brain expansion was rapid, but uneven across hominin species

with around 2.5 in chimpanzees). Neanderthals had slightly larger brains in absolute terms, but lower EQs, in principle, because they needed more neurons to control their larger bodies. According to this idea, whales and elephants possess enormous brains but modest EQs, as much of their neural capacity supports body regulation rather than higher cognition. The tiny mormyrid electric fish devotes about 3 per cent of its body mass to its brain – a proportion similar to humans – primarily for navigating its murky aquatic world through electrical signals rather than abstract thought.

HIGH PRICE OF INTELLIGENCE

A large brain may seem advantageous, but it comes at a steep cost. Human brains make up only about 2.5 per cent of our body mass yet consume roughly 20 per cent of the body's resting energy – around 13 watts out of 60.

Unlike muscles, the brain cannot be powered down and requires a constant supply of glucose and oxygen. For most species, it is thought that this cost far outweighs the benefits. The idea is that only when enhanced cognition offers a clear survival advantage can a large brain evolve.

This raises a central evolutionary question: why did large, and especially human-sized, brains evolve at all? Scientists have proposed several explanations, largely falling into three main categories. Ecological, social, and cultural brain hypotheses each highlight different selective pressures that may have driven the evolution of large brains and complex intelligence.

HYPOTHESES OF BRAIN EVOLUTION

The ecological brain hypothesis holds that intelligence evolved to meet environmental challenges, such as locating, remembering, and extracting difficult foods in complex or changing

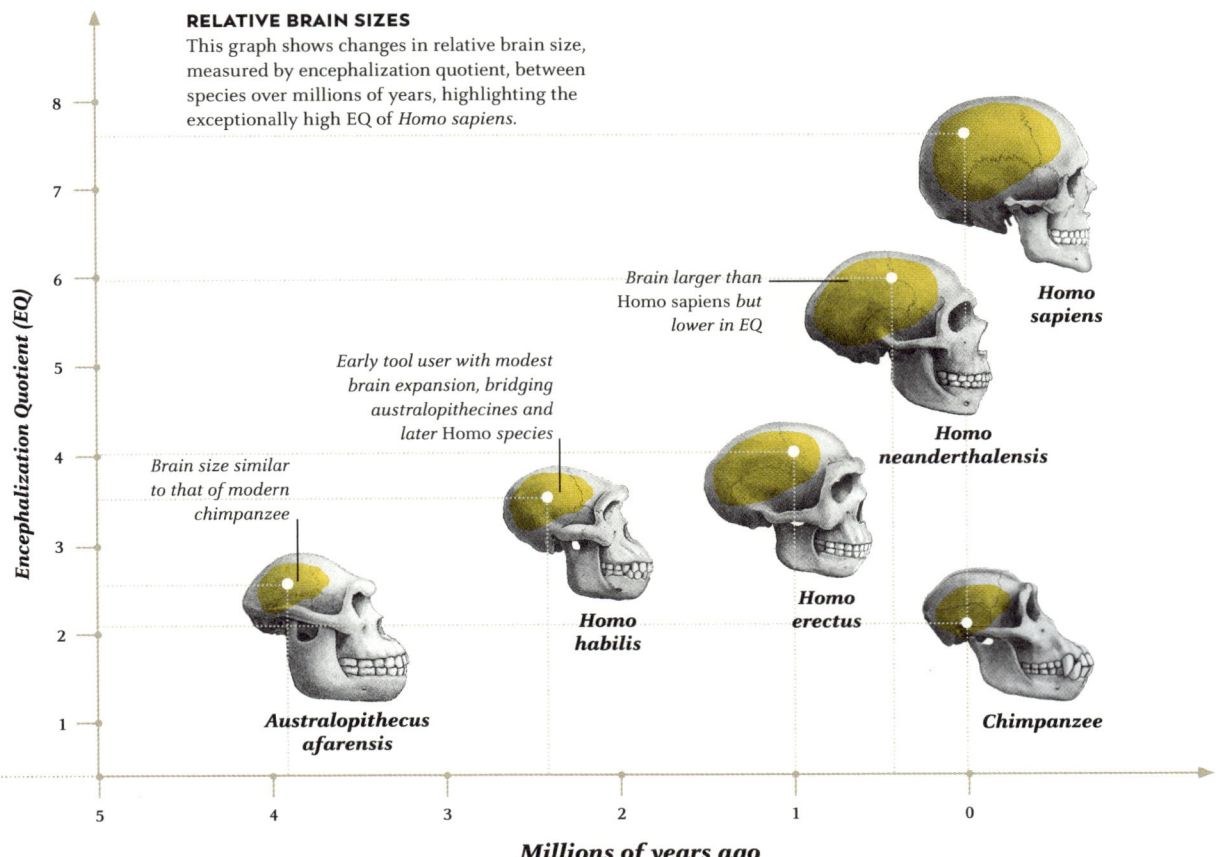

RELATIVE BRAIN SIZES
This graph shows changes in relative brain size, measured by encephalization quotient, between species over millions of years, highlighting the exceptionally high EQ of *Homo sapiens.*

Homo sapiens

Brain larger than Homo sapiens but lower in EQ

Homo neanderthalensis

Early tool user with modest brain expansion, bridging australopithecines and later Homo species

Homo erectus

Brain size similar to that of modern chimpanzee

Homo habilis

Australopithecus afarensis

Chimpanzee

Encephalization Quotient (EQ)

Millions of years ago

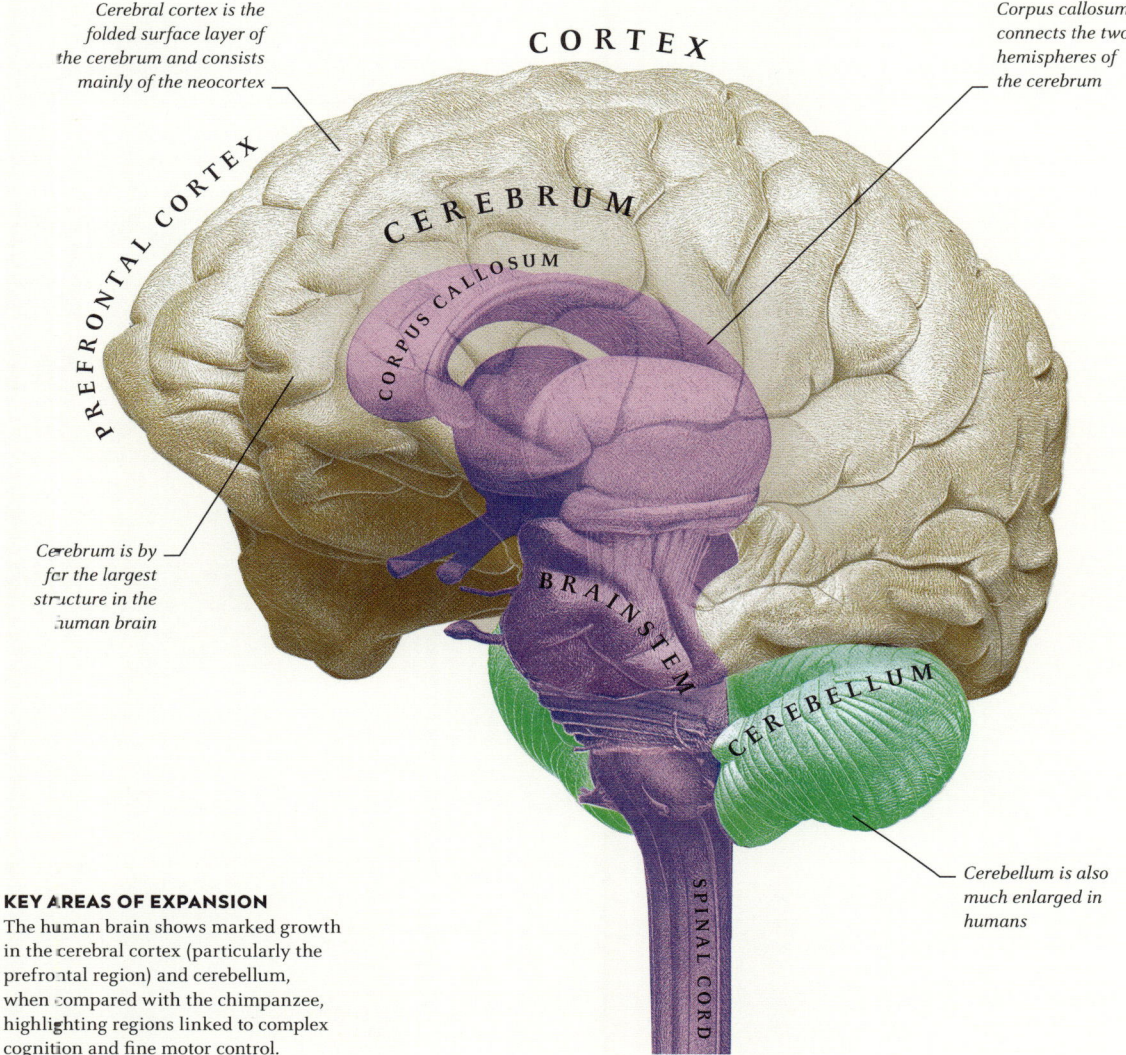

Cerebral cortex is the folded surface layer of the cerebrum and consists mainly of the neocortex

CORTEX

Corpus callosum connects the two hemispheres of the cerebrum

PREFRONTAL CORTEX

CEREBRUM

CORPUS CALLOSUM

Cerebrum is by far the largest structure in the human brain

BRAINSTEM

CEREBELLUM

Cerebellum is also much enlarged in humans

SPINAL CORD

KEY AREAS OF EXPANSION

The human brain shows marked growth in the cerebral cortex (particularly the prefrontal region) and cerebellum, when compared with the chimpanzee, highlighting regions linked to complex cognition and fine motor control.

habitats. Memory, planning, and technical skill conferred strong advantages. The "expensive tissue" idea adds that higher-quality diets, including cooked and animal foods, freed energy for brain growth over evolutionary time and generations (see pp.146–47). The social brain hypothesis centres on the cognitive demands of group living: recognizing individuals, tracking relationships, managing alliances, detecting deception, and resolving conflict. It proposes that the neocortex grows as typical group size increases, with humans maintaining more relationships than any other primate. In this view, navigating social complexity has been the main driver of brain expansion. Finally, the cultural brain hypothesis emphasizes the teaching and learning of skills and knowledge from other individuals as driving brain evolution (see pp.228–31).

Diet and environmental complexity may have driven the evolution of larger brains

TESTING THE THEORIES

Testing these different ideas is difficult because evolutionary events cannot be reproduced experimentally. We cannot rerun history to see what would have happened if early humans lacked fire or lived in smaller, more isolated groups. Instead, scientists use approaches such as "in-silico" evolution, in which virtual organisms

Brain evolution is modelled using simulations when experiments are impossible

evolve under controlled conditions. By varying environmental challenges, social structure, and learning ability, researchers can observe how brain size changes over thousands of simulated generations. These models suggest that brain expansion is most likely when individuals must find creative solutions when facing difficult, changing environments and can learn skills and knowledge from others, supporting both the ecological and cultural brain hypotheses.

PRIMATE BRAIN SPECTRUM

Homo sapiens is one of more than 500 primate species, all descended from a common ancestor about 66 million years ago. We occupy an extreme: our brain is roughly 800 times larger than a mouse lemur's, yet only about one-quarter the size of a humpback whale's. Much of this difference lies in the neocortex, the outer layer responsible for language, abstract reasoning, and cultural learning. Even closely related

STIMULATION SIMULATION

This graph shows a computer model, or simulation, of brain size and body size evolution as a result of the three different ways of finding food: through social learning (working as a team, competing with other groups), cultural learning (learning from one another), and ecological learning (individual problem-solving).

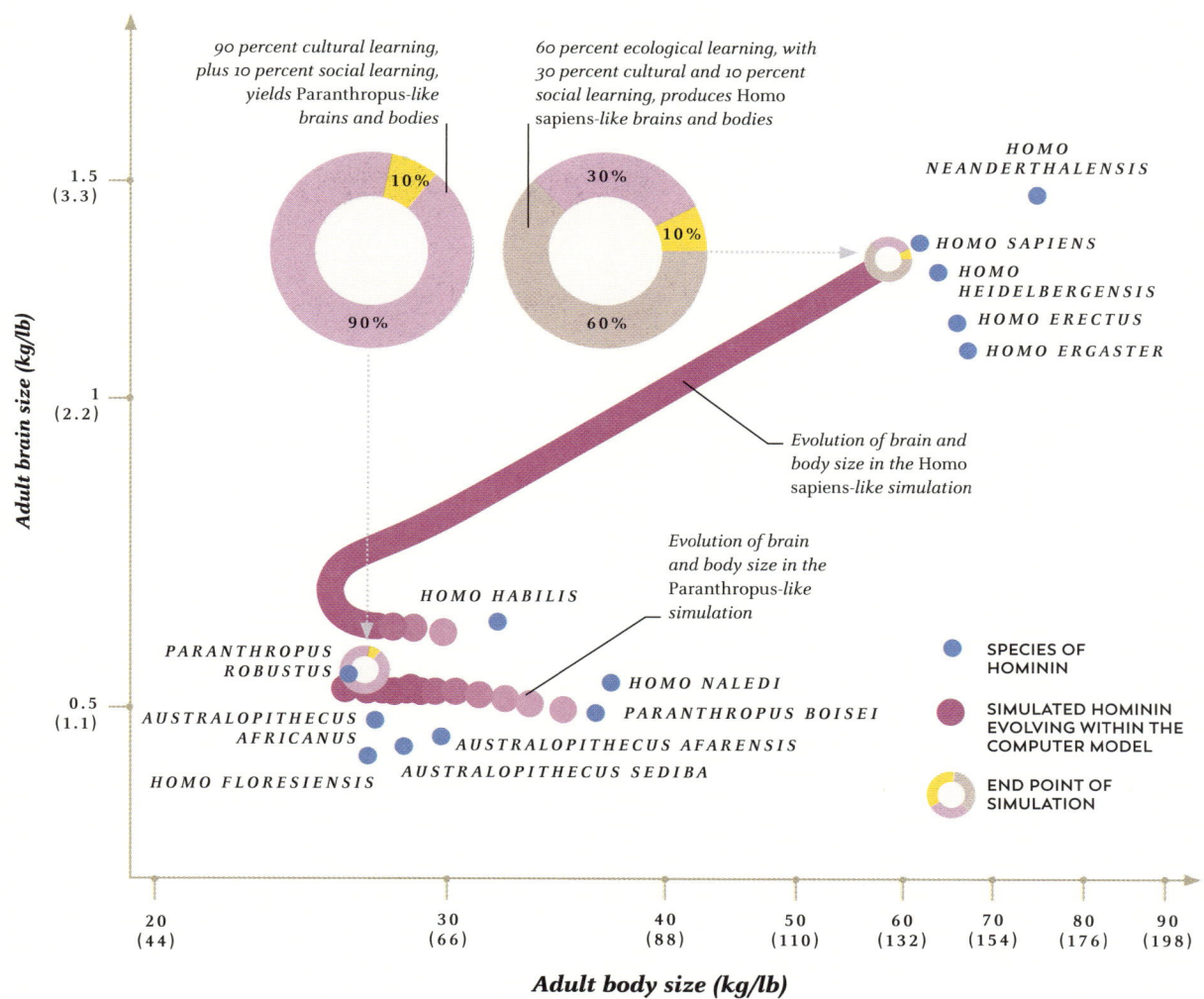

90 percent cultural learning, plus 10 percent social learning, yields Paranthropus-like brains and bodies

60 percent ecological learning, with 30 percent cultural and 10 percent social learning, produces Homo sapiens-like brains and bodies

Evolution of brain and body size in the Homo sapiens*-like simulation*

Evolution of brain and body size in the Paranthropus*-like simulation*

- SPECIES OF HOMININ
- SIMULATED HOMININ EVOLVING WITHIN THE COMPUTER MODEL
- END POINT OF SIMULATION

HOMO NEANDERTHALENSIS
HOMO SAPIENS
HOMO HEIDELBERGENSIS
HOMO ERECTUS
HOMO ERGASTER
HOMO HABILIS
PARANTHROPUS ROBUSTUS
HOMO NALEDI
PARANTHROPUS BOISEI
AUSTRALOPITHECUS AFRICANUS
AUSTRALOPITHECUS AFARENSIS
AUSTRALOPITHECUS SEDIBA
HOMO FLORESIENSIS

Adult brain size (kg/lb)

1.5 (3.3)
1 (2.2)
0.5 (1.1)

Adult body size (kg/lb)

20 (44) 30 (66) 40 (88) 50 (110) 60 (132) 70 (154) 80 (176) 90 (198)

225

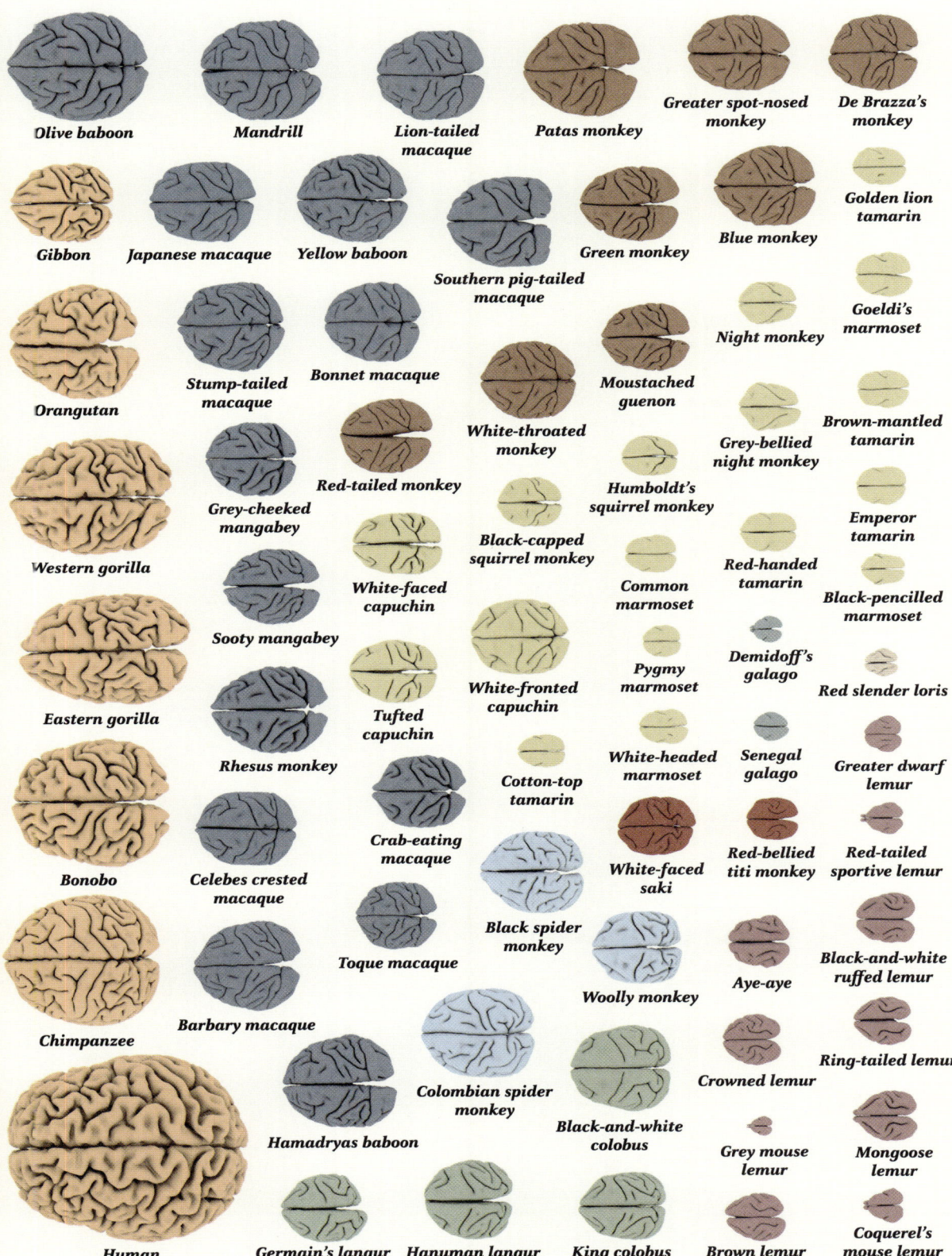

Olive baboon

Mandrill

Lion-tailed macaque

Patas monkey

Greater spot-nosed monkey

De Brazza's monkey

Gibbon

Japanese macaque

Yellow baboon

Southern pig-tailed macaque

Green monkey

Blue monkey

Golden lion tamarin

Orangutan

Stump-tailed macaque

Bonnet macaque

Moustached guenon

Night monkey

Goeldi's marmoset

White-throated monkey

Grey-bellied night monkey

Brown-mantled tamarin

Western gorilla

Red-tailed monkey

Humboldt's squirrel monkey

Emperor tamarin

Grey-cheeked mangabey

Black-capped squirrel monkey

Red-handed tamarin

Eastern gorilla

Sooty mangabey

White-faced capuchin

Common marmoset

Black-pencilled marmoset

Pygmy marmoset

Demidoff's galago

Red slender loris

Bonobo

Rhesus monkey

Tufted capuchin

White-fronted capuchin

White-headed marmoset

Senegal galago

Greater dwarf lemur

Crab-eating macaque

Cotton-top tamarin

Celebes crested macaque

White-faced saki

Red-bellied titi monkey

Red-tailed sportive lemur

Toque macaque

Black spider monkey

Aye-aye

Black-and-white ruffed lemur

Chimpanzee

Barbary macaque

Woolly monkey

Crowned lemur

Ring-tailed lemur

Colombian spider monkey

Hamadryas baboon

Black-and-white colobus

Grey mouse lemur

Mongoose lemur

Human

Germain's langur

Hanuman langur

King colobus

Brown lemur

Coquerel's mouse lemur

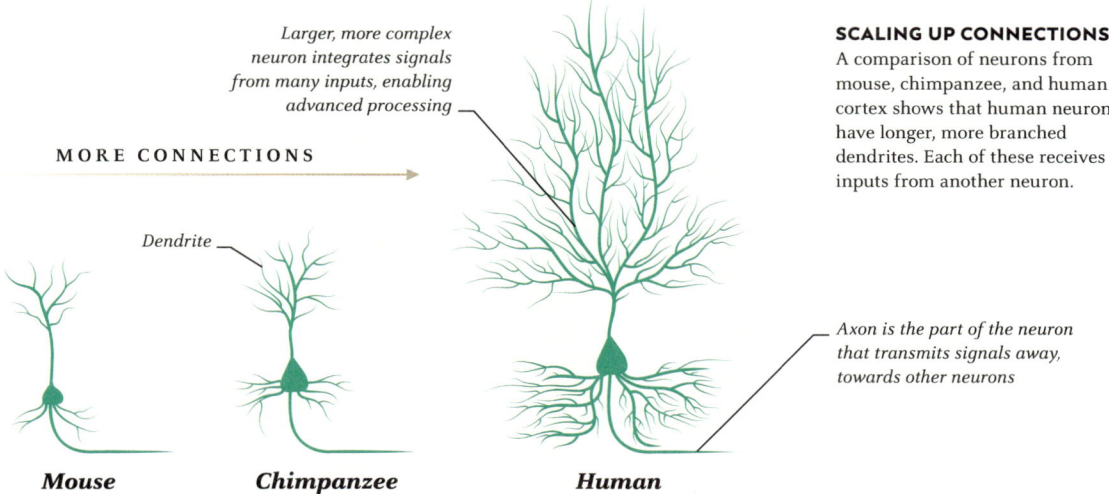

MORE CONNECTIONS

Larger, more complex neuron integrates signals from many inputs, enabling advanced processing

Dendrite

Mouse

Chimpanzee

Human

Axon is the part of the neuron that transmits signals away, towards other neurons

SCALING UP CONNECTIONS
A comparison of neurons from mouse, chimpanzee, and human cortex shows that human neurons have longer, more branched dendrites. Each of these receives inputs from another neuron.

primates can diverge dramatically: marmosets and capuchins, living in similar forests, evolved very different brain sizes in just 20 million years, while more distant species sometimes have similar structures.

PATTERNS BENEATH DIVERSITY

Despite this variation, brains follow predictable rules as they enlarge. Frontal regions expand more than other areas. White matter, which carries signals between brain regions, grows more than grey matter, which processes information, and the surface spreads and folds more than it thickens. Human brains follow these trends exactly.

The cerebellum, involved in movement as well as thinking and emotions, shows the same pattern. Although only about 10 per cent of human brain volume, it contains roughly 80 per cent of all neurons. Larger brains also have bigger, more complex neurons. A typical human

neuron forms about 8,000 connections, and the density of connections is fairly constant. Each neuron sends a single axon – the main, threadlike part carrying outgoing signals – to connect distant regions.

STABILITY AND FLEXIBILITY

Although humans have a remarkably large brain, its basic structure follows the same rules as other primates. Stable features provide consistency, while flexible ones allow variety and adaptability. Lifelong plasticity continually reshapes the brain through experience. The true brilliance of the human brain isn't that it breaks the rules, but that by following them, and pushing them to an extreme, it enables symbolic language, abstract thought, and a cumulative culture that builds across generations (see pp.228–31).

Human brains balance structural consistency with continuous change

CORTEX FOLDS
Across primate lineages, cortical expansion and folding follow predictable patterns, that are inevitable due to mechanical constraints. Once a cortex reaches a certain size, in an animal's life or over evolutionary time, it inevitably folds in a predicatable way.

- HOMINOIDEA
- COLOBINAE
- PAPIONINI
- CERCOPITHECINI
- ATELIDAE
- CEBIDAE
- GALAGONIDAE
- LORIDAE
- PITHECIIDAE
- LEMURIFORMES

Cultural evolution

Human evolution is not only written in our DNA but in our ideas. Culture – our ability to learn, teach, and innovate – has become humanity's most powerful evolutionary force.

Evolution is not solely biological; humans also evolve culturally, transmitting knowledge, behaviours, tools, beliefs, and skills through teaching and learning. This cultural evolution acts in parallel with genetic evolution, and the two constantly influence each other. Unlike genetic evolution, which depends on reproduction and unfolds over many generations, cultural evolution can occur both vertically, from parents to offspring, and horizontally, between peers. This flexible system allows ideas to spread and adapt far more quickly than genes, making cultural evolution orders of magnitude faster than biological evolution.

Cultural change happens with remarkable speed. Knowledge, technologies, and behaviours can shift within a single lifetime, allowing humans to adapt quickly to changing environments and

Culture spreads rapidly through teaching and learning, unlike slow genetic change

TYPES OF SELECTION
Natural selection occurs when individuals vary in heritable traits; some increase survival or reproduction and are passed on. Cultural selection occurs when individuals adopt diverse behaviours, and the most successful or prestigious are copied and spread.

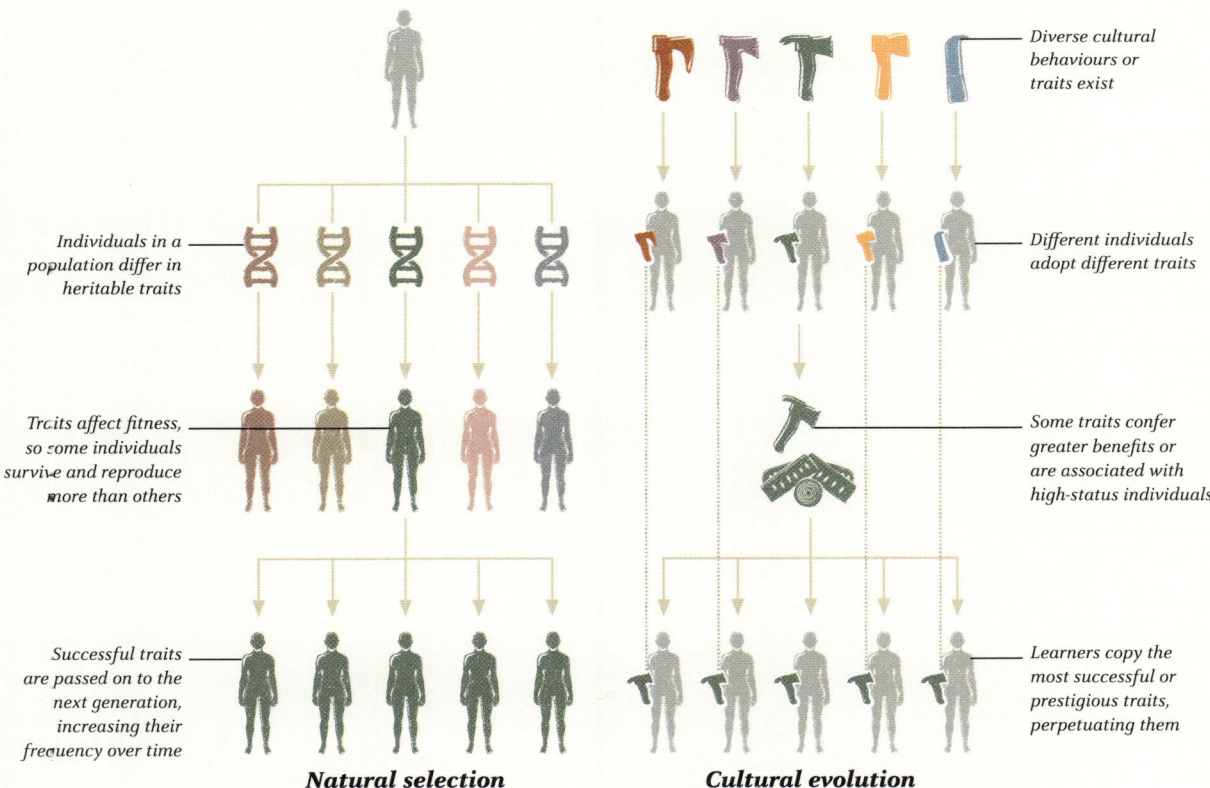

Individuals in a population differ in heritable traits

Traits affect fitness, so some individuals survive and reproduce more than others

Successful traits are passed on to the next generation, increasing their frequency over time

Natural selection

Diverse cultural behaviours or traits exist

Different individuals adopt different traits

Some traits confer greater benefits or are associated with high-status individuals

Learners copy the most successful or prestigious traits, perpetuating them

Cultural evolution

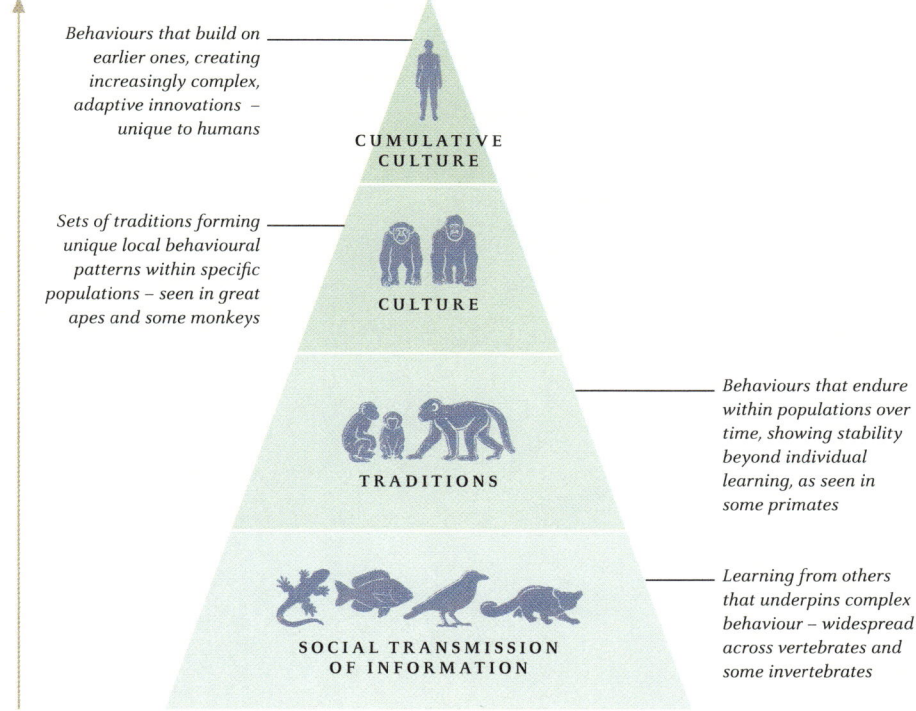

Behaviours that build on earlier ones, creating increasingly complex, adaptive innovations – unique to humans

CUMULATIVE CULTURE

Sets of traditions forming unique local behavioural patterns within specific populations – seen in great apes and some monkeys

CULTURE

Behaviours that endure within populations over time, showing stability beyond individual learning, as seen in some primates

TRADITIONS

Learning from others that underpins complex behaviour – widespread across vertebrates and some invertebrates

SOCIAL TRANSMISSION OF INFORMATION

Effectiveness or fidelity of information transmission

social conditions. Teaching, imitation, and communication let ideas spread rapidly, reshaping societies in decades rather than millennia. A new tool, belief, or social rule can be adopted by a community and alter daily life almost overnight. This ability to adapt quickly through shared learning is a cornerstone of our evolutionary success.

TEACHING AND LEARNING

Human teaching is deliberate, explains ideas clearly, and is more advanced than in other animals

Humans are exceptional teachers and learners – a rarity in the animal world. Many species engage in social learning, but active, intentional teaching, in which one individual deliberately instructs another, is exceedingly rare. Meerkats, for instance, help pups learn to handle scorpions by offering dead or disabled prey, but even this behaviour lacks the deliberate explanation and abstraction seen in humans.

Human teaching is distinctive because it combines explanation, demonstration, correction, and feedback with an understanding of the learner's perspective. Teachers adjust their approach to suit a pupil's level, conveying general principles as well as specific actions. This blend of empathy, communication, and flexibility greatly increases both the speed and accuracy of knowledge transmission, allowing complex information to be preserved and refined over time.

Human learning is equally sophisticated. We do not merely copy others but engage in emulation – understanding goals and intentions – and in active, instructional learning, in which we seek

FROM SOCIAL LEARNING TO CUMULATIVE CULTURE
This pyramid shows how clearly and effectively information is passed on, and how that affects the complexity of a culture. Most animals display basic social learning, some develop traditions and cultures, while humans achieve cumulative culture.

and process guidance. These processes depend on attention, memory, reasoning, and abstraction. From early childhood, humans display a powerful drive to teach and learn, enabling cultural knowledge to accumulate and evolve with remarkable fidelity.

THE RATCHET EFFECT

The accumulation of cultural progress produces a "ratchet effect". Once a useful idea or technique emerges, it can be maintained, improved, and built upon, rather than being lost or reinvented. Each generation stands on the shoulders of those before it, adding to a growing body of collective knowledge. Stone tools became metal tools, which in turn evolved into machines, computers, and spacecraft. Human progress, in every domain, from science to art, reflects this cumulative nature of culture.

By contrast, most animal cultures show little cumulative development. Chimpanzees may pass on termite-fishing methods or nut-cracking techniques, but these behaviours appear to change little over time. However, in humans cultural knowledge compounds as numbers and mobility allow innovations not only to arise, but to persist, spread, and become widely adopted.

Cultural knowledge builds cumulatively, driving continuous human innovation and progress

Each generation stands on the shoulders of those that came before it.

RECIPROCAL INFLUENCE

One hypothesis suggests that culture and biology shaped each other in a reinforcing feedback loop. Teaching and learning place major cognitive demands on both teachers and learners, possibly influencing human brain evolution. Effective teaching may require theory of mind, empathy, patience, and language, while learners need motivation and the ability to use knowledge flexibly. In this view, groups better at sharing information gained an advantage, creating a "cultural arms race". Over time, this feedback could have supported larger, more capable brains and, in turn, increasingly complex culture – a process known as gene-culture coevolution.

HOW LEARNING BIASES SHAPE CULTURE

These illustrations show how different biases guide what we learn from others: we copy what most people do (conformity), what's new or unusual (novelty), what respected people do (prestige), and what seems to work best (success).

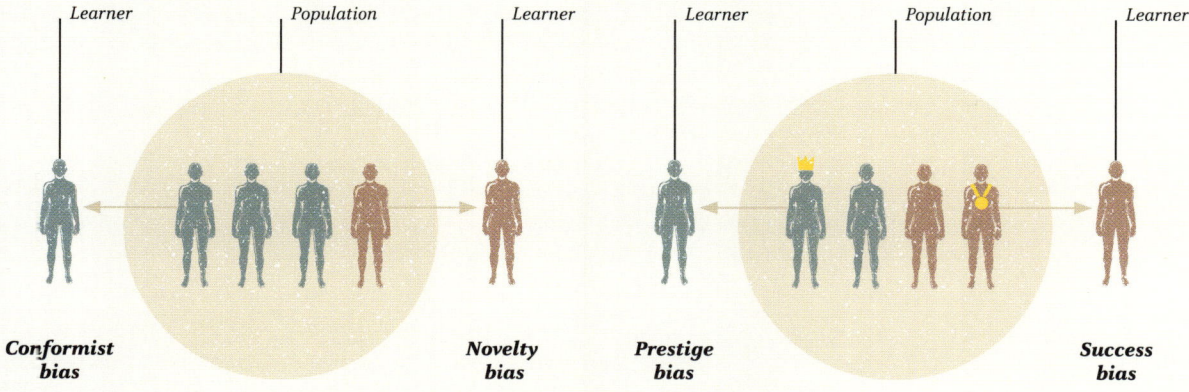

Learner | Population | Learner   Learner | Population | Learner

Conformist bias   **Novelty bias**   **Prestige bias**   **Success bias**

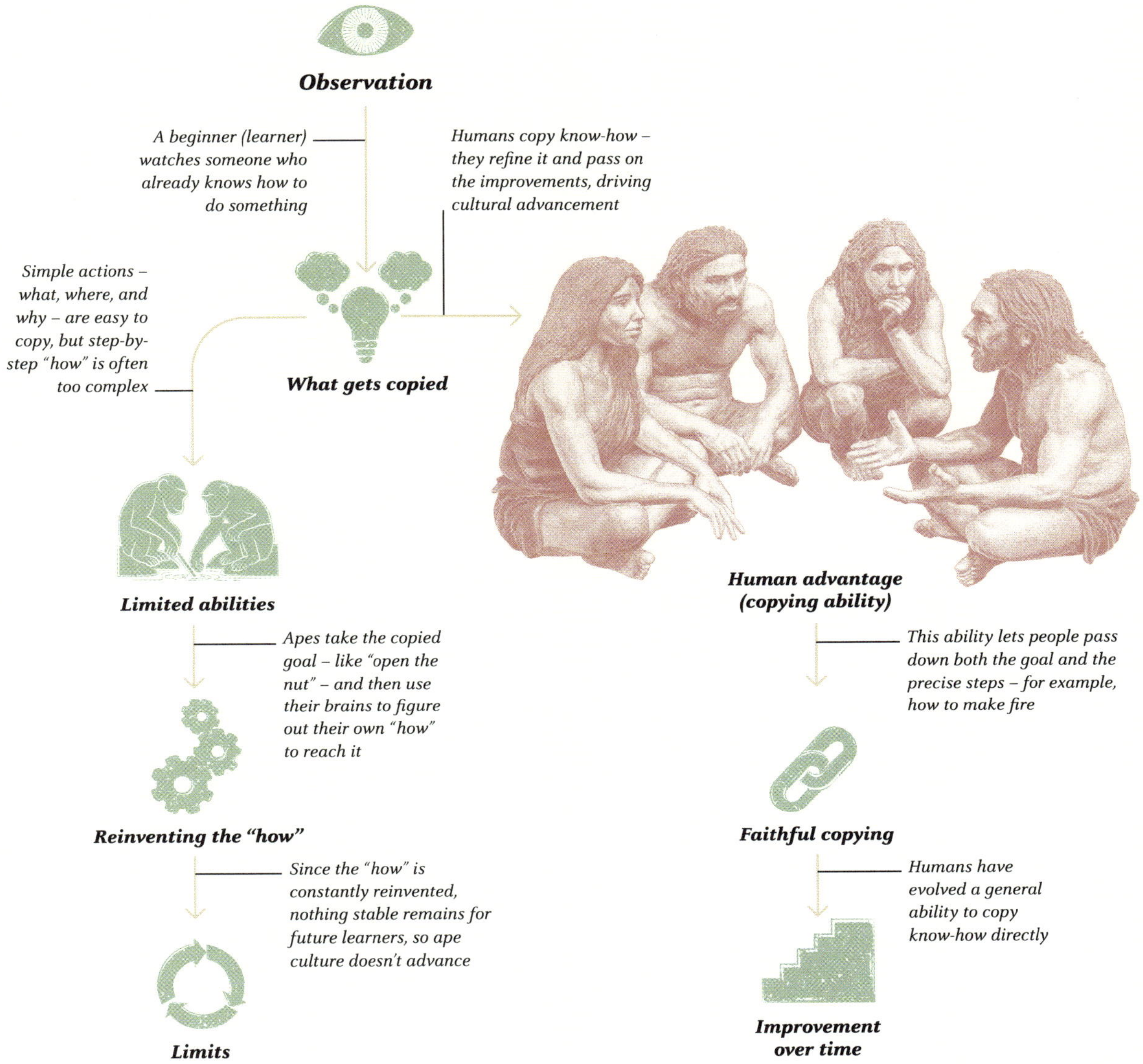

Observation

A beginner (learner) watches someone who already knows how to do something

Humans copy know-how – they refine it and pass on the improvements, driving cultural advancement

Simple actions – what, where, and why – are easy to copy, but step-by-step "how" is often too complex

What gets copied

Limited abilities

Apes take the copied goal – like "open the nut" – and then use their brains to figure out their own "how" to reach it

Human advantage (copying ability)

This ability lets people pass down both the goal and the precise steps – for example, how to make fire

Reinventing the "how"

Since the "how" is constantly reinvented, nothing stable remains for future learners, so ape culture doesn't advance

Faithful copying

Humans have evolved a general ability to copy know-how directly

Limits

Improvement over time

APES RESET, HUMANS BUILD

Apes pass down what tools are for, but each generation reinvents how to use them, so culture resets. Humans share purpose and method, letting knowledge accumulate. By copying and improving details, human culture builds across generations, creating rich cumulative traditions.

KEY TO SURVIVAL

Cultural transmission has always been vital for survival. Through stories, observation, and practice, people passed on essential knowledge about toolmaking, navigation, language, medicinal plants, food gathering, and social rules. This collective inheritance allowed humans to thrive in an extraordinary range of environments, from Ice Age tundras to tropical forests, far faster than biological adaptation alone could achieve. Culture gave us the ability to solve problems, share innovations, and preserve hard-won experiences.

Shared knowledge ensured survival, enabling rapid adaptation across diverse environments

CULTURAL EVOLUTION

Controlling the environment

Human evolution isn't just about expanding minds and social bonds – it's also about shaping our surroundings. Where other species must adapt to their environments, humans can modify their surroundings to suit themselves. This shift, from being shaped by nature to shaping it, defined our future.

What began with hunting innovations would ultimately culminate in agriculture and animal domestication. Early humans became "ecological engineers", managing ecosystems to serve their needs rather than just surviving in them. This transformation was not instinctive or isolated, as seen in some animals, it was cumulative, intentional, increasingly systemic, and ultimately very destructive.

MEGAFAUNA EXTINCTIONS

This graph shows how large mammal extinctions over the past 50,000 years varied by body size. At least 161 species vanished, with giant herbivores such as mammoths hardest hit – only 11 survive today, and all are still in sharp decline.

● EXTINCT
● TOTAL SPECIES

THE DAWN OF HUNTING

Early hominins were opportunistic foragers, consuming fruits, leaves, insects, and scavenged meat. Around 2 million years ago, *Homo erectus* is thought to have shifted to hunting, enabled by endurance running, accurate throwing, and larger brains that supported planning and teamwork.

By 500,000–200,000 years ago, evidence of big-game hunting grows. Stone spearheads in South Africa show high-velocity impact scars (see p.146) and spears in Germany – fire-hardened wooden weapons over 2 m (7 ft) long – would have been ideal for hunting deer (see p.143). This required cooperation, ambush planning, and knowledge of prey behaviour. Later, Neanderthals on Jersey, drove herds into ambushes, using terrain to control movement. These strategies reflect early environmental mastery: reshaping animal behaviour through coordinated, landscape-scale action.

Early humans evolved from scavengers to skilled hunters through tools and teamwork

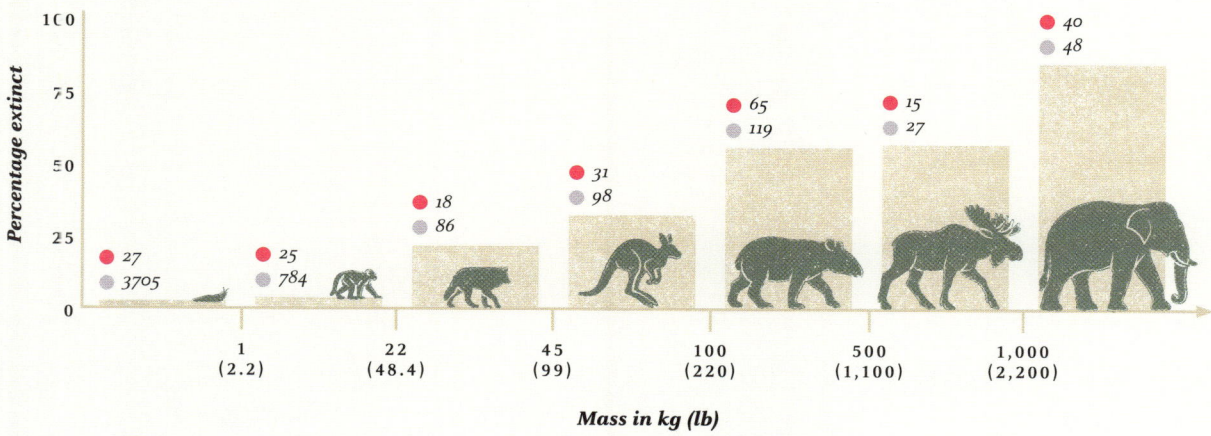

● 27	● 25		● 18	● 31	● 65	● 15	● 40
● 3705	● 784		● 86	● 98	● 119	● 27	● 48

Percentage extinct (y-axis: 0, 25, 50, 75, 100)

Mass in kg (lb): 1 (2.2), 22 (48.4), 45 (99), 100 (220), 500 (1,100), 1,000 (2,200)

Mass in kg (lb)

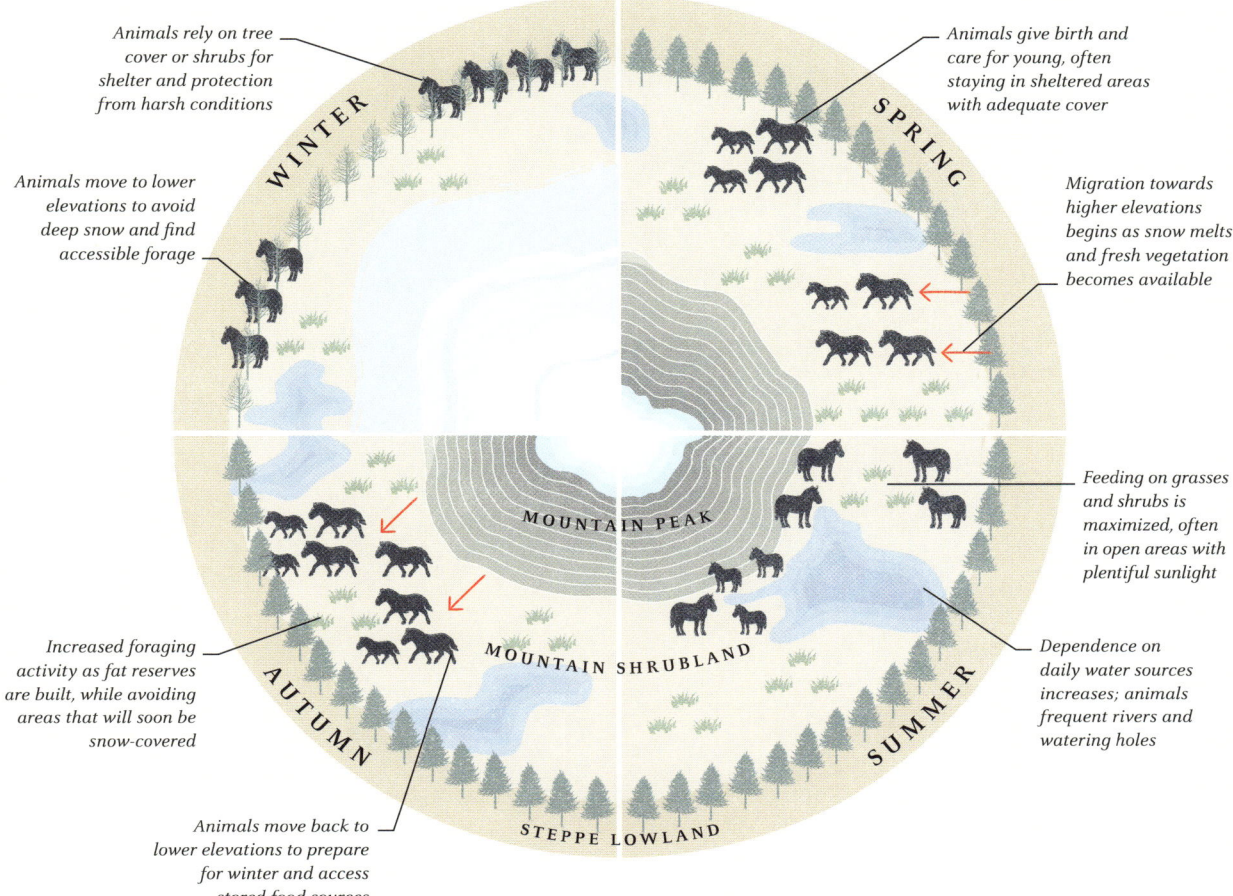

Animals rely on tree cover or shrubs for shelter and protection from harsh conditions

Animals give birth and care for young, often staying in sheltered areas with adequate cover

Animals move to lower elevations to avoid deep snow and find accessible forage

Migration towards higher elevations begins as snow melts and fresh vegetation becomes available

WINTER

SPRING

Feeding on grasses and shrubs is maximized, often in open areas with plentiful sunlight

MOUNTAIN PEAK

Increased foraging activity as fat reserves are built, while avoiding areas that will soon be snow-covered

MOUNTAIN SHRUBLAND

Dependence on daily water sources increases; animals frequent rivers and watering holes

AUTUMN

SUMMER

Animals move back to lower elevations to prepare for winter and access stored food sources

STEPPE LOWLAND

OVEREXPLOITATION OF PREY

Human predation appears to have been a contributing factor, along with climate change, in the extinction of many megafaunal species, including mammoths, woolly rhinoceroses, and cave bears. As humans spread into new regions, previously resilient populations were unable to withstand the combined pressure of persistent hunting and climate fluctuations. Archaeological and fossil records show that even low-intensity, repeated hunting over centuries could drive large, slow-reproducing species to extinction.

Although climate change at the end of the last ice age altered habitats and resources, many species had survived earlier environmental changes, indicating that human activity may have often been the decisive factor. The loss of these major herbivores reshaped ecosystems: forests expanded into grasslands, plant communities

Overhunting large animals reshaped ecosystems and led to extinctions, and shifts in vegetation

CYCLES OF SURVIVAL

This illustration shows the annual cycles of wild prey. By observing these patterns, prehistoric hunters anticipated migrations, followed animals across the landscape, and hunted them when conditions were most favourable. Such knowledge tied survival to close observation of animal behaviour and seasonal change.

shifted, and fuel loads increased, leading to more frequent wildfires. By removing keystone species, humans permanently transformed landscapes, altering ecological dynamics and the resource base on which later human societies depended.

USING FIRE

The earliest possible traces of fire control comes from Kenya and South Africa (about 1.5 MYA) and northern China (about 1.27 MYA), with more

HUNTING WITH DOGS

Rock art from Libya's Tadrart Acacus region, made about 8,000 years ago, shows humans hunting ibex and gazelles with domesticated dogs. Using dogs made hunting more effective, representing a major advancement in how people exploited resources.

1 *Hunters with bows and spears, shown in dynamic poses, pursue fleeing prey*

2 *Dogs depicted chasing and flanking prey together reflects trained, coordinated hunting*

3 *Slender ibex and gazelles with curved horns, rendered in profile amid sparse desert terrain*

widely accepted evidence by 1 million years ago at Wonderwerk Cave, South Africa, and 400,000 years ago at Beeches Pit, England. Once fire was reliably controlled, it quickly became a powerful tool for shaping landscapes. Humans cleared vegetation, created open hunting grounds, flushed prey, and reduced wildfire risks near settlements. Controlled burning also promoted useful plants, reflecting growing ecological knowledge.

Unlike some animals that use fire instinctively, humans applied it with foresight and planning, maintaining fires, transporting embers, and passing knowledge across generations. These practices reveal early intentional ecosystem management rooted in cultural learning.

ROOTS OF DOMESTICATION

Even before full domestication, humans influenced animals and plants in systematic, far-reaching ways. Between roughly 50,000 and 12,000 years ago, groups across Europe, Asia, and

Early fire use shows humans beginning to manage and shape environments

North America followed herds – reindeer, bison, and horses – along seasonal migration routes (see p.233). Archaeological sites such as Pincevent in France and Dolní Věstonice in the Czech Republic show people returned annually, building predictable hunting zones. In some cases, humans may have subtly shaped animal behaviour through indirect herding, landscape manipulation, or repeated fire. This laid the groundwork for domestication by establishing ongoing relationships with certain species.

Early environmental control hinted at the beginnings of domestication

Archaeological and genetic evidence shows early cooperation between humans and animals. The domestication of dogs – descended from wolves – may have occurred between 40,000 and 20,000 years ago (see pp.154–55). Dogs likely aided hunting and provided security, further deepening human interaction with animals.

Humans were also increasingly manipulating plants. At the Ohalo II site in Israel, dated to around 23,000 years ago, people harvested and ground wild cereals and legumes. They invented tools for the job, including stone sickles, grinding stones, and other food processing tools (see p.152). Storage pits – lined with stone or clay – began to appear at later hunter-gatherer sites, marking the beginnings of surplus food storage.

These practices enabled longer-term planning and risk management, allowing some communities to remain in one place for extended periods. As communities developed semi-sedentary lifestyles, moving around less over time, they became more deeply invested in actively managing their immediate environments. This growing dependence on local resources and increasingly detailed ecological knowledge laid the crucial foundation for the next major shift in human history: the intentional domestication of plants and animals.

CONTROLLING THE ENVIRONMENT

Adapting to new habitats

Humans evolved in Africa's tropical savannahs and woodlands but eventually spread worldwide, adapting to deserts, islands, mountains, tundra, and Arctic coasts. Thriving in such diverse habitats required not just endurance but also cooperation, foresight, and innovation – key traits that helped to make Homo sapiens *a successful global species.*

Hard stone generates sparks when it strikes pyrite nodule

Flint striker

Worn surface shows where mineral has been struck thousands of times

Iron pyrite nodule

*Horse hoof fungus (*Fomes fomentarius*) catches and holds sparks when dried*

SPARKING SURVIVAL

A replica fire-starting kit with flint and iron pyrite shows the simple yet powerful technology early humans used to generate sparks – an innovation that opened new possibilities for survival and adaptation.

Fungus tinder

CARRYING VESSELS

Replicas of birch bark carrying vessels demonstrate one way that humans transported tools, food, and materials. Lightweight and durable, these containers enabled the movement of essential resources across the landscape.

Stitched with lime bast – the inner bark fibres of the linden tree

As humans spread into new habitats, they adapted culturally, rather than biologically, to new challenges. They developed tools, strategies, and social systems to endure freezing winters, arid deserts, scarce resources, and unfamiliar predators. No other primate matches this adaptability. Early humans such as *Homo erectus*, which left Africa 1.9 million years ago, showed flexibility but probably remained tied to open woodland and coasts with year-round food, and climates like their east African home. With no evidence of shelters or clothing, we can only speculate how *Homo erectus* fared in colder climates. This first, modest migration laid the foundation for later, more sophisticated expansions of our species.

Cultural innovation, not biology, enabled humans' migrations beyond Africa

PLANNING FOR THE UNKNOWN

Later species pushed into far harsher environments that required foresight and long-term planning. Entering deserts or cold steppe zones was only possible with advance preparation: water-carrying containers, fire-starting tools, durable clothing materials, and cooperative social strategies. Archaeological finds from Jebel Faya in Arabia and Misliya Cave in the Levant show early humans traversing vast, arid landscapes between roughly 120,000 and 70,000 years ago. These journeys demanded detailed memory of landscapes, accurate prediction of seasonal conditions, careful group coordination, and the deliberate teaching of survival knowledge across multiple generations.

Around 65,000 years ago, lowered Ice Age sea levels formed Sahul – the combined landmass of Australia and New Guinea (see p.246). Reaching it from Southeast Asia required open-sea crossings of 60–100 km (37–62 miles), a feat implying advanced seafaring, cooperation, and imagination.

In Australia's deserts, Indigenous populations later encoded ecological and navigational knowledge into songlines – oral maps describing water sources, seasonal cycles, and rituals. Such cultural tools reveal how language and shared knowledge enabled survival in resource-scarce environments.

INNOVATION ON THE MOVE

Expansion into diverse habitats drove rapid technological change. The Acheulean handaxe (see pp.38–39) gave way to specialized regional toolkits. The Levallois technique (see p.40) produced sharp, standardized flakes, and by the Upper Palaeolithic (50,000–10,000 years ago), tools diversified dramatically to include long blades, projectile points, fishing gear, and bone implements. In colder regions, hide scrapers and bone needles became essential for making layered clothing.

Gravettian cultures of Ice Age Europe (30,000–20,000 years ago) combined figurative art with innovation, producing Gravette Points – small, backed blades suited to hunting big game on snowy steppes. In forested zones, hafted tools and projectiles improved hunting elusive prey. These technologies reflect foresight, material expertise, and knowledge-sharing across generations.

SURVIVING THE COLD

Physiologically, humans are ill-suited to cold – we lack insulation and are prone to frostbite. Yet through cooking, clothing, and shelter-building,

ADAPTING TO NEW HABITATS

REPLACEABLE SPEAR TIP

An antler spear tip designed to break on impact allowed early hunters to continue using their valuable wooden shafts. Reindeer antlers were plentiful on the mammoth steppe, but before entering the treeless wastes, hunters had to plan ahead and carry in wood from elsewhere.

Replica was made with Palaeolithic techniques

The spear tip is relatively weak and breaks before the shaft is in danger of damage

Antler spear tip is separated into flanges, which allow it to slot onto the wooden shaft

Wooden shaft ends in a wedge

Side view

Spear tip is made of reindeer antler

Shaft is made of a straight piece of imported silver birch

Front view

our ancestors flourished in frigid climates. Cooking unlocked more calories from roots and meat while neutralizing toxins in plants. Traces of controlled fire appear as early as 1 million years ago in different parts of the world. In Ice Age Eurasia, bone awls and needles reveal the creation of multi-layered garments protecting extremities. Gravettian groups also built semi-subterranean dwellings insulated with mammoth bones and hides. Culture, not biology, provided the insulation that allowed humans to master the cold.

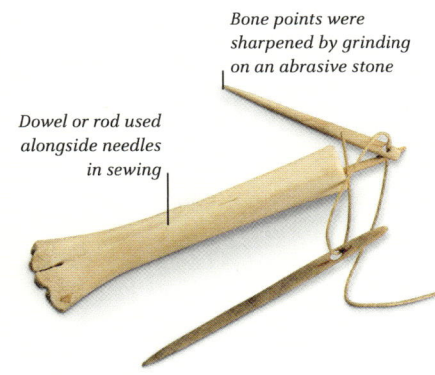

Bone points were sharpened by grinding on an abrasive stone

Dowel or rod used alongside needles in sewing

CARRYING CULTURE

Walking upright freed hands, fostering transport, trade, and portable technologies

Bipedalism freed human hands, enabling transport of infants, tools, and resources – an enormous advantage in environments where materials were scarce. Archaeological evidence shows obsidian and ochre traded across east Africa and southwest Asia. Portable toolkits – blade cores, resharpening flakes, fire-making kits – appear across dispersal sites.

On the mammoth steppe, wood suitable for spear shafts was rare, so hunters used replaceable tips to preserve shafts. The invention of hide bags, woven pouches, and bark containers expanded humans' ability to carry food, tools, and symbolic items. Humans carried culture with them, ensuring continuity in unfamiliar lands.

NEEDLE AND DOWEL
A replica Palaeolithic needle and dowel made of bone illustrate innovations in sewing and construction vital for making clothing, shelters, and other essentials in harsh climates.

INNOVATION AS SURVIVAL
Human dispersal was more than a geographic expansion; it marked a cognitive revolution that transformed our species from niche occupants into niche creators. Through cumulative knowledge and collective ingenuity, humans learned to shape their surroundings intentionally, creating fire, building shelters, and crafting tools. This capacity for cultural adaptation, rather than physical strength or speed, became our defining survival trait, enabling us to flourish in nearly every ecosystem on Earth.

MAMMOTH BONE SHELTER
This sequence shows how Ice Age hunters, living where wood was scarce, may have built durable, windproof homes from mammoth bones, tusks, and hides, providing warmth and stability in harsh glacial conditions.

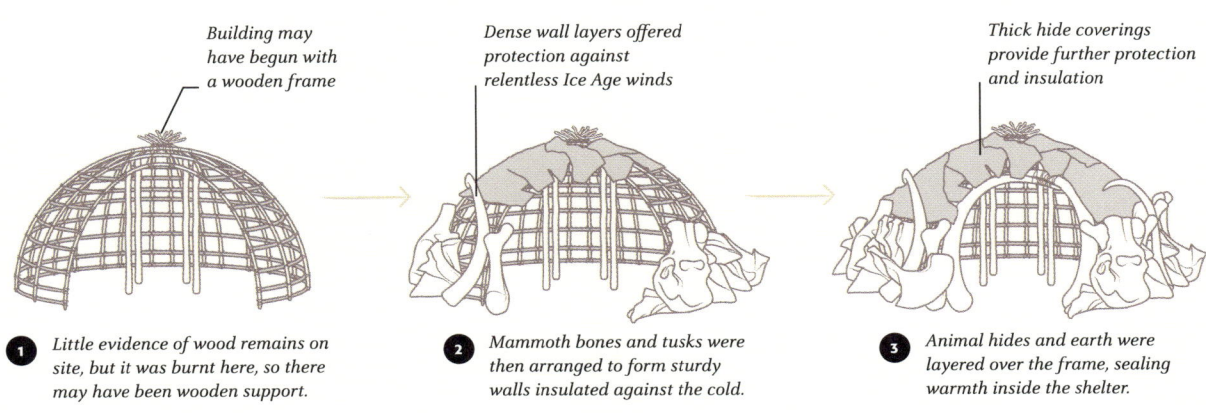

Building may have begun with a wooden frame

Dense wall layers offered protection against relentless Ice Age winds

Thick hide coverings provide further protection and insulation

1 *Little evidence of wood remains on site, but it was burnt here, so there may have been wooden support.*

2 *Mammoth bones and tusks were then arranged to form sturdy walls insulated against the cold.*

3 *Animal hides and earth were layered over the frame, sealing warmth inside the shelter.*

ADAPTING TO NEW HABITATS

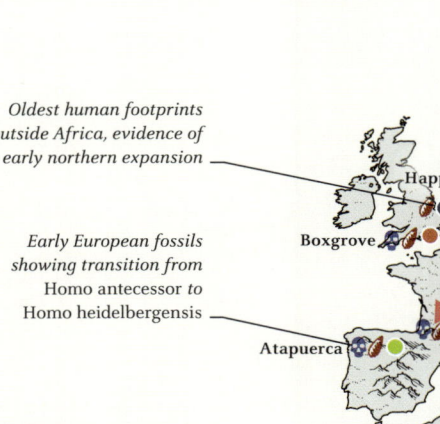

Oldest human footprints outside Africa, evidence of early northern expansion

Happisburgh

EUROPE

Mauer **1.2 MYA**

Boxgrove

Isernia
La Pineta

Early European fossils showing transition from Homo antecessor to Homo heidelbergensis

Arago

Dmanisi

Petralona

Atapuerca

Ceprano

Kocabaş

Ubeidiya

One defining trait of humans is how geographically widespread we are as a species. But long before Homo sapiens *spread across the world, several earlier species of humans dispersed widely beyond their presumed origins in Africa.*

AFRICA

MORE THAN 1.8 MYA

Buia

Daka

Koobi Fora

Lake Turkana

Worldwide dispersal

Olorgesailie

Oldupai

Key site with early human fossils and some of the oldest stone tools

Early humans evolved in Africa before gradually dispersing across the world

From early *Homo erectus* to the rise of *Homo sapiens*, our story has been shaped by adaptation, migration, and evolution across diverse habitats. Early hominins were confined to Africa. The first known – *Sahelanthropus*, *Orrorin*, and *Ardipithecus* (see pp.12–13) – are known from isolated African sites. By 4.3 million years ago (MYA), *Australopithecus* had spread across eastern, northern, and southern Africa, but none ventured beyond the continent.

OUT OF AFRICA

Homo erectus (known in Africa as *Homo ergaster*) seems to have emerged in Africa about 2 MYA and was the first hominin to expand into other regions. Larger brains, humanlike anatomy, and prolonged childhood suited it for mobility and adaptation. Fossil sites from Ubeidiya in the Levant to Dmanisi in Georgia and Riwat in Pakistan trace early dispersal routes. By 1.7 MYA, *Homo erectus* reached China and Java. Evidence of possible fire use at

Xihoudu in China, about 1.27 MYA, marks another milestone. Western Europe was reached by 1.2 MYA, with fossils found in Spain.

At Gran Dolina in Atapuerca, Spain (1.2–0.8 MYA), fossils attributed to *Homo antecessor* reveal a mosaic of traits: primitive jaws and teeth with a more modern face. Likely descended from early Eurasian *Homo erectus*, this species may have evolved into later European lineages such as *Homo heidelbergensis*, though some see it as an extinct side branch.

Nihewan Zhoukoudian

1.7–1.3 MYA

Xihoudu

Lantian

Yunxian Nanjing

Riwat

Narmada

Site of the "Peking Man" fossils, with associated stone tools and evidence of fire use

1.8 MYA

EQUATOR

Mojokerto

Sangiran

Trinil

"Java Man" fossils found alongside tools

AUSTRALIA

EARLY HOMININ MIGRATION

This map traces the earliest hominin migrations out of Africa, beginning around 2 million years ago. It highlights key routes through Eurasia, marking significant archaeological sites and migration patterns that reveal how early humans gradually dispersed, adapted, and evolved across diverse landscapes over thousands of generations.

- *HOMO ERGASTER* SITE
- *HOMO GEORGICUS* SITE
- *HOMO ERECTUS* SITE
- *HOMO ANTECESSOR* SITE
- *HOMO HEIDELBERGENSIS* SITE
- SITE OF UNKNOWN SPECIES OF *HOMO*
- FOSSILS
- TOOLS
- JOURNEY

DIVERSITY AND ISOLATION

Regional populations of hominins evolved distinct traits shaped by local environments and isolation, highlighting the importance of geography and time. Island populations underwent dramatic changes. On Flores in Indonesia, *Homo floresiensis* (190,000–50,000 years ago, YA) stood about a metre tall, with a brain the size of a chimpanzee's – evidence of island dwarfism. In the Philippines, *Homo luzonensis* (about 67,000 YA) displayed a unique mix of primitive and newly evolved features. These species show that early humans could reach remote islands and that some isolated populations survived long after *Homo sapiens* arose in Africa.

NEW SPECIES LEAVE AFRICA

In Africa, *Homo heidelbergensis* likely evolved from *Homo ergaster* (the African form of *Homo erectus*) after a population bottleneck 900,000–800,000 YA. With a larger brain, robust build, and Acheulean tools (see pp.38–39), it dispersed

EURASIA'S HOMININ LANDSCAPE

This map shows the approximate distribution of early human species across Eurasia around 100,000 YA. As modern *Homo sapiens* began migrating out of Africa, they encountered these early humans, such as Neanderthals in Europe and Denisovans in Asia, setting the stage for interaction, competition, and interbreeding.

■ NEANDERTHALS
■ DENISOVANS
■ *HOMO ERECTUS*
■ *HOMO SAPIENS*

FOSSIL FINDS

This map also marks major sites where ancient hominin fossils from this period have been discovered.

● NEANDERTHALS
● DENISOVANS
● *HOMO ERECTUS*
● *HOMO FLORESIENSIS*
● *HOMO LUZONENSIS*
● ARCHAIC *HOMO SAPIENS* (OLDER THAN 100,000 YEARS)
● *HOMO NALEDI*
● *HOMO SAPIENS*

Classic Neanderthal burial site revealing social behaviour and cultural practices

Among the oldest known Homo sapiens *fossils, dating back around 195,000 years*

EUROPE

Vindija Cave
Mezmaiskaya Cave
La Ferrassie
Shanidar Cave
Misliya Cave
Qafzeh Cave
Jebel Irhoud
Grotte des Pigeons
Skull Cave

AFRICA

Herto
Omo Kibish
Panga Ya Saidi
Rising Star Cave
Ga-Mohana Hill North Rockshelter
Border Cave
Diepkloof Rockshelter
Pinnacle Point
Blombos Cave
Klasies River Mouth

into Europe by 700,000 YA, leaving fossils at Boxgrove (England), Arago (France), and Mauer (Germany). Some groups were skilled hunters of large animals such as elephants and rhinos. *Homo heidelbergensis*, or something very similar to it, may have included the common ancestors of Neanderthals (*Homo neanderthalensis*) in Europe and Asia, Denisovans (*Homo longi*) in Asia, and *Homo sapiens* in Africa, though the species itself is not well defined.

ARCHAIC COUSINS

Denisovans were first identified from DNA in a juvenile finger bone from Denisova Cave in Siberia. Genetic evidence links them to modern Melanesians and Aboriginal Australians, showing ancient interbreeding. They likely lived between 285,000 and 25,000 YA and interbred with both Neanderthals and *Homo sapiens* (see pp.212–15). Fossil evidence is scarce but growing: in China, the Xiahe jaw from the Tibetan Plateau (about

160,000 YA) and the Harbin skull from the northeast (about 146,000 YA) suggest Denisovans were widespread and regionally adapted, with robust, cold-tolerant features distinct from both Neanderthals and *Homo sapiens*.

Neanderthals were also stocky, cold-adapted humans with large brains, Mousterian tools, and burial practices. They were the first hominins to occupy much of Europe continuously during the Middle Paleolithic, following earlier species such as *Homo antecessor* and *Homo heidelbergensis*.

Neanderthals evolved in Europe, adapting to cold and burying their dead

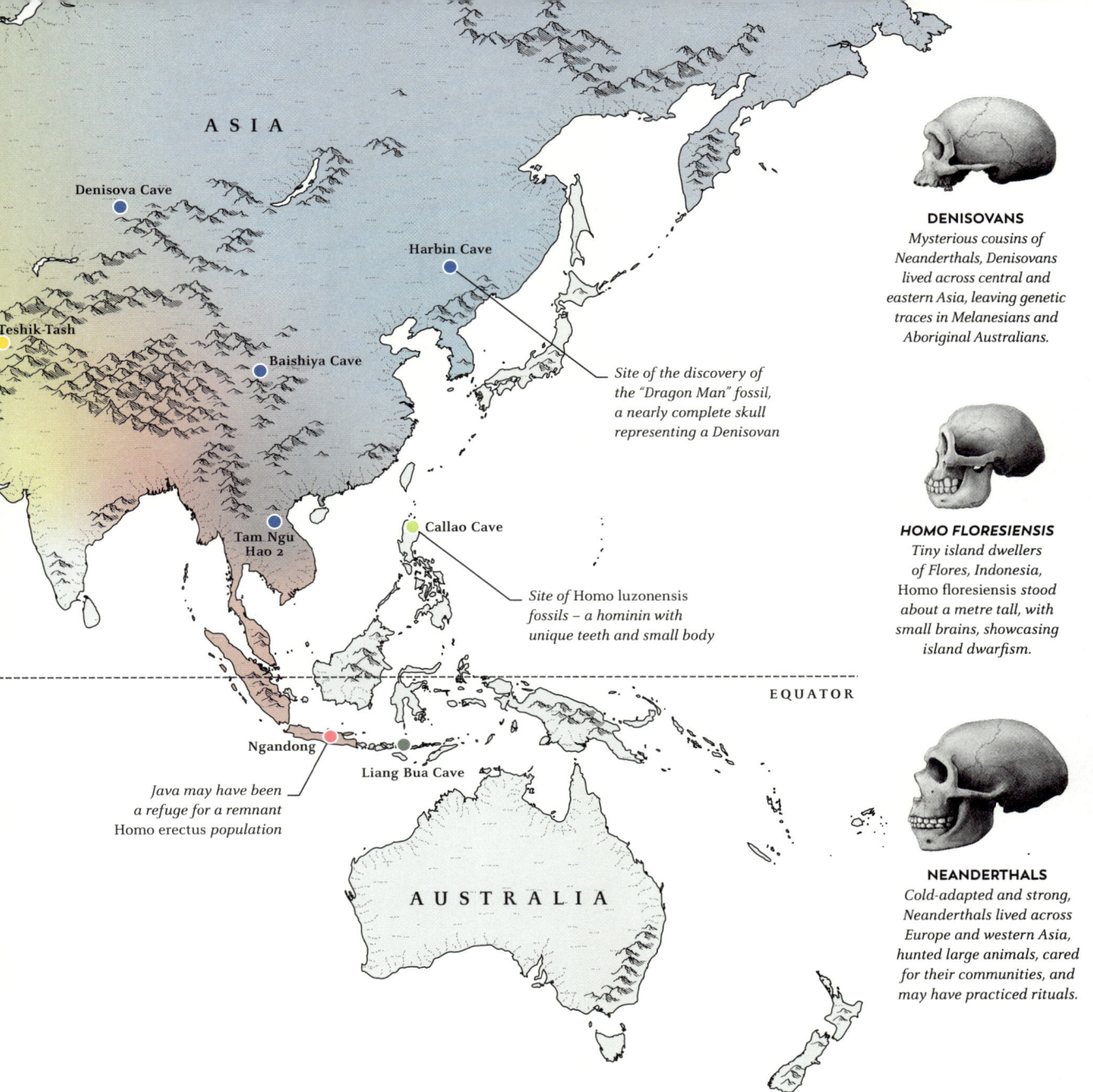

DENISOVANS
Mysterious cousins of Neanderthals, Denisovans lived across central and eastern Asia, leaving genetic traces in Melanesians and Aboriginal Australians.

HOMO FLORESIENSIS
Tiny island dwellers of Flores, Indonesia, Homo floresiensis stood about a metre tall, with small brains, showcasing island dwarfism.

NEANDERTHALS
Cold-adapted and strong, Neanderthals lived across Europe and western Asia, hunted large animals, cared for their communities, and may have practiced rituals.

ASIA

Denisova Cave

Harbin Cave

Teshik-Tash

Baishiya Cave

Site of the discovery of the "Dragon Man" fossil, a nearly complete skull representing a Denisovan

Tam Ngu Hao 2

Callao Cave

Site of Homo luzonensis *fossils – a hominin with unique teeth and small body*

EQUATOR

Ngandong

Liang Bua Cave

Java may have been a refuge for a remnant Homo erectus *population*

AUSTRALIA

A CROWDED WORLD

Between 300,000 and 100,000 YA, several human species coexisted. Neanderthals lived in Europe and western Asia; Denisovans in central and eastern Asia; *Homo erectus* persisted in parts of Asia; *Homo floresiensis* and *Homo luzonensis* survived on islands in Southeast Asia; and *Homo naledi* occupied caves in South Africa. Meanwhile, *Homo sapiens* was evolving in Africa and nearby Asia. This diverse human mosaic gradually shifted as *Homo sapiens* expanded out of Africa, first in limited forays, then with a concerted move between 70,000 and 60,000 YA (see pp.270–71), interacting with other human species and even interbreeding with them (see p.215). Over time, our species became the last surviving branch of a once-crowded evolutionary tree (see pp.18–19), carrying within us the genetic echoes of those earlier humans.

Consciousness

Consciousness lies at the heart of what it means to be alive and aware. Exploring its origins reveals how awareness, emotion, and imagination evolved across species and through early human history, shaping the richness of inner experience.

Consciousness is the awareness of oneself, one's thoughts, and the external world, though it remains remarkably difficult to define precisely. It raises enduring questions: is it the same as thinking, does it depend on language, and can animals truly be conscious?

Most researchers now view consciousness as a dynamic spectrum rather than an all-or-nothing trait. They distinguish between primary consciousness – the basic sensory and emotional awareness shared with many species – and higher consciousness, which involves self-reflection, imagination, deliberate reasoning, and understanding the thoughts and intentions of others. This layered model helps us explore how consciousness may have evolved gradually through increasing brain complexity, social cooperation, and symbolic communication, rather than appearing suddenly and fully formed.

Consciousness likely evolved gradually, from basic awareness to complex self-reflection

CLUES FROM ANIMALS

Modern animals offer valuable insights into the roots of consciousness. Chimpanzees display empathy, use tools, and even mourn their dead. Elephants have been seen gently touching the bones of deceased companions, while dolphins use signature whistles that function like names and cooperate to help injured peers. Even crows

GLIMPSES OF CONSCIOUSNESS
Behaviours such as grief, empathy, and social bonding suggest that elements of consciousness, such as emotion, foresight, and memory, may extend beyond humans.

An elephant mourns a lost relative, demonstrating empathy, social bonding, and emotional intelligence

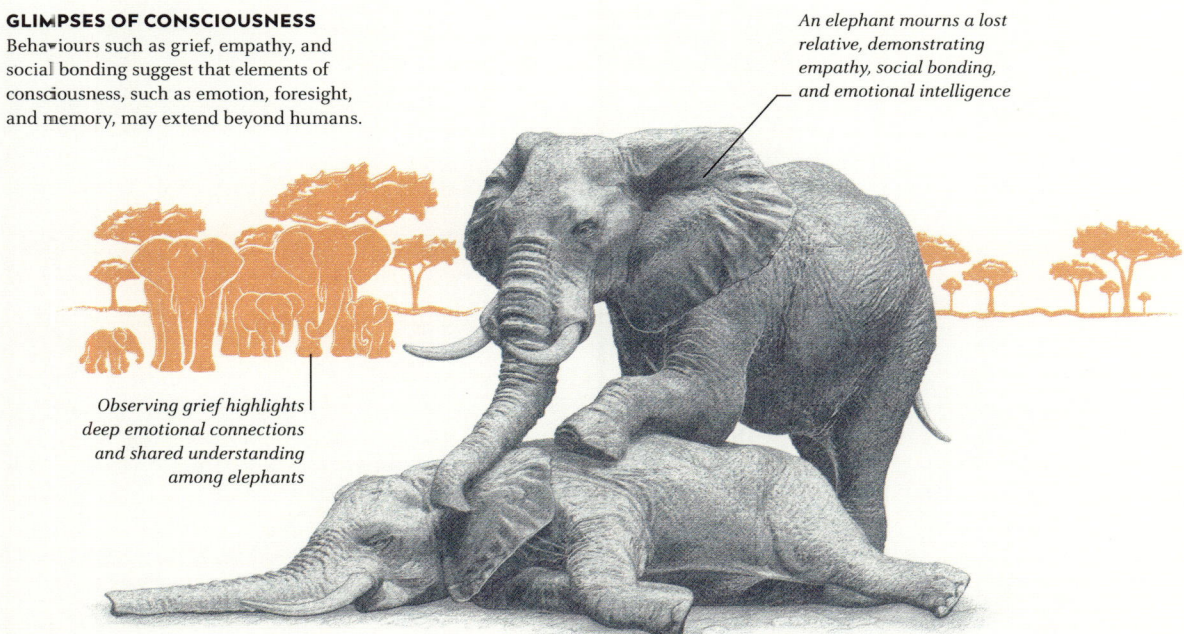

Observing grief highlights deep emotional connections and shared understanding among elephants

Mourning the dead

RITUAL AND AWARENESS

The Sunghir burials, Russia, about 34,000 years ago, reveal elaborate rituals, including thousands of ivory beads and red ochre. Such practices indicate symbolic thought, social awareness, and complex world views.

and ravens show striking intelligence, solving problems, caching food for later, and fashioning tools from sticks and leaves.

These behaviours, empathy, memory, planning, and self-recognition, suggest that the fundamental building blocks of consciousness are not uniquely human but part of a shared evolutionary heritage among intelligent species.

EVIDENCE FROM EARLY HUMANS

Archaeological discoveries can provide clues about when and how consciousness deepened in our ancestors and hominin relatives. The survival of disabled individuals such as a *Homo georgicus* at Dmanisi, Georgia, who lived well into maturity despite losing all his teeth, implies

Evidence of care suggests early humans exhibited empathy and social support

compassion and caregiving. Similarly, the Neanderthal known as Shanidar 1, found in Iraq, lived for decades with severe injuries, showing sustained social support within his group.

Burial practices offer early evidence of symbolic thought and theory of mind – an awareness that others possess inner experiences, preferences, and social identities. Deliberate body placement and the inclusion of red ochre, tools, ornaments, or animal remains suggest not

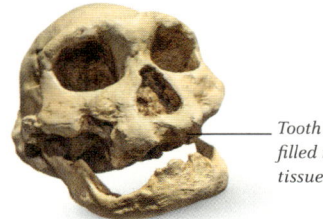

Tooth sockets filled in with bone tissue over years

EARLY EMPATHY

A *Homo georgicus* fossil reveals survival after tooth loss, suggesting that this individual was a member of a supportive group that cared for him, implying empathy.

THEORIES OF CONSCIOUSNESS

A range of theories explain how the brain produces awareness, each giving a different view on how our experiences become part of our conscious mind.

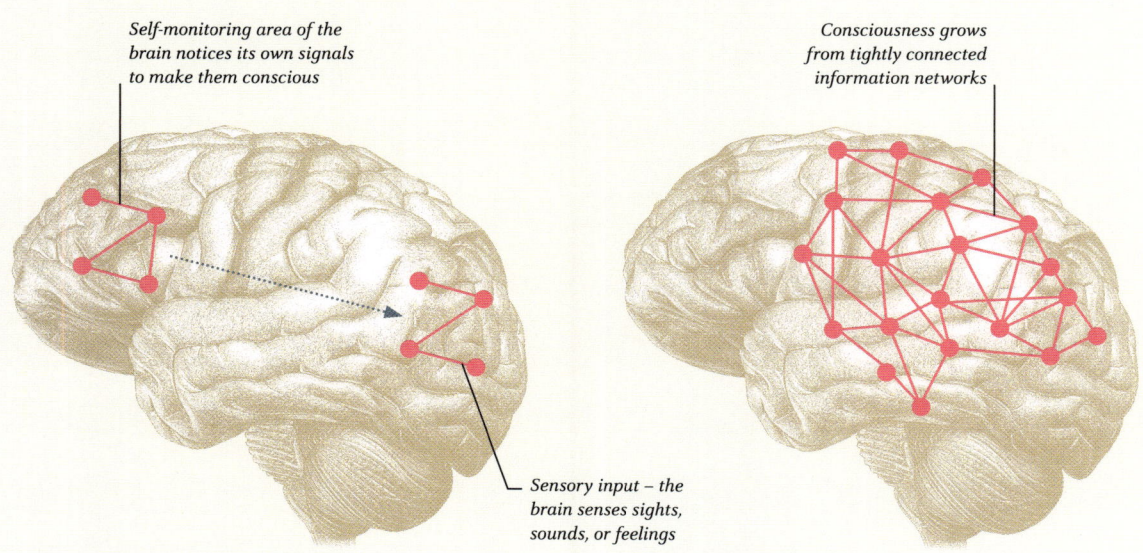

Self-monitoring area of the brain notices its own signals to make them conscious

Consciousness grows from tightly connected information networks

Sensory input – the brain senses sights, sounds, or feelings

HIGHER-ORDER THEORIES

We become aware of experiences when the brain notices its own activity, turning basic feelings or sights into conscious thoughts.

INTEGRATED INFORMATION THEORY

Consciousness arises from how deeply information is connected within a system. The more integrated and unified the network, the richer and more complex the experience.

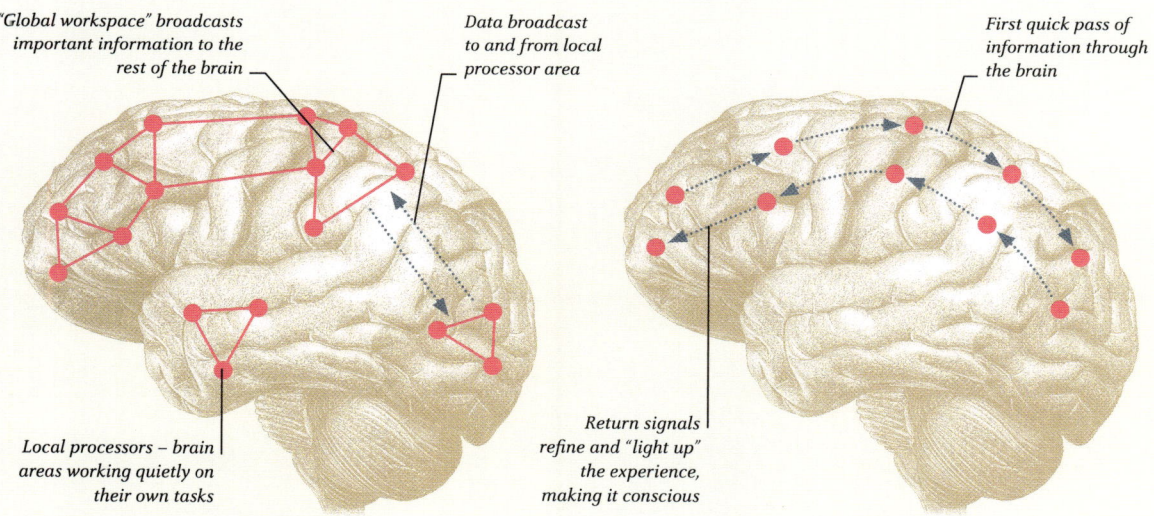

"Global workspace" broadcasts important information to the rest of the brain

Data broadcast to and from local processor area

First quick pass of information through the brain

Local processors – brain areas working quietly on their own tasks

Return signals refine and "light up" the experience, making it conscious

GLOBAL WORKSPACE THEORIES

Consciousness happens when important information is broadcast across the brain, letting many areas share it and work on it together.

RECURRENT PROCESSING THEORY

Consciousness arises when signals loop back and forth between brain areas, creating a stronger, more detailed picture of what we are experiencing.

PERCEPTION IN FOCUS
The Rubin's vase illusion exemplifies how
consciousness negotiates competing
sensory information, highlighting that
perception is an active, interpretive process
rather than a passive reflection of reality.

Opposing faces?   *A vase?*

just ritual, but an empathetic recognition of another mind and its continuing place in a community's social memory.

The debated "flower burial" at Shanidar Cave, Iraq, where pollen concentrations were thought to indicate blossoms placed around the body, might have contributed evidence of concern for the dead. Sadly, this evidence has been thrown into doubt, since action by bees or rodents could have caused the pollen build-up.

Other behaviour, such as the practice of art and adornment, reflect imagination and shared social meaning. Planning and cooperation in hunting, food preparation, and child-rearing demonstrate communication, foresight, and empathy – all hallmarks of a conscious inner life.

THEORIES OF CONSCIOUSNESS

Consciousness remains one of science's most challenging mysteries, and researchers have developed several influential theories to explain how the brain produces awareness and integrates experience.

The Global Workspace Theory proposes that consciousness functions like a spotlight, broadcasting selected information across the brain. The Integrated Information Theory suggests that the richness of experience depends on how much information is bound together into a unified whole. Higher-order Theories argue that a mental state becomes conscious only when the brain represents it – such as knowing that one is in pain rather than simply feeling it. The Recurrent Processing Theory emphasizes feedback loops between brain regions as the source of awareness.

Aspects of everyday perception illustrate these ideas. In the Rubin vase illusion – where one can see either a vase or two faces – the same sensory input produces different experiences. While the Global Workspace Theory describes shifting attention between alternatives, the Recurrent Processing Theory highlights feedback circuits stabilizing one perception over another. Although none of these models fully explains why brain activity gives rise to subjective experience, they frame the mystery in testable ways and guide research into both modern and ancient minds.

THE RISE OF INNER LIFE

Consciousness likely evolved gradually, shaped by environmental pressures, tool use, and the demands of increasingly complex social living. The social brain hypothesis (see p.225) proposes that expanding brains enabled our ancestors to track relationships, interpret intentions, and navigate alliances – cognitive skills that underpinned empathy, reflection, and self-awareness. According to this idea, these capacities supported the emergence of imagination, foresight, and moral reasoning, allowing humans to plan, cooperate, and create shared cultural meaning.

The need to be conscious of other minds may have been an evolutionary driver of brain expansion

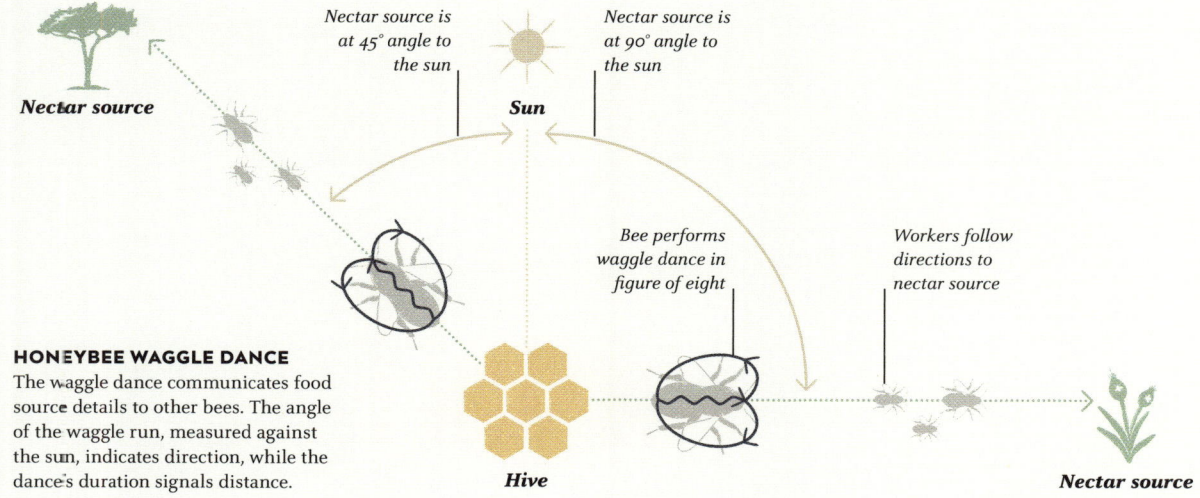

HONEYBEE WAGGLE DANCE
The waggle dance communicates food source details to other bees. The angle of the waggle run, measured against the sun, indicates direction, while the dance's duration signals distance.

Diagram labels: *Nectar source* · *Nectar source is at 45° angle to the sun* · *Nectar source is at 90° angle to the sun* · **Sun** · *Bee performs waggle dance in figure of eight* · *Workers follow directions to nectar source* · **Hive** · *Nectar source*

Unlocking language

Communication is widespread across the living world. At its simplest, in the form of chemical signals, it may be as ancient as life itself. But the flexibility and open-ended creativity of human language sets it apart from all other animal communication systems.

We are a talkative species. An average English speaker utters about 16,000 words every day, most of it in conversations with others. Language allows us not only to name objects in the world but also to reflect on experiences, imagine hypothetical scenarios, and communicate about the past and the future – things that are not present in the here and now.

Other organisms, of course, also communicate, often in remarkable ways. Bacteria can sense chemical signals in their environment; honeybees perform elaborate waggle dances to guide hive-mates to distant flowers; cuttlefish create intricate visual displays; and vervet monkeys emit distinct alarm calls for different predators. Yet despite this diversity, most animal communication within a species shows little variation. Even songbirds, whose regional "dialects" might seem like languages, differ only slightly across populations – more akin to the differences between British and American English than to entirely separate languages such as English and Navajo.

Animal communication shows little variation, even across regional "dialects"

"The polka-dotted tiger kissed the striped humpback whale."

WHAT MAKES LANGUAGE UNIQUE?

Human language's structure allows infinite, creative, and flexible expressions

At its core, a language is a highly complex system of interlocking parts: sounds or gestures combine to form words, and these words can be arranged into sentences. This hierarchical, combinatorial structure gives human language remarkable open-ended productivity, allowing us to generate an essentially infinite number of meaningful expressions. Consider a sentence no one has ever uttered before: *"The polka-dotted tiger kissed the striped humpback whale"*. Despite its novelty and absurdity, you immediately understand it – a testament to language's flexibility and its power to convey any idea, however fanciful or abstract.

Learning language is one of the most intricate tasks humans undertake. By adulthood, most people have acquired tens of thousands of words, each with subtle shades of meaning. Children must not only build their vocabulary,

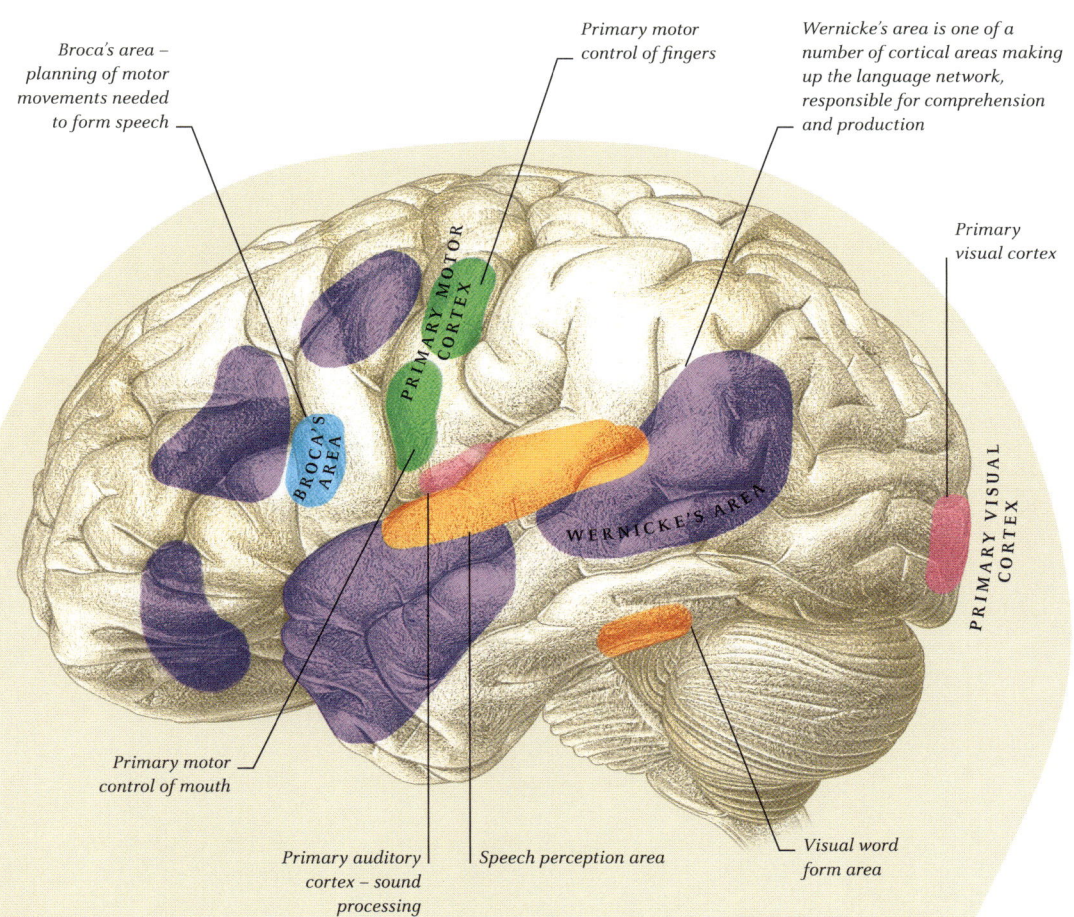

Broca's area – planning of motor movements needed to form speech

Primary motor control of fingers

Wernicke's area is one of a number of cortical areas making up the language network, responsible for comprehension and production

Primary visual cortex

PRIMARY MOTOR CORTEX

BROCA'S AREA

WERNICKE'S AREA

PRIMARY VISUAL CORTEX

Primary motor control of mouth

Primary auditory cortex – sound processing

Speech perception area

Visual word form area

BRAIN REGIONS USED IN LANGUAGE

Parts of the brain work together to support language production and comprehension. Though multifunctional, and active in other species, the coordination of these brain regions in humans enables speaking, understanding, reading, and writing.

- ● LOW-LEVEL MOTOR CONTROL
- ● MOTOR PLANNING
- ● PERCEPTION
- ● LANGUAGE
- ● LOW-LEVEL SENSORY PROCESSING

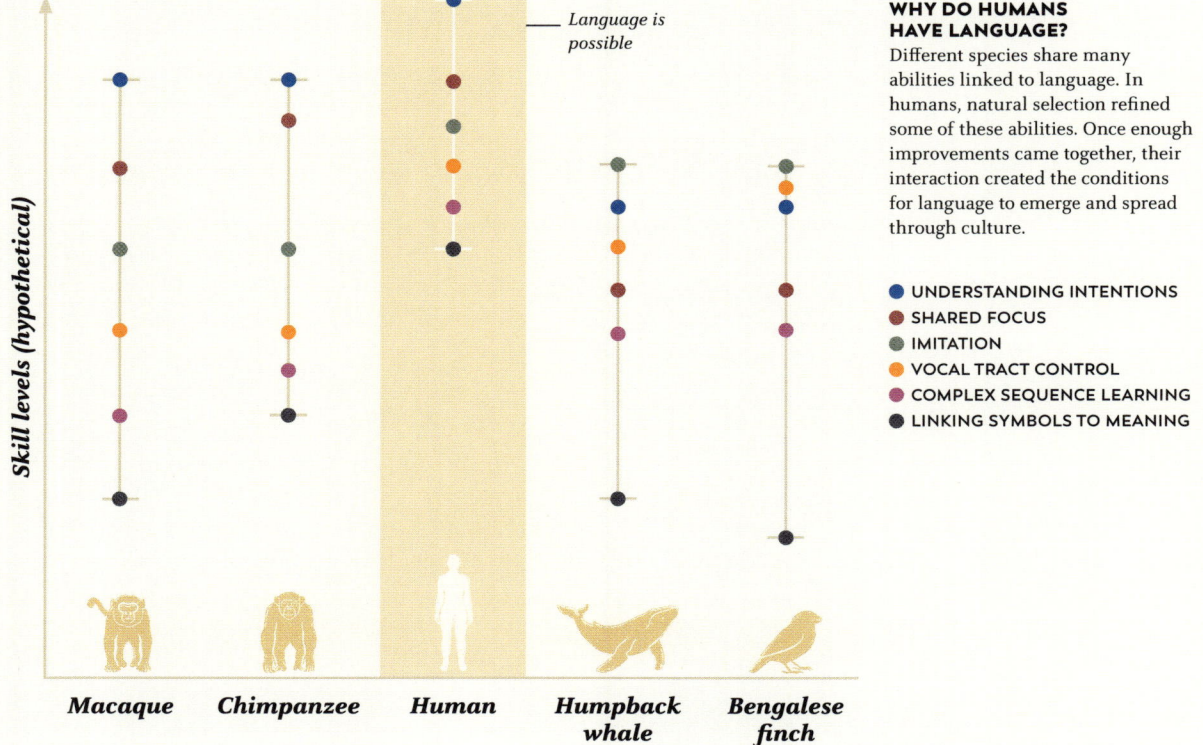

Language is possible

WHY DO HUMANS HAVE LANGUAGE?

Different species share many abilities linked to language. In humans, natural selection refined some of these abilities. Once enough improvements came together, their interaction created the conditions for language to emerge and spread through culture.

- ● UNDERSTANDING INTENTIONS
- ● SHARED FOCUS
- ● IMITATION
- ● VOCAL TRACT CONTROL
- ● COMPLEX SEQUENCE LEARNING
- ● LINKING SYMBOLS TO MEANING

Skill levels (hypothetical)

Macaque · **Chimpanzee** · **Human** · **Humpback whale** · **Bengalese finch**

but also learn to assemble these elements into meaningful structures, mastering grammar, syntax, and idiomatic expressions. In some languages, like English or Chinese, this involves combining words into sentences. In others, such as Yupik – an indigenous language spoken in parts of Alaska and Siberia – a single long word can convey what would require an entire English sentence. For example, the word *tuntussuqatarniksaitengqiggtuq* roughly translates as "he had not yet said again that he was going to hunt reindeer".

Given this complexity, it's unsurprising that children take years to master language fully, often reaching mature abilities only after puberty, even though they communicate effectively earlier. Along the way, they also learn social and cultural nuances – how tone, context, and gesture shape meaning – highlighting that language is as much a social tool as a cognitive one.

Language mastery takes years; children also learn social and cultural cues

EVOLUTIONARY ORIGINS OF LANGUAGE

How humans acquired the gift of language has fascinated scholars for millennia. Unlike bones or tools, language does not fossilize, leaving only indirect clues in the archaeological record. Many early artefacts – clothing, baskets, leather – simply decay over time, making it difficult to determine when symbolic communication first emerged or how sophisticated it might have been.

Current research suggests that the human brain evolved to be "language-ready" gradually over hundreds of thousands of years. Evidence indicates that both Neanderthals and Denisovans may have possessed some linguistic abilities, hinting that the roots of language extend beyond our own species. Language likely emerged through a virtuous cycle: a larger brain enabled more complex social interactions, which created pressure for more effective communication, driving further brain development, and so on over many generations.

Language itself did not arise from a single new ability. Rather, it piggybacked on existing neural mechanisms, such as the capacity to process sequences, link arbitrary sounds to meanings, and to understand others' communicative intentions. While other animals exhibit these skills to varying degrees, humans are unique in combining them into a fully

The increased use of video-conferencing has significantly reshaped American Sign Language. Users have adapted signs for smaller screens, adjusted their body positioning, and clarified gestures for clearer, more effective communication.

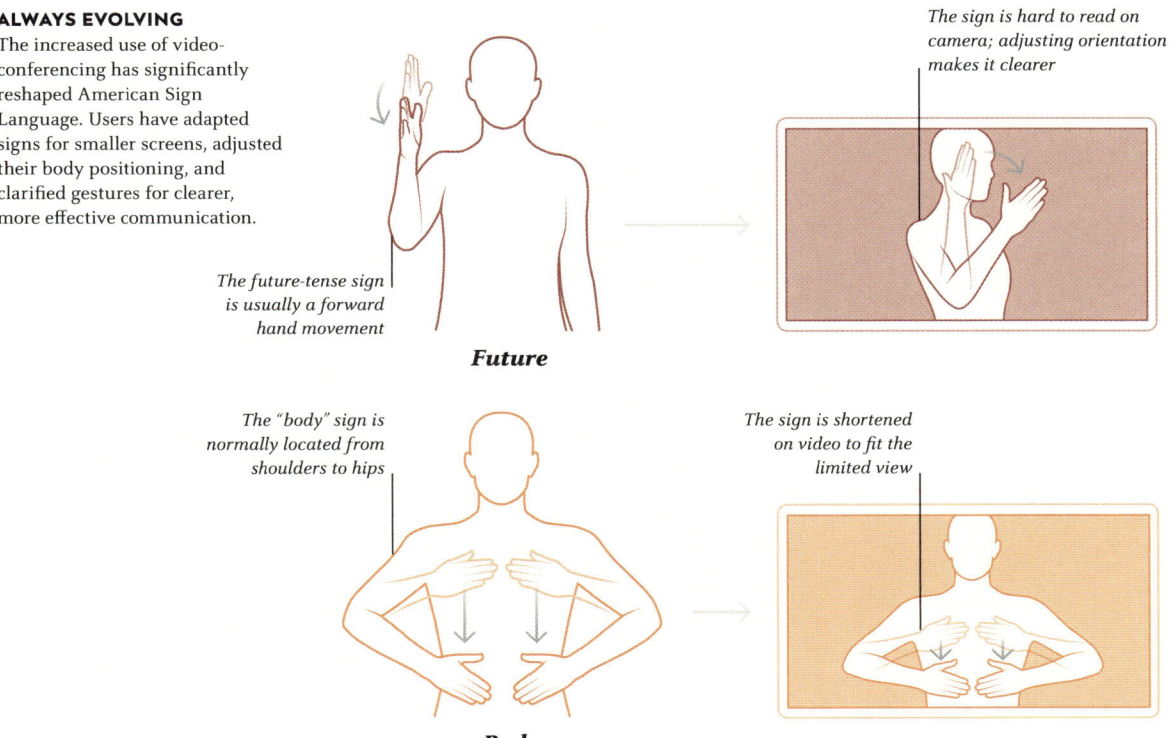

The future-tense sign is usually a forward hand movement

The sign is hard to read on camera; adjusting orientation makes it clearer

Future

The "body" sign is normally located from shoulders to hips

The sign is shortened on video to fit the limited view

Body

functional linguistic system. This allows us not only to convey information, but also to share abstract ideas, negotiate social relationships, and create rich symbolic worlds that extend far beyond immediate experience.

LANGUAGE AND HUMAN CULTURE

The advent of language was a turning point in human evolution. It enabled the transmission of knowledge across generations, such as navigating terrain and planning expeditions. Language also allows for coordination and cooperation on scales far beyond any other species, fuelling cumulative cultural evolution (see pp.228–29).

Over millennia, this process gave rise to complex societies, advanced technologies, science, religions, and social institutions. Beyond practical knowledge, language also enabled humans to create art, myths, and rituals, fostering shared identities and social cohesion. Storytelling, poetry, and songs allowed cultures to preserve history, pass on moral values, and explore imagination in ways that shaped civilizations. Without language, our social structures might resemble those of other primates far more closely than the intricate, symbolic, and deeply interconnected societies we live in today.

ONGOING EVOLUTION

Language is never static. It evolves constantly, reflecting changes in culture, technology, and society. Modern phenomena such as internet slang, texting abbreviations, and emojis are simply the latest chapters in a long story of linguistic change – often met with disapproval by older generations. Studying these shifts not only helps us understand the evolution of language itself but also sheds light on what it means to be human.

From chemical signals in bacteria to emojis on our smartphones, communication surrounds us. Yet human language remains unique: flexible, creative, and endlessly productive. It is both a mirror and a driver of our humanity – a gift that has shaped who we are and continues to shape who we will become.

Language constantly evolves, reflecting culture, technology, and society

The power of storytelling

Once upon a time, we began to tell stories. Long before they were written down, stories were passed through legend, poetry, and song. Wherever there is human life, there is storytelling. The phrase "Homo narrans" captures how narrative defines and distinguishes our species.

Because storytelling was originally only oral, we cannot know exactly when it began. But cave paintings from more than 50,000 years ago seem to depict a hunting story, suggesting ancient roots. More evidence of antiquity comes from similar stories that are widely dispersed. One myth, which exists globally, depicts the Big Dipper as a hunting scene in the night sky – with the three stars of the handle as hunters and the dipper as their quarry. Researchers suspect all versions of this story to have a shared origin that is at least 15,000 years old, when certain migration events took place.

WHY DO WE TELL STORIES?

We may never know when storytelling evolved, but the fact that it exists at all is a puzzle. Creating narratives may be a fundamental part of human cognition. In a 1944 study, researchers showed people a simple animation of two triangles and a circle. After viewing, almost every viewer described a story involving characters, emotions, and intentions – despite the fact that they were just abstract shapes.

It's been suggested that stories help us survive. They can act like databases, storing useful knowledge. Supporting this, studies show survival-related information is more memorable. People were more likely to recall fake urban legends when they involved spiders in a woman's hair or steroids in chickens causing illness. In some communities, such as the Agta of the Philippines, storytellers have higher social prestige – pointing to an evolutionary advantage not just in hearing stories, but in telling them.

THE MORAL OF THE STORY

To tell a story requires cooperation: sharing and listening. Some scientists propose our brains evolved to handle complex social lives, and that

Stories convey survival knowledge and boost social status, suggesting deep evolutionary importance.

OBJECTS WITH CHARACTER

In the Heider-Simmel experiment, viewers watched simple shapes move around a rectangle, yet nearly everyone imagined a dramatic story. This classic study reveals how our brains instinctively assign motives, emotions, and relationships, even to lifeless objects.

Threatened "character"

Protector

Bully

1 *Most viewers saw characters: a "bully", a "protector", and someone in danger.*

ANCIENT NARRATIVE

Cave art from Sulawesi, Indonesia, dating to 44,000 years ago, depicts buffalo pursued by hybrid human-animal figures wielding spears and ropes. This striking scene may represent the earliest known narrative ever recorded.

storytelling helped us manage it. Our ability to imagine fictional characters is linked to "theory of mind" – our capacity to imagine what others think or feel. In this view, storytelling is like a social flight simulator. We can explore emotions, intentions, and relationships – all without real-world consequences.

As one Navajo storyteller explained, "If my children hear coyote stories, they will grow up to be good people, [otherwise] they will turn out bad." Many stories punish characters who break social rules. As audiences, we feel satisfaction when those who misbehave get what they deserve. That emotional payoff helps reinforce cooperative behaviour in the real world.

Storytelling trains social understanding, teaches morals, and reinforces cooperation

WHY DO SOME STORIES STICK?

Not all stories survive. Some are told repeatedly, while others disappear. The Brothers Grimm, famous 19th-century folktale collectors, gathered

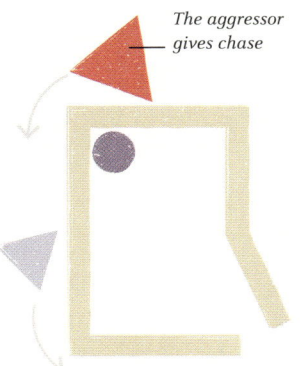

The aggressor gives chase

2 *The large triangle's movements were seen as aggressive. Viewers described a conflict unfolding.*

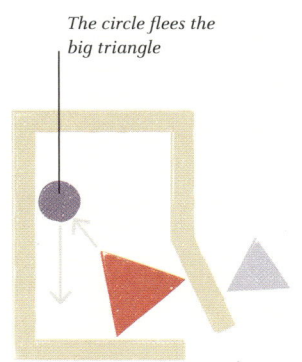

The circle flees the big triangle

3 *With the help of the small triangle, the circle tries to evade the big triangle.*

The big triangle is left frustrated

4 *The smaller triangle and circle worked together to hide or flee, suggesting teamwork.*

WANDJINA FIGURE

For Aboriginal communities, Wandjina figures embody powerful Dreaming stories. Painted on rock over thousands of years, they preserve law, culture, and history – storytelling that carries wisdom across generations.

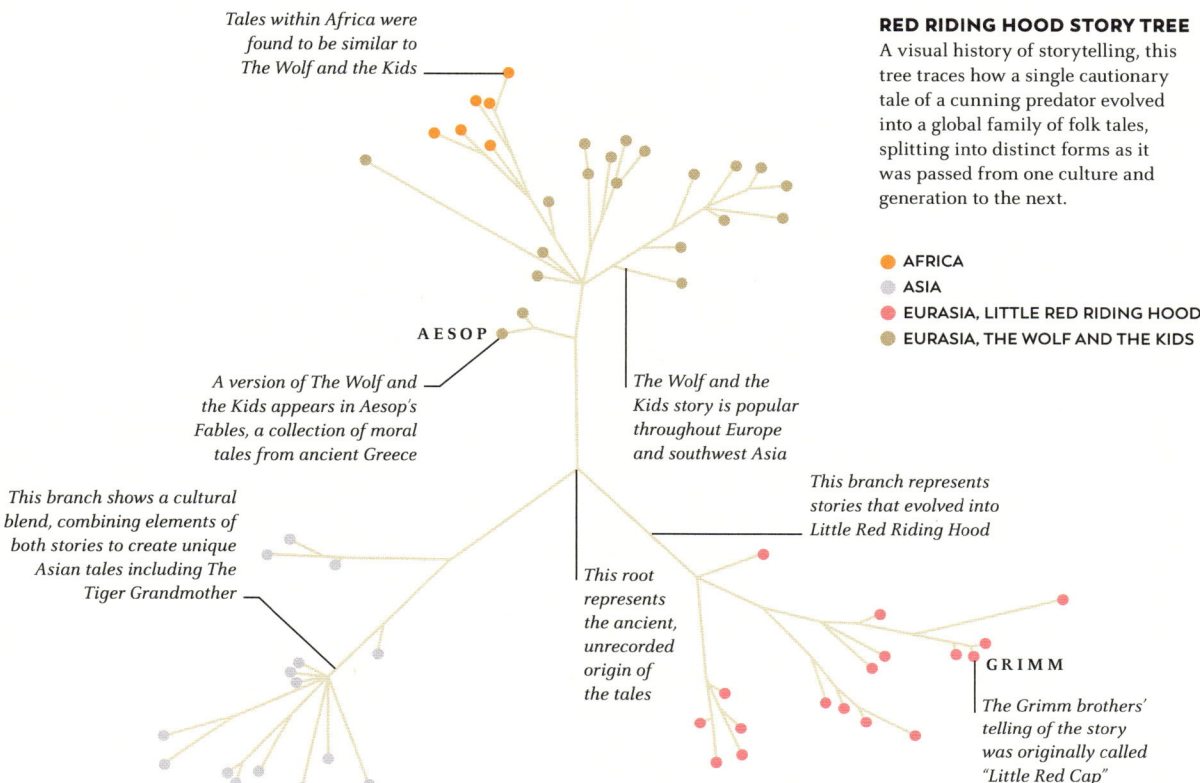

Tales within Africa were
found to be similar to
The Wolf and the Kids

RED RIDING HOOD STORY TREE
A visual history of storytelling, this
tree traces how a single cautionary
tale of a cunning predator evolved
into a global family of folk tales,
splitting into distinct forms as it
was passed from one culture and
generation to the next.

● AFRICA
● ASIA
● EURASIA, LITTLE RED RIDING HOOD
● EURASIA, THE WOLF AND THE KIDS

A E S O P

A version of The Wolf and
the Kids appears in Aesop's
Fables, a collection of moral
tales from ancient Greece

The Wolf and the
Kids story is popular
throughout Europe
and southwest Asia

This branch shows a cultural
blend, combining elements of
both stories to create unique
Asian tales including The
Tiger Grandmother

This branch represents
stories that evolved into
Little Red Riding Hood

This root
represents
the ancient,
unrecorded
origin of
the tales

G R I M M

The Grimm brothers'
telling of the story
was originally called
"Little Red Cap"

stories from around their town in Germany –
such as Cinderella, Little Red Riding Hood, and
Rapunzel. While some remain household
names, others have faded. Researchers call this
"content bias": some types of stories are more
likely to be remembered and retold.

Stories that evoke disgust – like many
urban legends – tend to be more memorable.
As noted earlier, survival-related information
sticks in memory more easily; information
about disease, danger, or parasites holds
survival value. We also naturally gravitate
toward gossip and social drama. Stories about
who did what to whom spread more easily
when told as narratives.

We also remember stories that are
"minimally counter-intuitive". That is, they
bend expectations just enough to be interesting,
without becoming too strange. A ghost that
walks through walls violates our understanding
of reality, but is still graspable. Too many
violations, and the story becomes hard to follow.
This helps explain the popularity of tales with
teleporting wizards or talking animals – they
are fantastical, but not incomprehensible.

*Stories last when
they balance the
familiar with
small, graspable
violations of
expectation*

The Grimm brothers believed that their stories
weren't only German – they reflected themes
shared across cultures. Today, scientists use
phylogenetic analysis, a technique borrowed
from evolutionary biology. In phylogenetic
reconstruction, they use short, repeatable story
elements, instead of genes, to study how stories
change and spread. This has revealed surprising
links. Little Red Riding Hood, for instance,
shares traits with The Wolf and the Kids, a story
that may trace back to Aesop's fables in ancient
Greece, around 620–564 BCE. Studying how
stories evolve helps us understand what makes
them last through time and across cultures.

Storytelling is all around us: in books, films,
songs, conversations, and the tales we hear as
children. The human urge to tell stories is both
mysterious and deeply rooted. It helps us store
knowledge, navigate relationships, and connect
communities. We continue to gather – watching,
reading, and listening – to something as simple,
and as powerful, as a story.

THE POWER OF STORYTELLING

Art is often seen as a hallmark of humanity, and its roots stretch deep into our evolutionary past. From carvings to painted caves, the earliest traces of expression suggest that creativity and imagination were vital to survival, communication, and identity across ancient societies.

Origins of human creativity

ARTISTIC TIMELINES

Timelines stretching back 500,000 years feature some of prehistory's most significant works of art, showing mobile art's portable expression versus parietal art's enduring rock surfaces.

— **MOBILE ART**
— **ROCK (PARIETAL) ART**

c. 64,000 YA
La Pasiega paintings, if date is accepted, show Neanderthal symbolic expression

c. 500,000 YA
Engraved abstract markings on shell, Trinil, Indonesia

500 000 YA

c. 77,000 YA
Ochre engraved with early graphic patterns, Blombos Cave, South Africa

c. 500,000–300,000 YA
Venus of Tan-Tan, Morocco, controversial quartzite figurine claimed to have pigment traces

c. 60,000 YA
Ostrich shells bear geometric engravings, Diepkloof, South Africa

c. 400,000–350,000 YA
Line-incised bone, Bilzingsleben, Germany

Marks show deliberate design, not accidental or functional use

Red ochre was also ground and used as a pigment

c. 270,000 YA
Twin Rivers ochre, Zambia, indicates early body decoration

c. 100,000–92,000 YA
Skhul and Qafzeh burials, Israel, show ritual and symbolism

Blombos Cave ochre slab

100 000 YA

250 000 YA

c. 130,000 YA
Eagle talon jewellery is among earliest European ornamentation

c. 200,000 YA
Preserved handprints and footprints, Quesang, Tibet, China

c. 142,000 YA
Bizmoune Cave shell beads, Morocco, show early adornment

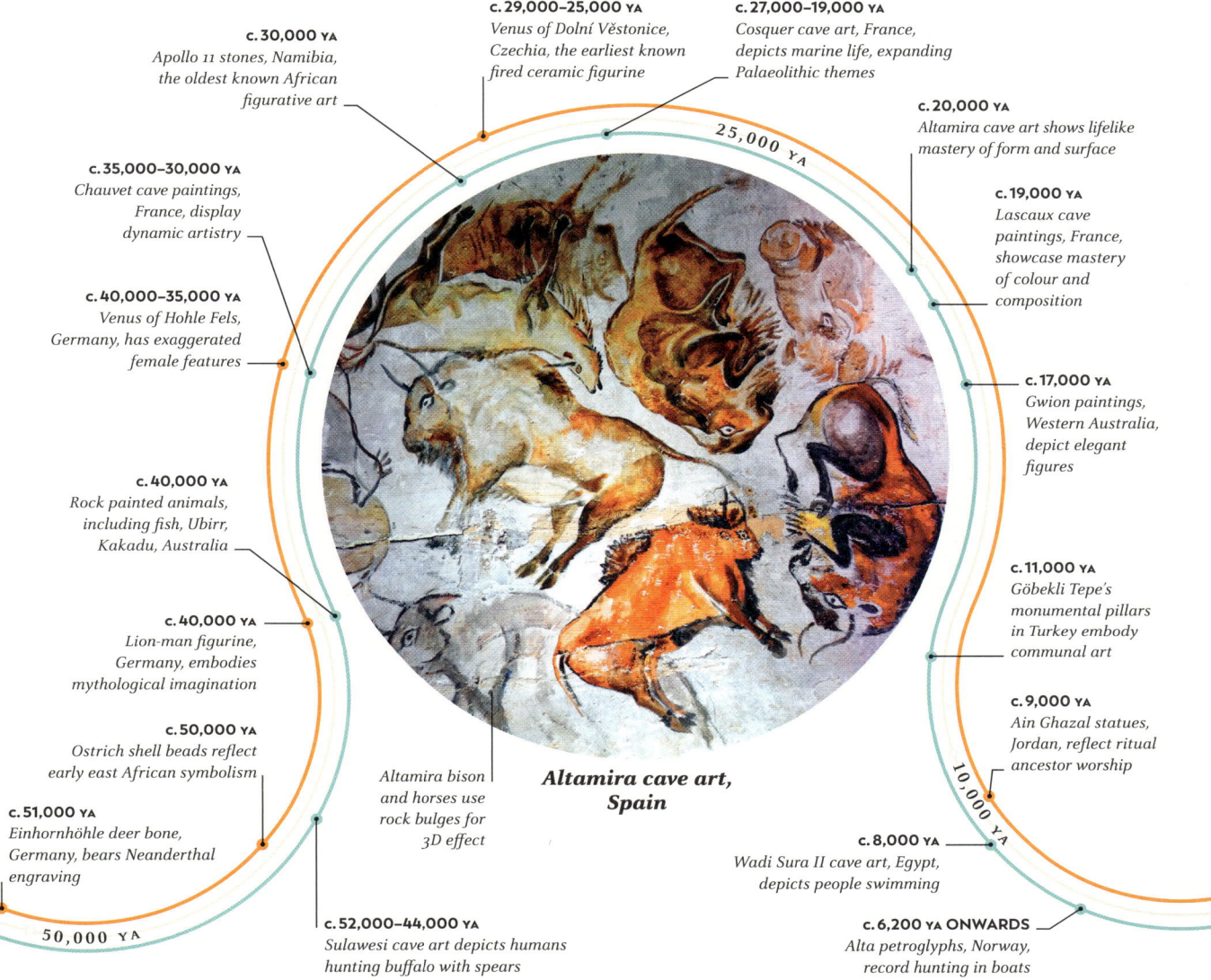

c. 30,000 YA
Apollo 11 stones, Namibia, the oldest known African figurative art

c. 29,000–25,000 YA
Venus of Dolní Věstonice, Czechia, the earliest known fired ceramic figurine

c. 27,000–19,000 YA
Cosquer cave art, France, depicts marine life, expanding Palaeolithic themes

c. 20,000 YA
Altamira cave art shows lifelike mastery of form and surface

c. 35,000–30,000 YA
Chauvet cave paintings, France, display dynamic artistry

c. 19,000 YA
Lascaux cave paintings, France, showcase mastery of colour and composition

c. 40,000–35,000 YA
Venus of Hohle Fels, Germany, has exaggerated female features

c. 17,000 YA
Gwion paintings, Western Australia, depict elegant figures

c. 40,000 YA
Rock painted animals, including fish, Ubirr, Kakadu, Australia

c. 11,000 YA
Göbekli Tepe's monumental pillars in Turkey embody communal art

c. 40,000 YA
Lion-man figurine, Germany, embodies mythological imagination

c. 9,000 YA
Ain Ghazal statues, Jordan, reflect ritual ancestor worship

c. 50,000 YA
Ostrich shell beads reflect early east African symbolism

c. 51,000 YA
Einhornhöhle deer bone, Germany, bears Neanderthal engraving

Altamira cave art, Spain

Altamira bison and horses use rock bulges for 3D effect

c. 8,000 YA
Wadi Sura II cave art, Egypt, depicts people swimming

c. 52,000–44,000 YA
Sulawesi cave art depicts humans hunting buffalo with spears

c. 6,200 YA ONWARDS
Alta petroglyphs, Norway, record hunting in boats

25,000 YA

10,000 YA

50,000 YA

The story of art begins long before *Homo sapiens*. At Trinil, Indonesia, a mussel shell engraved with zigzag lines some 500,000 years ago – probably by *Homo erectus* – provides a tantalizing clue to the origins of symbolic thought. Though modest, these marks are deliberate and unlikely to be accidents of wear. They suggest an early sensitivity to pattern and a recognition that objects could carry meaning beyond function.

Other traces support the idea of deep roots. Early humans collected pigments such as red ochre and applied them in ways that served no immediate practical purpose. Pinnacle Point in South Africa, for example, provides evidence of pigment use dating to around 164,000 years ago. Modified shells, bones, and stones – scratched or patterned in ways unlikely to be random – hint at early aesthetic awareness. Whether these acts were ritualistic or simply exploratory, they reveal an emerging capacity for symbolic thought. By the time of early *Homo sapiens*, this latent sensitivity had become a creative force that reshaped objects, spaces, and social interactions.

Another tentative example comes from hand and foot impressions in Tibet, China, dated between 226,000 and 169,000 years ago. We

Early humans' markings reveal emerging symbolic thought and aesthetic awareness

257

cannot know the intention behind these marks, and there is no clear evidence that their makers aimed to communicate or imagined future observers. They may simply reflect curiosity or playfulness. However, it is possible that these impressions were a simple gesture of presence, a way of leaving a trace, linking them to later human acts of marking space and identity. The species responsible is unknown, though Denisovans are a likely candidate.

Early art reflects reverence, aesthetic awareness, and thoughtful contemplation

Some archaeologists argue that an artistic awareness may have developed alongside the first sparks of spirituality. The attention given to certain objects – shells, stones, pigments – reflects careful selection and handling of materials, suggesting that beauty had begun to hold social or emotional value. In this sense, the earliest art was not just communication, but contemplation: a way of pausing to see the world differently.

ANATOMY OF CREATIVITY

Art arose from imagination, but it depended on the body and brain evolving together. Hands refined through toolmaking gained precision grips that allowed engraving, sculpting, and painting with extraordinary control. At the same time, expansion of the prefrontal cortex supported abstraction, planning, and symbolic reasoning, while visual and emotional brain networks reinforced the pleasure of shaping meaningful forms.

These anatomical and neural developments made artistic behaviour adaptive: representing ideas in tangible form enhanced communication, strengthened social learning, and signalled intelligence and skill. The physical act of creation – from carving a figurine to tracing pigment on stone – not only expressed thought but helped refine the cognitive capacities that made such thought possible.

Durable material had ideal surface for engraving

Repeated lines form geometric pattern

OSTRICH EGGSHELL FRAGMENTS
Dating to around 60,000 years ago, these decorated water containers from Diepkloof Rock Shelter, South Africa, bear repeated geometric motifs, possibly marking group identity or ownership.

Recent research has suggested that making art engaged brain areas linked to empathy and foresight, helping early humans cooperate and build social trust. Shared creative activity may have encouraged group cohesion before language was fully developed.

THE FIRST ARTISTS

Early Homo sapiens *used art to express identity and ritual*

By around 100,000 years ago, *Homo sapiens* were producing objects with clear symbolic intent. At Blombos Cave in South Africa, engraved ochre pieces and perforated shell beads show repeated motifs over long periods. At Border Cave, the continued use of pigments and ornaments indicates that decoration had become part of social and ritual life.

Burials at Skhul and Qafzeh in the Levant included ochre and grave goods, providing evidence of symbolic practice. At Diepkloof Rock Shelter, South Africa, engraved ostrich eggshell containers may have served as group identifiers. These examples show that humans were assigning meaning to objects, reflecting growing social and cognitive complexity.

Some researchers view these finds as part of a gradual development, while others argue for a sharp increase in symbolic behaviour between 70,000 and 50,000 years ago. In either case, art reinforced social bonds and identity. Decorative items such as beads, pigments, and engravings likely marked membership or status and helped maintain group cohesion.

Art also served as a form of memory. Designs and symbols preserved information about myths, territories, and shared knowledge, allowing early humans to record and transmit ideas across generations, and to maintain continuity within their communities.

EARLIEST CAVE ART
Around 52,000 years old, these hand stencils and images of buffalo, in a cave in Sulawesi, Indonesia, rank among the world's earliest known figurative artworks. They suggest a desire to communicate, record experiences, and leave a mark.

ORIGINS OF HUMAN CREATIVITY

EARLY MASTERPIECES

The flourishing of visual art is vividly evident in cave paintings and carvings across the globe. In Sulawesi, Indonesia, hand stencils and animal images around 52,000 years old predate Europe's celebrated sites. A hunting scene depicts part-human, part-animal figures tracking warty pigs, arguably the earliest narrative image (see p.252). These stencilled hands and animal depictions serve as human signatures – a tangible connection to the people who created them tens of thousands of years ago. Through them, we glimpse not just hunters and gatherers, but storytellers with a capacity for imagination, symbolic thinking, and cultural memory.

In Europe, these early experiments blossomed into some of the most ambitious artworks of the prehistoric world. The paintings in Chauvet Cave, France, dating to between 37,000 and 28,000 years ago, show a refinement that is astonishing for their time. With skilled shading, naturalistic movement, and vivid depictions of 14 different species, they demonstrate that the impulse to create was already matched by remarkable technical ability.

Far from being simple marks, Chauvet's images are full compositions. The ancient artists masterfully used the contours of the rock to shape their figures, layered animals to suggest depth, and captured them in dynamic poses

Chauvet Cave paintings display naturalistic animals, motion, shading, and early human creativity

that convey motion and narrative. These sophisticated choices reveal a mind capable of imagining worlds beyond the immediate present. In the flicker of torchlight, lions would have seemed to move, rhinos to charge, and mammoths to advance – the cave itself became a living stage.

These dramatic scenes, executed in charcoal, red ochre, and engraved stone, exploit every available surface and crevice. Each brushstroke and engraving hints at storytelling, ritual, and humanity's earliest conscious effort to convey meaning visually. In the flicker of firelight, the images would have seemed alive, creating an early setting for collective storytelling.

LIVING STONE

Dynamic scenes in Chauvet Cave show lions, rhinos, mammoths, and a horse in motion, their layered forms following the rock's contours to create depth, rhythm, and a vivid, timeless sense of life and movement.

1 *The artists scraped areas of the surface of the stone wall before painting figures*

2 *Series of woolly rhinoceroses, which went extinct in the area 17–15,000 years ago*

3 *A single horse features in this scene; horses are common in other parts of the cave*

4 *A few bison appear behind the lions – bison still exist in Europe today in tiny refuge populations*

5 *The irregular shape of the wall enhances the sense of motion, making the animals surge*

6 *Overlapping lion portraits imply movement of the pride towards the bison*

Spiral bands on the head may represent braided hair or a woven cap

Chauvet shows early humans projecting imagination into shared cultural spaces

This ability – to picture the unseen and to shape imagined scenes – marks a significant shift in human cognitive development. Chauvet is evidence of a mind capable of projecting inner visions outward into shared cultural spaces. Viewed this way, the cave becomes more than a prehistoric gallery: it is a monument to the creative power of the mind and a milestone in our evolutionary history, when art became integral to what defines humanity.

Later masterpieces built on this foundation. Lascaux's walls, in France, pulse with bulls, stags, and horses, while Altamira's bison, in Spain, are painted on curved ceilings that create depth and realism. These sites show early humans combined technical skill with imagination, conveying storytelling, emotion, and the impulse to make the unseen visible.

Material choice enabled experimentation with three-dimensional human representation

Traces of red ochre suggest ritual or symbolic significance

Curved forms and smooth surfaces show careful modelling skill

WILLENDORF VENUS

The Venus of Willendorf, carved more than 25,000 years ago, is thought to represent fertility and survival, and shows how early *Homo sapiens* emphasized the human form in their art.

DOLNÍ VĚSTONICE VENUS

Made from fired clay nearly 30,000 years ago, the Dolní Věstonice Venus is the world's oldest known ceramic object, highlighting human innovation in material technology.

OBJECTS OF MEANING

Portable art flourished alongside cave painting across different regions. Figures such as the "Venus" figurines from Willendorf, Dolní Věstonice, and Brassempouy show varied ways of representing the human form – from exaggerated features linked to fertility to individualized faces suggesting the beginnings of portraiture. The Lion-Man of Hohlenstein-Stadel, combining human and animal traits, reflects early imaginative thinking and creativity. Although their purpose remains uncertain – whether as fertility symbols, ritual objects, or toys – these artefacts demonstrate conceptual skill and cognitive adaptability.

Portable figurines reveal symbolic diversity and early mythic imagination

Together, both cave paintings and portable objects form a shared global heritage: early humans universally experimented with form, narrative, and symbolism, transforming raw materials into expressions of thought, identity, and imagination. This creative capacity to make the unseen visible marks the enduring signature of humanity.

FUNCTIONS AND COMPLEXITY

Art served many roles in early human societies. It transmitted knowledge, recorded hunting strategies and moral lessons, expressed group identity, and signalled status, skill, or social roles through personal ornaments, decorated tools, and shaped objects. In non-literate societies, images may have acted as living libraries, preserving important information. Handprints on cave walls, for example, asserted presence and continuity across generations, linking people to their history.

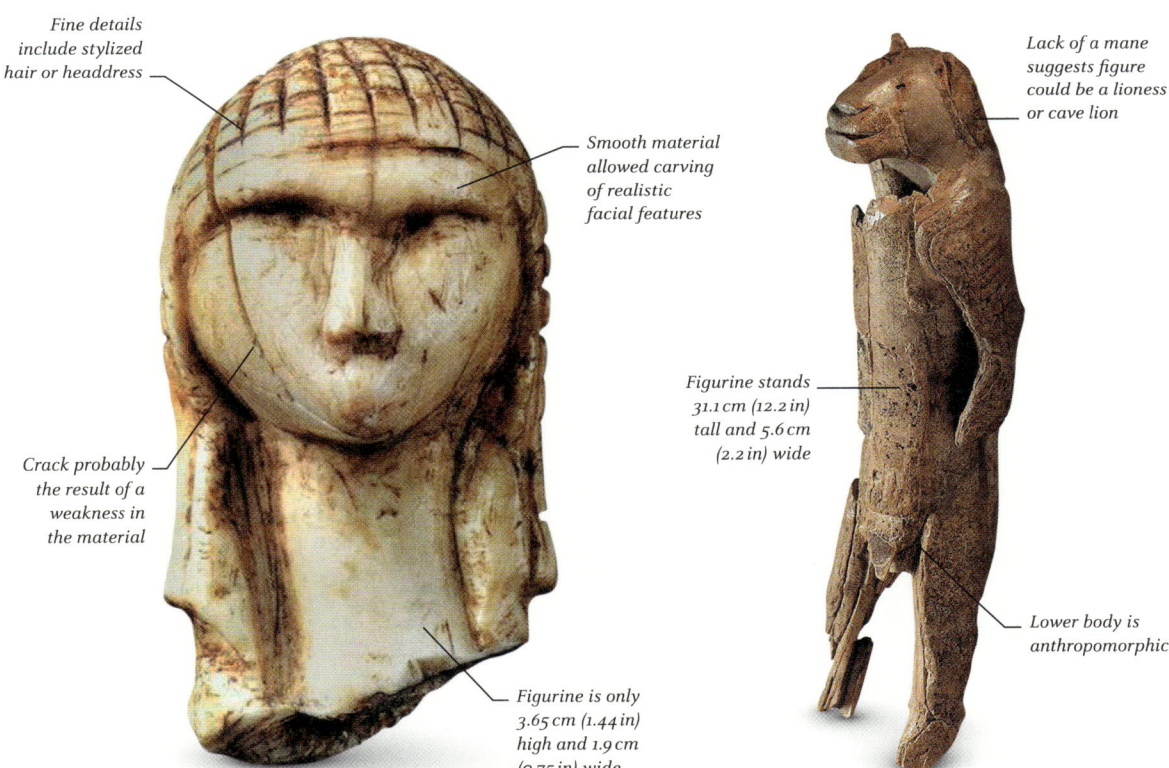

Fine details include stylized hair or headdress

Smooth material allowed carving of realistic facial features

Crack probably the result of a weakness in the material

Figurine is only 3.65 cm (1.44 in) high and 1.9 cm (0.75 in) wide

Lack of a mane suggests figure could be a lioness or cave lion

Figurine stands 31.1 cm (12.2 in) tall and 5.6 cm (2.2 in) wide

Lower body is anthropomorphic

VENUS OF BRASSEMPOUY

The Venus of Brassempouy, finely carved from mammoth ivory around 25,000 years ago, is one of the earliest realistic human faces, revealing remarkable Ice Age artistry and accurate representation.

LION-MAN OF HOHLENSTEIN-STADEL

Carved from mammoth ivory 40,000 years ago, this figurine blends human and animal traits, an early example of therioanthropic imagery, suggesting transformation between human and lion.

INDIAN CAVE ART
Bhimbetka Rock Shelters, spanning over 700,000 years of human history, preserve vivid cave paintings. This image, likely created between 4,500 and 2,000 years ago, shows men on horses wielding weapons, possibly on a hunting expedition.

The merging of imagination, ritual, and social cognition is evident in complex works like the Lion-Man, which illustrates the coevolution of mind, culture, and artistic expression. As societies developed, so did their art. Mobile hunter-gatherers focused on depictions of wild animals. Herders and early farmers incorporated humans, livestock, and cultivated landscapes.

Monumental sites such as Göbekli Tepe (Türkiye), Nabta Playa (Egypt), and Callanish (Scotland, see pp.286–87) demonstrate art's role in ritual and communal life. Australian rock art, for example, mapped ancestral beings onto stars, weaving cosmology into daily existence. Across continents, art mediated the relationship between society, nature, and the cosmos.

As tools of connection, art objects facilitated cultural exchange. Pigments like red ochre, marine shells from the Indian Ocean, and carved figures moved across hundreds of kilometres, showing creativity was a shared, non-isolated endeavour. These transmissions likely reinforced cultural alliances and helped spread ideas,

I WAS HERE
In Cueva de los Manos, Argentina, vivid handprints spread across the canyon walls – a 9,000-year-old testament to early humans' creativity and their desire to leave lasting traces of their presence.

making art an early form of diplomacy and dialogue. Not all art was purely practical or ritualistic; some pieces existed for aesthetic or playful reasons. Whether ceremonial, symbolic, or exploratory, art universally reflects imagination at work and stimulated cognitive traits such as curiosity, innovation, and empathy.

ART'S LEGACY
From the delicate early engravings on shells to monumental stone structures, art traces a continuous, unbroken thread through human history. It has been the timeless mechanism by which humanity preserved myths, asserted

Some art was playful or aesthetic, fostering curiosity and empathy

group identities, structured essential rituals, and celebrated inherent beauty. Its legacy is both a profound archaeological record and a vibrant, living tradition.

Living traditions with roots in the ancient past – such as Australian songlines – maintain symbolic systems that connect the deep past and the dynamic present. Modern artists still draw inspiration from this prehistoric imagery, while global institutions protect and conserve certain fragile, irreplaceable sites.

In today's world, the creative impulse remains constant, even as the medium evolves through technology. Contemporary forms,

Ancient traditions inspire art; global efforts protect fragile cultural sites

including digital installations, virtual reality, and community murals, extend the same primal instinct that once guided ochre-stained hands: to leave a mark, to share a vision, and to make inner meaning visible.

Making art has been part of human life for hundreds of thousands of years. From engraved shells and ochre pieces to cave paintings and figurines, it records imagination, thought, and social connection, linking people across time and generations.

The musical brain

Music is universal, accompanying celebrations, rituals, and daily life across cultures. It stirs emotions, strengthens bonds, and carries traditions, yet its evolutionary purpose remains uncertain. From ancient flutes to digital playlists, music has shaped humanity for millennia, revealing shared biology and cultural creativity.

PALAEOLITHIC FLUTE

This flute from the Hohle Fels cave in southwestern Germany, dating to around 40,000–35,000 years ago, is one of the oldest known musical instruments. It features five finger holes and a V-shaped mouthpiece, demonstrating remarkable craftsmanship.

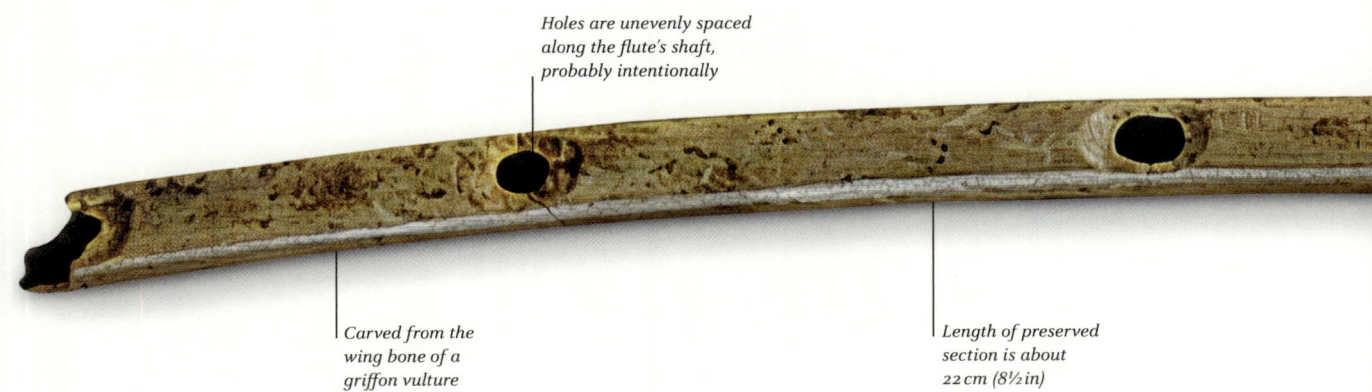

Holes are unevenly spaced along the flute's shaft, probably intentionally

Carved from the wing bone of a griffon vulture

Length of preserved section is about 22 cm (8½ in)

All over the world, human communities regularly gather to make and listen to music. Often accompanying important social functions such as festivals, weddings, and funerals, music can evoke strong emotional responses, sometimes moving us to tears. However, music has no obvious evolutionary benefits in survival and reproduction, which led Charles Darwin to describe it as one of the "most mysterious" human behaviours. Why, then, is music such an important part of the human experience?

Music's evolutionary advantages remain intriguingly unclear

MUSIC'S ORIGINS

Pinpointing the origin of music is difficult because it leaves few physical traces. The oldest instruments, such as flutes from the Hohle Fels cave in Germany, date to 43,000–35,000 years ago, but the story likely begins earlier. Modern humans evolved in Africa around 300,000–200,000 years ago, and music likely arose with them. Early instruments may have been made of wood, skin, or other perishable materials that do not survive – just as many hunter-gatherer groups today still use simple, handcrafted, biodegradable instruments passed down through generations.

Still, some ancient sound-producing objects survive. Simple bone-carved idiophones – instruments producing sound through the vibration of the material itself, such as clappers, rattles, or struck bones – appear as early as 40,000–35,000 years ago and are among the earliest evidence. Bullroarers – flat pieces of wood spun on a cord to create a low drone – were in use by at least 20,000 years ago, seemingly for ritual or long-distance communication. Horns made from the shells of sea snails, such as conches, capable of resonant tones, appear in contexts dating back 18,000–12,000 years ago. Together, these artefacts point to a long-standing human impulse to shape sound, suggesting music, in some form, was already part of early human life.

Clues from early human art suggest a strong creative culture long before durable instruments. Beaded jewellery from at least 142,000 years ago, geometric carvings from 77,000 years ago, and cave paintings from over 52,000 years ago all point to symbolic thought and expression. Likely, music – like visual art and storytelling – was already woven into the fabric of human life.

MUSIC IN MIND AND BODY

Musical ability often seems to run in families, from the Bachs of Europe to the Marleys of Jamaica. In some societies, there are even hereditary castes of musicians such as griots in west Africa or the Manganiar of India, for whom music is believed to be "in the blood". But just like language, musical skill is shaped by both genes and culture.

There is no single "music gene" – in fact, rhythm perception alone involves at least 69 genetic variants. Everyone has some capacity for music, but learning and environment play a major role. Much like language, the brain is primed to learn music, but what kind of music we learn depends on our exposure, cultural surroundings, and our individual developmental experiences. Music is a multi-sensory process, involving brain areas linked to sound,

Musical talent combines inherited traits with cultural learning and environmental influences

V-shaped mouthpiece where musician blew air into tube

Flute's external diameter is about 1 cm (⅖ in)

SOUNDS OF THE ICE AGE
These remarkable prehistoric instruments highlight the diversity of early human sound-making practices, reflecting creativity, ritual, communication, and the deep roots of our musical traditions.

Carved notches were scraped to produce sound

IDIOPHONE
This notched reindeer bone from Mas d'Azil, France, dating to 15,000–10,000 years ago, is believed to have been used as a musical idiophone.

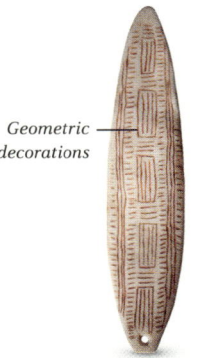

Geometric decorations

BULLROARER
This bullroarer, dating to 15,000–12,000 years ago, was carved from bone and decorated with finely incised lines and geometric patterns.

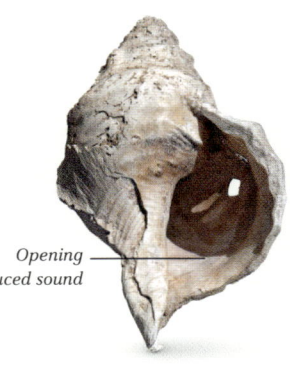

Opening produced sound

MARINE SHELL HORN
This horn, disovered in France, dates to around 18,000 years ago. It is crafted from a sea snail's shell and carefully modified to produce sound.

THE MUSICAL BRAIN

movement, memory, and emotion. When we listen, we might feel our heart race, shiver, or feel compelled to dance. Music literally moves us – and dancing with others in rhythm can help us feel closer and more connected.

MUSIC'S FUNCTION

Darwin thought music had little practical use. Yet today we know that music offers many benefits – both emotional and social. It reduces stress, promotes group bonding, and can even release endorphins, creating powerful feelings of euphoria and joy when people sing, play, or dance together.

Music may have evolved to strengthen social ties, especially in the small, highly cooperative communities of early humans. Across cultures, music often plays a role in group activities, helping people coordinate, communicate, and remember. For example, in Gaelic-speaking Scottish communities, songs traditionally accompanied tasks such as waulking (beating cloth) and rowing, helping synchronize movements and ease the pains of repetitive labour. Songs are also powerful tools for transmitting cultural

Music fosters cooperation, transmits knowledge, and bonds communities

WIRED FOR MUSIC

Music engages many brain regions at once, linking hearing, movement, memory, emotion, and attention. This connectivity explains why rhythm moves us, melodies linger, and shared musical experiences unite people across time and cultures.

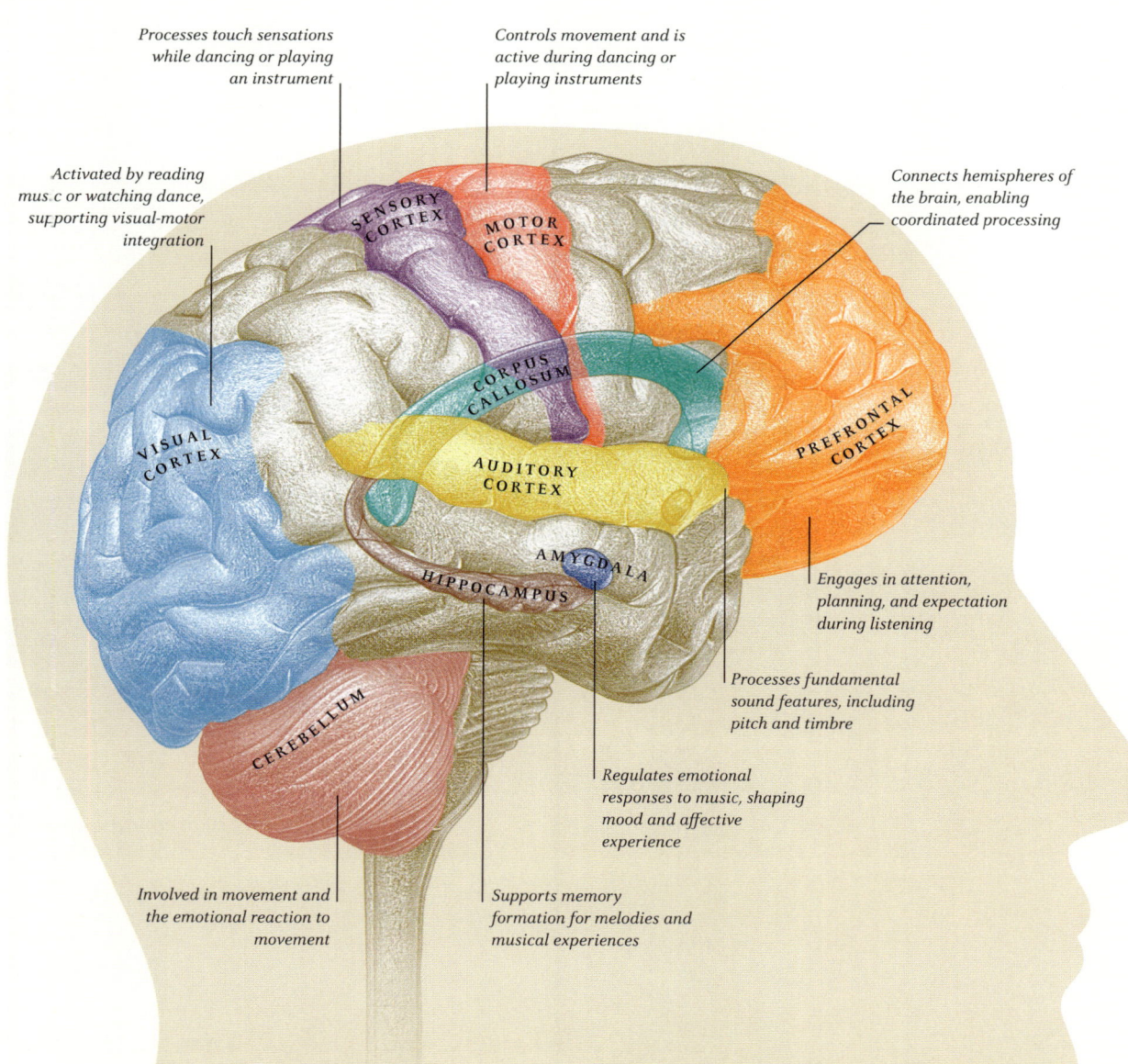

Processes touch sensations while dancing or playing an instrument

Controls movement and is active during dancing or playing instruments

Activated by reading music or watching dance, supporting visual-motor integration

Connects hemispheres of the brain, enabling coordinated processing

SENSORY CORTEX

MOTOR CORTEX

CORPUS CALLOSUM

VISUAL CORTEX

AUDITORY CORTEX

PREFRONTAL CORTEX

AMYGDALA

HIPPOCAMPUS

CEREBELLUM

Engages in attention, planning, and expectation during listening

Processes fundamental sound features, including pitch and timbre

Regulates emotional responses to music, shaping mood and affective experience

Involved in movement and the emotional reaction to movement

Supports memory formation for melodies and musical experiences

ANCIENT RITUAL
These 8,000–10,000-year-old paintings in Magura Cave, Bulgaria, depict human figures in rhythmic motion, likely engaged in a ritual dance. The scenes suggest that music and movement were already central to communal life.

Humans uniquely create music with symbolic, emotional, and cultural meaning

knowledge, such as navigation routes, ancestral stories, or hunting strategies. Music soothes infants, inspires rituals, and cements group identity. These social advantages may have helped musical humans survive and thrive.

Some scientists argue that music could simply be a by-product of language, without a direct evolutionary function. But if that were the case, it is hard to explain why humans across all cultures invest so much time and energy in making music – often at the expense of more obviously useful tasks.

IS MUSIC UNIQUELY HUMAN?

According to most definitions, humans are the only truly musical species. Some animals – like birds, gibbons, and whales – produce elaborate songs, but these usually serve clear functions such as attracting mates or defending territory. Human music, by contrast, often has no immediate function and carries complex symbolic and emotional meaning.

However, our musical capacity likely evolved from abilities shared with other species. Some birds and whales have culturally transmitted song traditions with regional "dialects". A few animals, like cockatoos, can keep time with a beat or imitate sounds – abilities rare in nature. Yet these traits are not especially strong in our closest relatives. Chimpanzees, for instance, struggle with rhythm and vocal learning.

Human musicality may have developed from traits that evolved independently in distant species, rather than being inherited directly from our closest primate ancestors.

ANCIENT AND EVER-EVOLVING

Our understanding of music's evolution has deepened over time. We now know that music has likely been central to human life for hundreds of thousands of years, playing a key role in social bonding, communication, and cultural transmission. Yet, music remains one of the most mysterious of human behaviours.

What is clear is that music has changed dramatically over time. For most of human history, music was created and shared within small groups, passed on orally through generations. Today, many people engage with music through vast digital libraries and algorithmically curated playlists. People are often to be found listening to music in private, through headphones, but live music remains enduringly popular, and music festivals bring vast numbers of people together. Music remains as important for us today as it ever was – allowing us to express ourselves and connect with each other in ways that cannot easily be put into words.

Modern humans

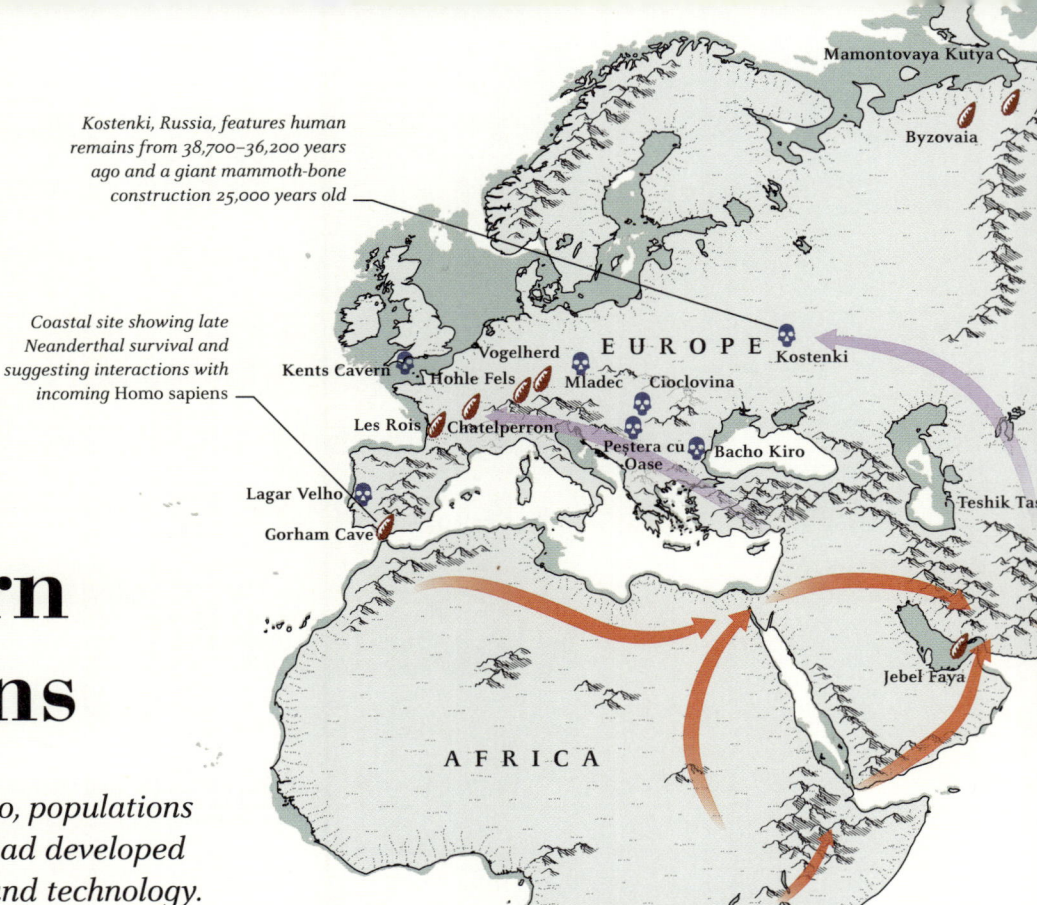

Kostenki, Russia, features human remains from 38,700–36,200 years ago and a giant mammoth-bone construction 25,000 years old

Coastal site showing late Neanderthal survival and suggesting interactions with incoming Homo sapiens

Mamontovaya Kutya

Byzovaia

Kostenki

EUROPE

Vogelherd
Kents Cavern
Hohle Fels   Mladec   Cioclovina
Les Rois   Chatelperron
Pestera cu   Bacho Kiro
Oase
Lagar Velho
Gorham Cave

Teshik Tash

Jebel Faya

AFRICA

EQUATOR

By 70,000 years ago, populations of Homo sapiens *had developed advanced culture and technology. As some spread out of Africa, their new abilities promoted global expansion – a journey traced by genetics, fossils, and archaeology.*

Modern humans – *Homo sapiens* – evolved in Africa between 300,000 and 200,000 years ago. The first migration out of Africa occurred around 120,000 years ago, reaching as far as China, but the major dispersal came about 70,000 years ago, when anatomically modern humans emerged from east Africa – more slender-bodied and fully modern in form. Climate shifts, population pressures, and technological innovations likely drove this expansion. Genetic evidence, including mitochondrial DNA and Y-chromosome studies, supports a single major migration from which all non-African populations descend. The successful routes included a northern path through the Levant and a southern coastal route along the Arabian Peninsula into south Asia, from where humans spread into Europe, Asia, Australasia, and later the Americas.

As modern humans spread across Eurasia, they met and interbred with other species such as Neanderthals and Denisovans. Today, all non-African genomes carry about 1–2 per cent Neanderthal DNA, while Denisovan ancestry is most pronounced in Melanesian and some Southeast Asian groups. These gene flows influenced immunity, altitude and cold

Climate, genetics, and innovation combined to drive humanity's great global expansion

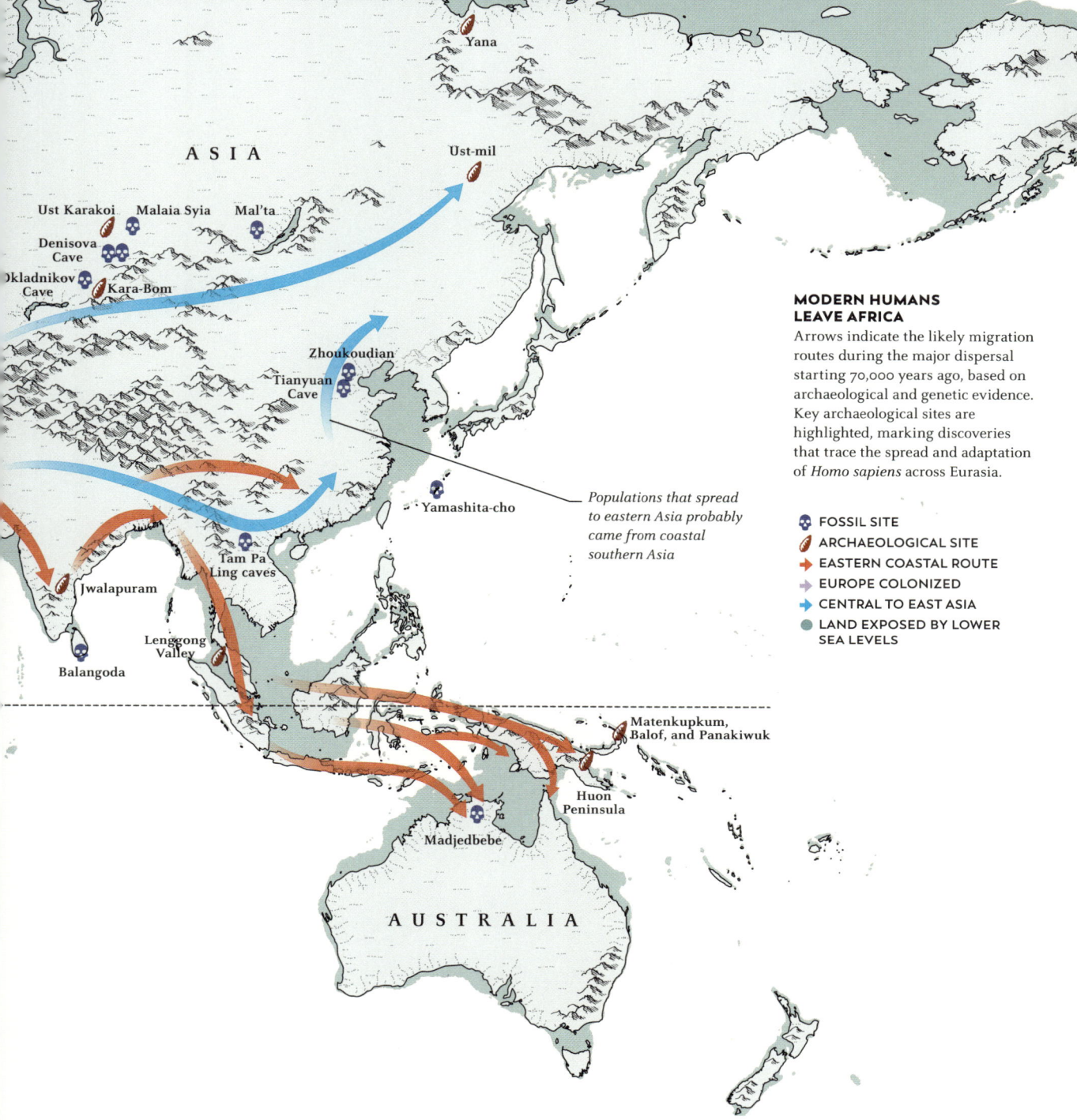

ASIA

Yana

Ust-mil

Ust Karakoi Malaia Syia Mal'ta

Denisova
Cave

Okladnikov
Cave Kara-Bom

Zhoukoudian

Tianyuan
Cave

Yamashita-cho

Jwalapuram

Tam Pa
Ling caves

Lenggong
Valley

Balangoda

Matenkupkum,
Balof, and Panakiwuk

Huon
Peninsula

Madjedbebe

AUSTRALIA

*Populations that spread
to eastern Asia probably
came from coastal
southern Asia*

MODERN HUMANS LEAVE AFRICA

Arrows indicate the likely migration routes during the major dispersal starting 70,000 years ago, based on archaeological and genetic evidence. Key archaeological sites are highlighted, marking discoveries that trace the spread and adaptation of *Homo sapiens* across Eurasia.

- 💀 FOSSIL SITE
- 🔴 ARCHAEOLOGICAL SITE
- → EASTERN COASTAL ROUTE
- → EUROPE COLONIZED
- → CENTRAL TO EAST ASIA
- ● LAND EXPOSED BY LOWER SEA LEVELS

tolerance, and skin pigmentation. This discovery, made in the 21st century, reshaped our understanding of human evolution, from a linear tree to a complex network of interwoven lineages that reflected deep complexity. Wherever *Homo sapiens* arrived, other human species eventually vanished, likely due to displacement and assimilation.

EASTWARD EXPANSION

From the Arabian Peninsula, modern humans rapidly expanded eastward, moving through Asia and even crossing the sea to Australia. Sites such as Jwalapuram in India and Madjedbebe in northern Australia provide some of the earliest archaeological evidence of modern humans outside Africa. Signs of early

Humans spread east adapting, innovating, and interbreeding along the way

MODERN HUMANS

ROUTES THROUGH THE AMERICAS

The arrows on this map trace the probable migration routes of humans as they spread through the Americas, based on archaeological and genetic evidence. Highlighted sites mark significant discoveries of tools and cultural remains that reveal how humans adapted to diverse environments across North and South America.

🥜 ARCHAEOLOGICAL SITE
➤ ASIAN ORIGINS
➤ FOUNDER AMERICANS
➤ NORTH AMERICAN EXPANSION
➤ PENETRATING CENTRAL AMERICA
➤ COLONIZING SOUTH AMERICA
⬤ LAND EXPOSED BY LOWER SEA LEVELS

Swan Point

Bluefish Cave and Old Crow River

Nenana Upward Sun River

NORTH AMERICA

During the Ice Age, ocean levels were low enough to expose a connection of land, known as Beringia between Asia and North America

Triquet island

Anzick

Manis mastodon

La Sena and Lovewell

Paisley Cave

Meadowcroft Rockshelter

Topper

Page-Ladson

One of the oldest sites including non-Clovis tools and a range of plants gathered for food

Arlington Springs

Clovis

Gault

Ixtapan

Evidence of transition from hunter-gatherer to early farming settlements

Taima-Taima

EQUATOR

SOUTH AMERICA

Serra da Capivara

Santa Elina

Human habitation with living floor, hearth, and horses

Lagoa Santa

Quebrada Santa Julia

Piedra Musea

Fell's Cave

symbolic behaviour and complex culture is found in rock art older than 45,000 years on the Indonesian islands. The arrival of humans in Australia by at least 65,000 years ago demonstrates remarkable seafaring skills, highlighting their ingenuity and innovation as they colonized challenging environments.

WEST INTO EUROPE

Modern humans entered Europe around 45,000 years ago, coexisting with Neanderthals for several millennia, as evidenced by sites such as Bacho Kiro in Bulgaria and Oase Cave in Romania. Although Neanderthals were well adapted to the cold climates of Ice Age Europe, *Homo sapiens* – with broader social networks and more advanced technology – proved more

Modern humans outcompeted Neanderthals through cooperation, and innovation in harsh climates

resilient. Neanderthals eventually disappeared, likely through a combination of competition, climate stress, and assimilation. The stronger social ties and higher population density of *Homo sapiens* may have further accelerated cultural exchange and innovation, ultimately tipping the balance in their favour.

POPULATING THE AMERICAS

The Americas were the final continents to be widely settled by humans. Current genetic and archaeological evidence indicates that people reached Beringia and the northern Yukon by at least 24,000 years ago and experienced a prolonged "Beringian standstill" after splitting from Siberian groups around 36,000 years ago. From around 17,000 years ago, populations expanded southward into the rest of North and South America via inland and coastal routes. Numerous well-documented pre-Clovis sites now confirm that human presence predates the once-assumed primacy of the Clovis culture, a culture

known for its distinctive stone points, and as people spread, they adapted to environments ranging from Arctic tundra to Amazon rainforest.

FINAL EXPANSION

The last phase of human expansion, beginning around 5,000 years ago, unfolded across the vast Pacific Ocean. Archaeological evidence points to the Lapita culture – a prehistoric people of the western Pacific – as the pioneers of this migration. They spoke an early Austronesian language and their voyages carried them across thousands of kilometres of open sea, reaching as far as Tonga, Samoa, Hawaii, Easter Island, and New Zealand. The Lapita people's distinctive dentate-stamped pottery, made by pressing toothed tools into clay, provides a key archaeological marker tracing where and when they spread throughout the Pacific.

Lapita seafarers launched the final human migrations, settling islands across the Pacific

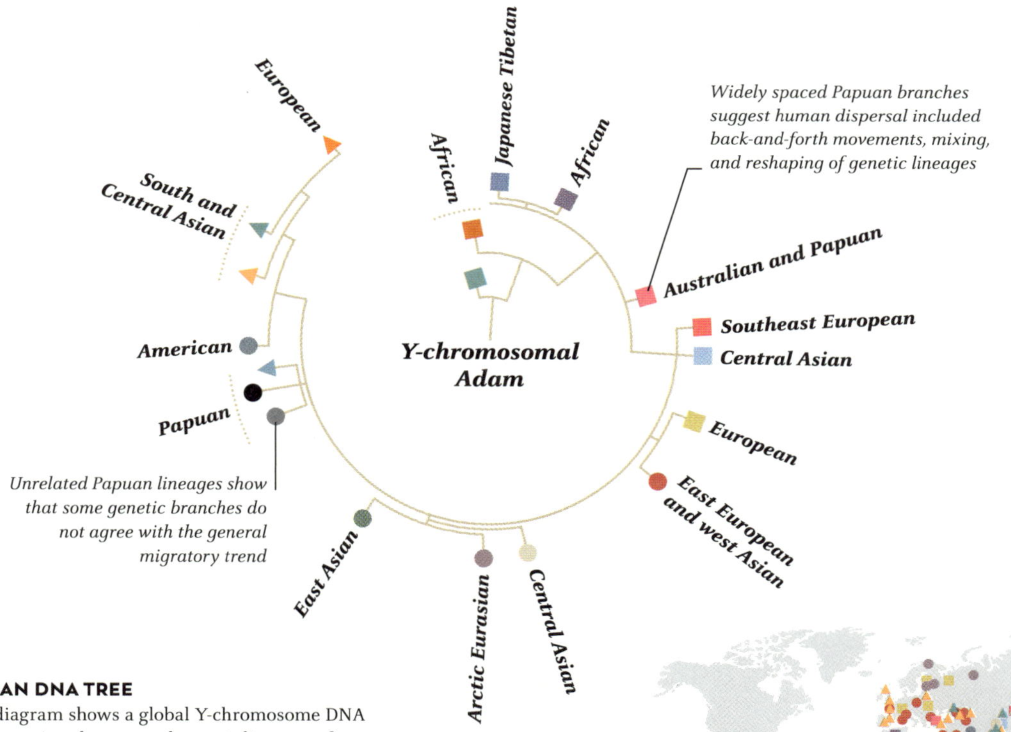

Y-chromosomal Adam

Widely spaced Papuan branches suggest human dispersal included back-and-forth movements, mixing, and reshaping of genetic lineages

Unrelated Papuan lineages show that some genetic branches do not agree with the general migratory trend

- European
- Japanese Tibetan
- African
- African
- Australian and Papuan
- Southeast European
- Central Asian
- European
- East European and west Asian
- Central Asian
- Arctic Eurasian
- East Asian
- Papuan
- American
- South and Central Asian

HUMAN DNA TREE

This diagram shows a global Y-chromosome DNA tree, mapping the paternal genetic lineages of modern humans. Based on mutations passed from father to child, the tree traces all Y-chromosome DNA back to a common ancestor – "Y-chromosomal Adam" – and reveals how human populations diverged and spread across the world over time.

Ancestral group locations

273

Some of the earliest signs that humans regularly turned to the sea for food appear with Neanderthals around 150,000 years ago (YA) at Bajondillo Cave, Spain, where they left mussel shells. In fact, about 30 coastal sites along the northern Mediterranean and Portugal's Atlantic coast are attributed to Neanderthals. At around the same time, 164,000 YA, *Homo sapiens* at Pinnacle Point on South Africa's southern coast had also expanded their diet to include marine foods. Middens – dense ancient refuse piles in distinct layers – suggest that they relied even more on seafood than Neanderthals.

Shellfish collection required working with tides and likely lunar cycles, as the lowest spring tides, offered the most productive harvesting. Coastal people didn't abandon their other means of subsistence – seafood would have supplemented

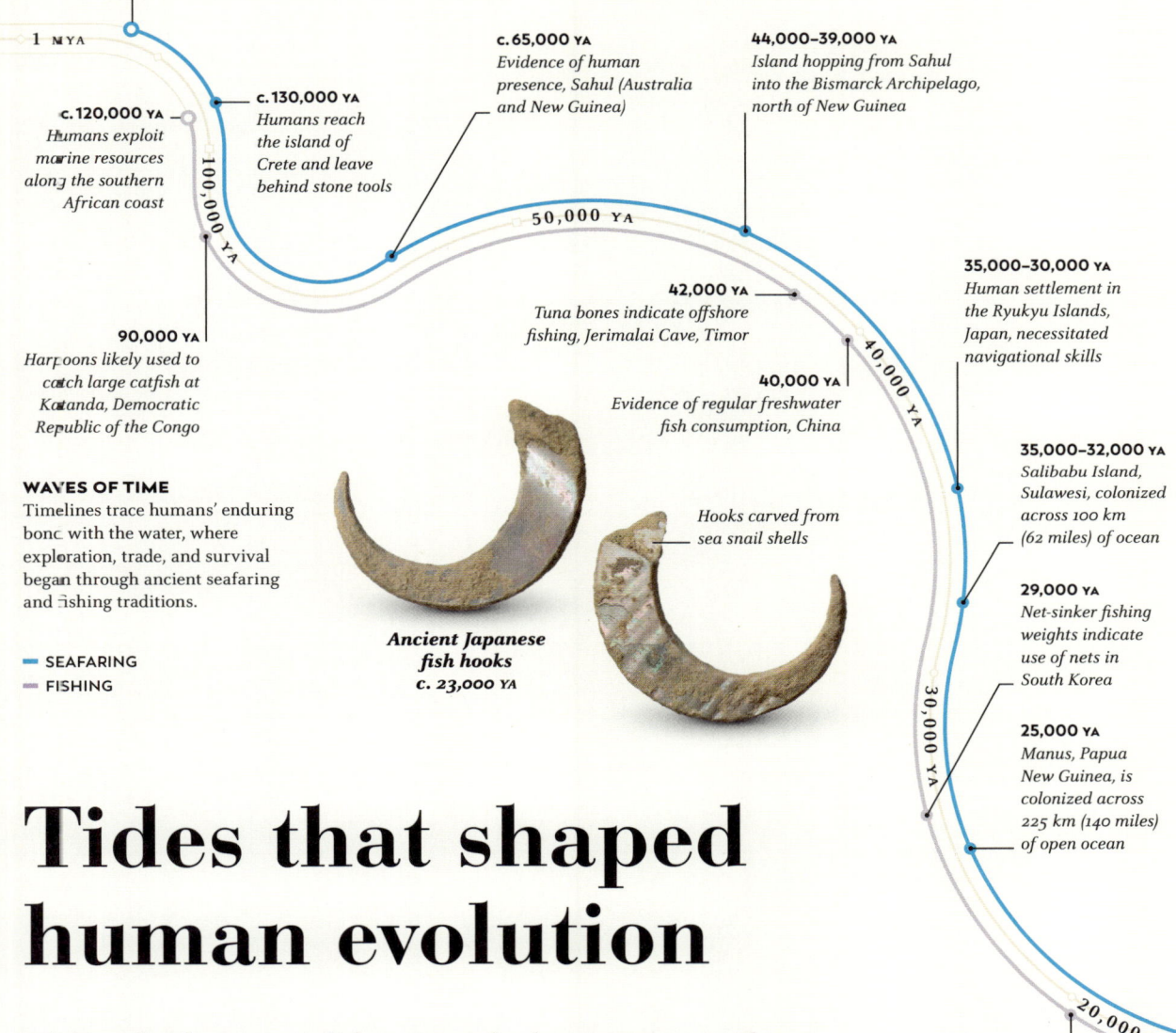

c. 777,000–631,000 YA
Luzon, Philippines, is colonized via an unknown route

1 MYA

c. 120,000 YA
Humans exploit marine resources along the southern African coast

c. 130,000 YA
Humans reach the island of Crete and leave behind stone tools

100,000 YA

c. 65,000 YA
Evidence of human presence, Sahul (Australia and New Guinea)

44,000–39,000 YA
Island hopping from Sahul into the Bismarck Archipelago, north of New Guinea

50,000 YA

90,000 YA
Harpoons likely used to catch large catfish at Katanda, Democratic Republic of the Congo

42,000 YA
Tuna bones indicate offshore fishing, Jerimalai Cave, Timor

40,000 YA
Evidence of regular freshwater fish consumption, China

40,000 YA

35,000–30,000 YA
Human settlement in the Ryukyu Islands, Japan, necessitated navigational skills

35,000–32,000 YA
Salibabu Island, Sulawesi, colonized across 100 km (62 miles) of ocean

29,000 YA
Net-sinker fishing weights indicate use of nets in South Korea

25,000 YA
Manus, Papua New Guinea, is colonized across 225 km (140 miles) of open ocean

30,000 YA

20,000 YA

23,000–16,000 YA
Fish hooks left in Jerimalai Cave, Timor

WAVES OF TIME
Timelines trace humans' enduring bond with the water, where exploration, trade, and survival began through ancient seafaring and fishing traditions.

— SEAFARING
— FISHING

Ancient Japanese fish hooks
c. 23,000 YA

Hooks carved from sea snail shells

Tides that shaped human evolution

The crash of waves and the scent of salt connect us to deep evolutionary roots. Early hominins who understood tides and coastlines expanded their diets, resilience, and reach, beginning humanity's enduring relationship with the sea.

terrestrial hunting and provided a dependable fallback, possibly aiding *Homo sapiens'* early expansion out of Africa.

EARLY ISLAND MIGRATION

Early humans traversed some of Indonesia long before Homo sapiens *appeared*

Compelling evidence shows that early humans could reach distant islands. On Sulawesi, artefacts over 1 million years old indicate people travelled from mainland Southeast Asia – a crossing of 50 km (30 miles) or more. Their identity remains uncertain – possibly *Homo erectus* or an even earlier species. Human species did reach Luzon in the Philippines and the island of Flores, Indonesia, and *Homo floresiensis* does indeed show some traits more archaic than *Homo erectus*. The most plausible explanation for these early dispersals is accidental rafting on vegetation mats dislodged

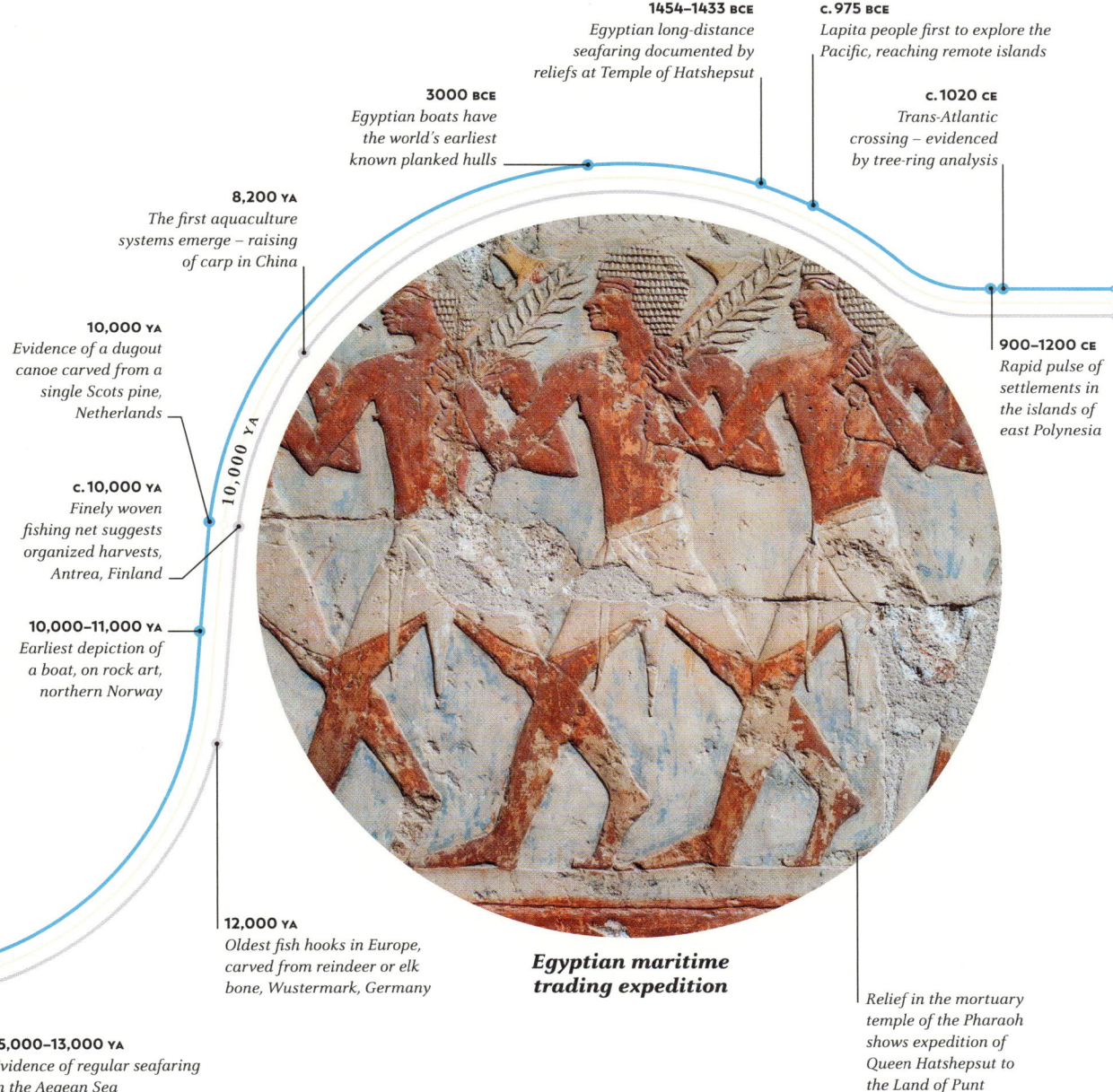

1454–1433 BCE
Egyptian long-distance seafaring documented by reliefs at Temple of Hatshepsut

c. 975 BCE
Lapita people first to explore the Pacific, reaching remote islands

3000 BCE
Egyptian boats have the world's earliest known planked hulls

c. 1020 CE
Trans-Atlantic crossing – evidenced by tree-ring analysis

8,200 YA
The first aquaculture systems emerge – raising of carp in China

10,000 YA
Evidence of a dugout canoe carved from a single Scots pine, Netherlands

10,000 YA

c. 10,000 YA
Finely woven fishing net suggests organized harvests, Antrea, Finland

10,000–11,000 YA
Earliest depiction of a boat, on rock art, northern Norway

900–1200 CE
Rapid pulse of settlements in the islands of east Polynesia

12,000 YA
Oldest fish hooks in Europe, carved from reindeer or elk bone, Wustermark, Germany

Egyptian maritime trading expedition

Relief in the mortuary temple of the Pharaoh shows expedition of Queen Hatshepsut to the Land of Punt

15,000–13,000 YA
Evidence of regular seafaring in the Aegean Sea

TIDES THAT SHAPED HUMAN EVOLUTION

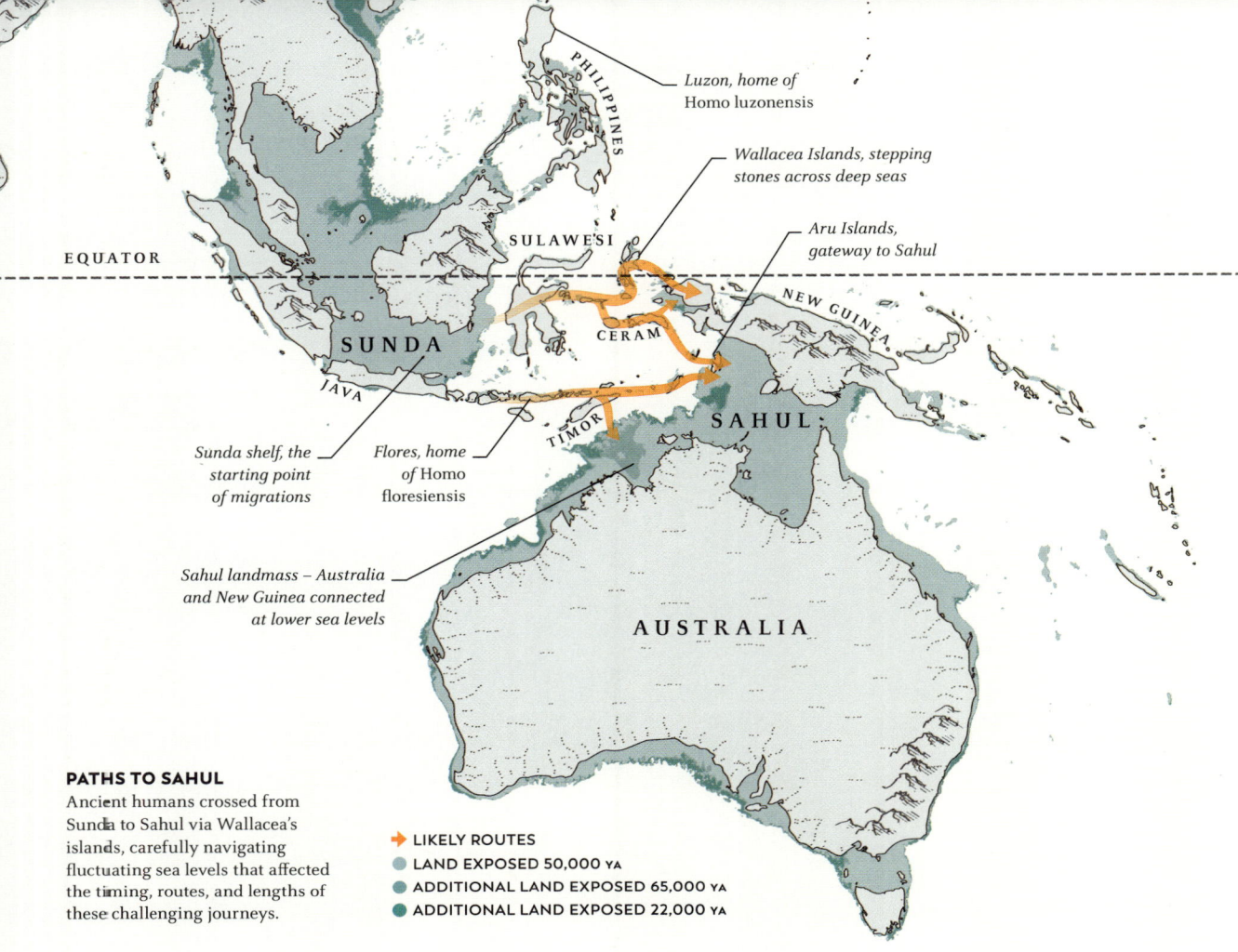

PATHS TO SAHUL
Ancient humans crossed from Sunda to Sahul via Wallacea's islands, carefully navigating fluctuating sea levels that affected the timing, routes, and lengths of these challenging journeys.

Luzon, home of Homo luzonensis

Wallacea Islands, stepping stones across deep seas

Aru Islands, gateway to Sahul

PHILIPPINES

SULAWESI

NEW GUINEA

CERAM

EQUATOR

SUNDA

JAVA

TIMOR

SAHUL

FLORES

Sunda shelf, the starting point of migrations

Flores, home of Homo floresiensis

Sahul landmass – Australia and New Guinea connected at lower sea levels

AUSTRALIA

→ LIKELY ROUTES
● LAND EXPOSED 50,000 YA
● ADDITIONAL LAND EXPOSED 65,000 YA
● ADDITIONAL LAND EXPOSED 22,000 YA

during cyclones or tsunamis. These rafts could have drifted with prevailing currents, moving through the Philippines towards Sulawesi and eventually, Flores. This would have allowed small, isolated populations to cross otherwise impassable sea barriers. Indonesia's tectonically active volcanic arc makes earthquakes, eruptions, and tsunamis frequent, providing repeated natural dispersal mechanisms.

THE FIRST SEAFARING VOYAGES

At one moment in the Pleistocene, people first launched boats intentionally to cross the sea to reach lands beyond the horizon. It may have been *Homo sapiens* that first did so, and once they mastered long-distance seafaring, the ocean became a vital resource. In Timor, by 42,000 years ago, people practiced deep-sea fishing, catching powerful oceanic predators, such as tuna, trevally, and sharks with seacraft and fishing equipment that is completely unknown

to us. Fish hooks only appear in the archaeological record around 23,000 YA, but they must have been in use much earlier.

By the late Pleistocene, people were also using sea routes to transport exotic raw materials, such as obsidian – a volcanic glass used to make sharp tools. Its dispersal across Wallacea by 16,000 YA, along with standardized shell beads, reveals some of the world's earliest extensive maritime exchange networks. Similar networks developed elsewhere after the peak of the Ice Age, including in Japan, the Mediterranean, and the islands of Papua New Guinea.

MARITIME SKILL AND EXPANSION

Experimental voyages show that skilled paddlers could cross treacherous waters such as the 110-km (68-mile) strait between Taiwan and Yonaguni Island – the shortest sea route to Japan. These reconstructions suggest that late Pleistocene voyagers were capable of deliberate crossings

Skilled paddlers transformed oceans into migration and resource pathways

using dugout canoes and averaging over a metre per second. Such feats required coordination, leadership, teamwork, and navigation guided by stars, swells, currents, and birds. By this time, humans had become highly competent seafarers, turning the ocean from a barrier into a bridge for migration, fishing, and expansion.

Homo sapiens likely used fleets to move viable groups, colonizing new lands. These societies fished and gathered coastal resources while evolving complex logistical, technological, and social strategies for survival and settlement.

Fleets enabled colonization, resource use, and complex societal development by humans

INNOVATION AND FOOD SECURITY

Building on maritime expertise, societies developed early aquaculture. The 9,000-year-old Antrea net, found in Finland, reflects organized harvesting. Fish farming appeared in China, around 8,000 YA, suggesting fish domestication preceded that of many land animals. In Victoria, Australia, the Gunditjmara built stone-lined eel traps about 6,600 YA, modifying habitats to secure reliable food. These innovations show an evolving relationship with aquatic resources, blending environmental knowledge and cultural ingenuity.

From shellfish collection to deep-sea fishing, inter-island voyaging, and aquaculture, humans transformed the ocean from a barrier into a critical resource. These developments enhanced food security, enabled migration and trade, and laid the foundation for complex maritime societies and global human expansion.

ANCIENT NORTHERN SEAFARERS

Norwegian rock art at Alta, dating back 7,000 years, depicts boats up to 4 m (16 ft) long, filled with archers and fishers casting nets – offering a vivid glimpse into prehistoric seafaring.

1 *The confident stance of the hunters suggests the boats were stable*

2 *One hunter stands with a raised bow, while another casts a net*

TIDES THAT SHAPED HUMAN EVOLUTION

Over thousands of years, humans gradually moved from mobile foraging to life in fixed communities. This uneven transition saw changes in food production, social organization, and humanity's impact on the landscape.

Settling down

The gradual change from foraging to farming was the most profound global transition in human history since the control of fire, seeing human populations shifting from mobile hunting and gathering to settled, food-producing, agricultural lifestyles. Often called the Neolithic Revolution, this transformation was incremental rather than revolutionary, and far from linear. Over the last 12,000 years, it emerged independently in multiple regions, each shaped by local flora, fauna, environment, and social contexts. Agriculture gradually replaced foraging in many areas, fundamentally altering how people obtained food,

The Neolithic Revolution was in fact a gradual change occurring in different ways in unconnected places and times

FIRST SETTLERS
As the Ice Age ended, landscapes grew richer and more predictable. Some nomadic bands kept roaming, but others settled in fertile areas for longer, forming the first semi-permanent communities.

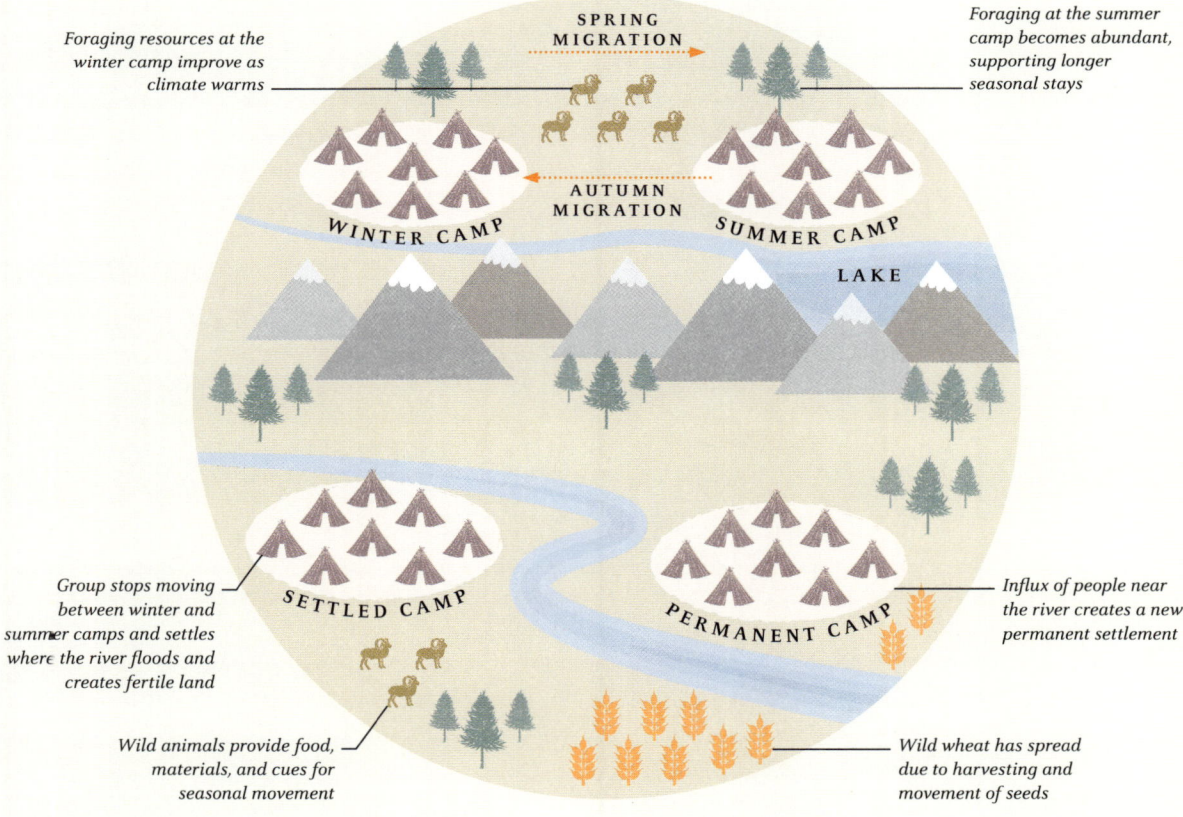

Foraging resources at the winter camp improve as climate warms

Foraging at the summer camp becomes abundant, supporting longer seasonal stays

SPRING MIGRATION

AUTUMN MIGRATION

WINTER CAMP

SUMMER CAMP

LAKE

Group stops moving between winter and summer camps and settles where the river floods and creates fertile land

SETTLED CAMP

PERMANENT CAMP

Influx of people near the river creates a new permanent settlement

Wild animals provide food, materials, and cues for seasonal movement

Wild wheat has spread due to harvesting and movement of seeds

organized communities, and interacted with the landscape. As populations grew, seasonal camps gave way to permanent hamlets and eventually villages, which featured increasingly complex forms of cooperation, ritual, and communal labour.

EARLY SEDENTARY COMMUNITIES

At the end of the Ice Age, warmer and wetter climates created more stable environments with more plentiful food resources. In many regions, this allowed groups to remain in one area for extended periods while still practising hunter-gatherer lifestyles. These populations are often described as affluent foragers – people who settled in areas of natural abundance, such as coasts, rivers, lakes, and forests, where diverse seasonal foods supported larger and more stable communities. With less pressure to move constantly and "pack light", they invested in specialized tools, including fishing gear, ground-stone axes, and finely crafted blades. This period also saw the emergence of pottery, used for cooking, storing surpluses, and processing plant foods.

Affluent foragers began settling in favourable areas long before farming began

In southwesternmost Asia, long-term camps gradually developed into true sedentary villages. The earliest evidence for sustained farming appears in the Fertile Crescent, covering parts of modern-day Iraq, Syria, and Türkiye, around 10,500 years ago, where people cultivated wild cereals, such as barley and emmer

GROWING COMMUNITIES

Rising populations anchored communities to the land. Villages with permanent buildings and defences expanded as people cultivated wild cereals and penned animals to ensure reliable food supplies.

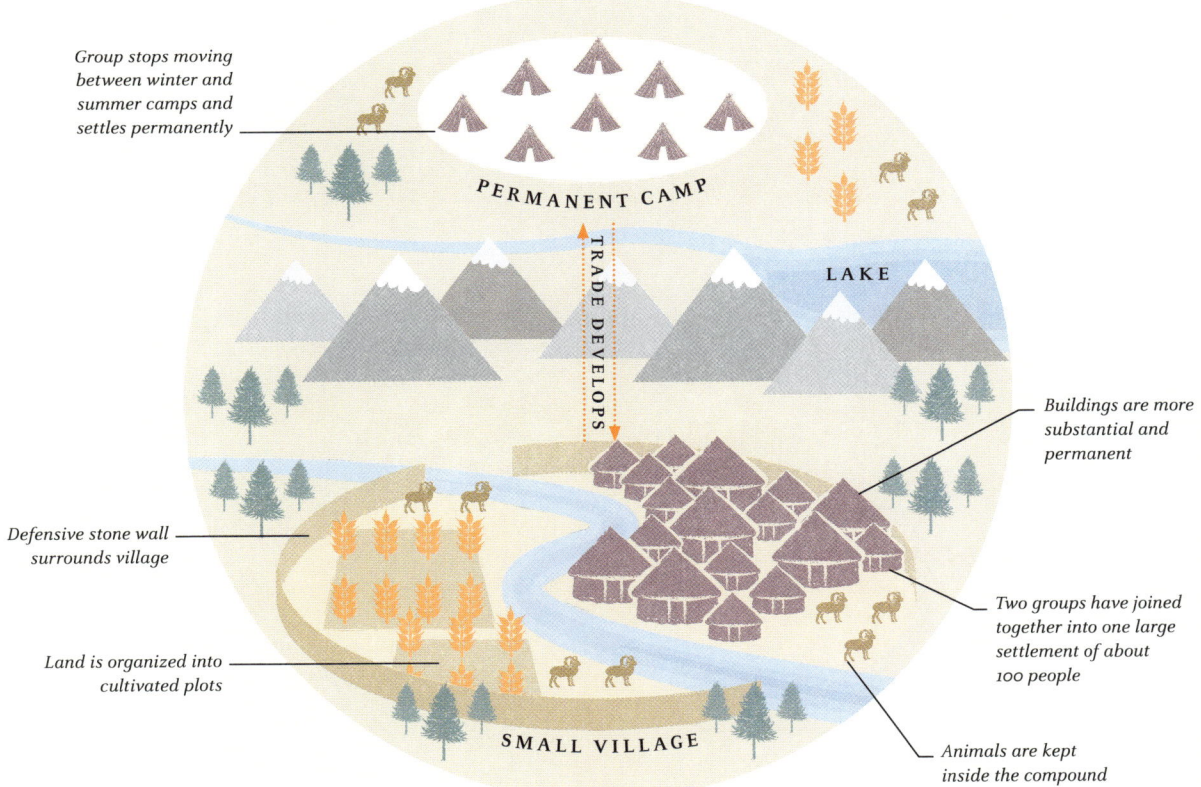

Group stops moving between winter and summer camps and settles permanently

PERMANENT CAMP

TRADE DEVELOPS

LAKE

Buildings are more substantial and permanent

Defensive stone wall surrounds village

Land is organized into cultivated plots

SMALL VILLAGE

Two groups have joined together into one large settlement of about 100 people

Animals are kept inside the compound

SETTLING DOWN

EARLY COMMUNAL LIVING

Göbekli Tepe is among the world's earliest built landscapes, where towering stone circles stood alongside everyday activity, offering a rare glimpse into early community life more than 11,000 years ago.

wheat, and domesticating sheep and goats. Sites such as Jericho and 'Ain Ghazal reveal stone houses, storage buildings, communal structures, and below-ground burials – early indicators of organized living and a deepening attachment to place. Many of these communities adopted technologies suited to settled life, including grinding stones for grain, plastered floors, and increasingly elaborate ritual objects.

Further north, Çatalhöyük in present-day Türkiye demonstrates a densely packed settlement of mudbrick houses accessed from the roof, featuring shared courtyards, vibrant wall paintings, and rituals integrated into daily domestic life. Nearby, Göbekli Tepe, renowned for its monumental stone circles, also contains evidence of domestic structures, cereal processing, water management, and an expanding toolkit for everyday tasks. Collectively, these sites illustrate that ritual gatherings, food production, and permanent settlement evolved together, rather than along a simple, linear path.

Farming, permanent settlements, and ritual life, all seem to have evolved together

Vessels found in graves are often painted in red and black

Large urn

Alternating handles

Earthenware vessel

Decorative painted bands characteristic of the culture of the Yellow River basin

Curvilinear design typical of the Majiayao phase (3200–2650 BCE)

Ceramic decorated vase

Spouted vessel

Beyond the Fertile Crescent, agriculture developed independently in multiple regions, each drawing on local plants, animals, and cultural traditions. In east Asia, communities along the Yellow River and Yangtze River cultivated rice by around 11,000 years ago. Pottery for storage and cooking, along with polished stone tools, facilitated land clearing and efficient food production.

MORE INDEPENDENT AGRICULTURAL REVOLUTIONS

Agriculture developed independently in several regions based on local flora and fauna

In the Americas, farming also arose in several centres. Mesoamerican communities domesticated the staple crops that would later support major civilizations, while Andean populations adapted agriculture to high-altitude environments. Africa, southern Asia, Southeast Asia, and North America underwent their own agricultural transformations around 5,000 years ago. African farmers pioneered crops such as sorghum, pearl millet, and yams, often alongside early cattle herding. In southern Asia, people combined wheat, barley, and pulses with increasingly

NEOLITHIC CHINESE CERAMICS
These ceramics, dating from the 5th–3rd millennium BCE, were products of the culture developed around farming in China.

281

REGIONAL AGRICULTURE
This map shows where farming first emerged in each region and illustrates how, when faced with similar conditions and resources, humans independently developed comparable innovations.

● THE AMERICAS
● AFRICA / EUROPE / ASIA
● OCEANIA
● THE PACIFIC ISLANDS

NORTH AMERICA

Northeast America 2000–1000 BCE: Plants including sunflowers, sumpweed, and goosefoot domesticated, later integrated with maize

Mesoamerica, 3000–2000 BCE: Farmers grew maize, squash, and beans and raised turkeys and dogs for meat

ATLANTIC OCEAN

Amazonia, 3000–2000 BCE: Early cultivation of cassava, arrowroot, squash, maize, peanuts, cotton, and fruit trees

EQUATOR

PACIFIC OCEAN

The Andes, 3000–2000 BCE: Quinoa, potatoes, and beans were grown; llamas and alpacas were domesticated from wild guanacos and vicuñas

SOUTH AMERICA

sophisticated irrigation. Southeast Asian groups domesticated taro, yams, and bananas. In eastern North America, Indigenous peoples developed a diverse farming–foraging economy.

Across these regions, the rise of agriculture was accompanied by technological innovation. Pottery stored harvests, ground-stone axes and adzes cleared forests, sickles harvested grain, and woven baskets transported crops. These tools reveal that agriculture was more than a dietary shift – it transformed labour, technology, and the rhythm of daily life.

CONSEQUENCES FOR AGRICULTURAL SOCIETIES

Farming profoundly reshaped human health, biology, society, and the environment. Diets high in carbohydrates but low in nutrients caused cavities, shorter stature, and signs of nutritional stress. Settled life and close contact with domesticated animals spread zoonotic diseases – smallpox, measles, influenza, tuberculosis, typhoid, and plague – that thrived in dense settlements and periodically caused devastating epidemics.

Living close together in settlements led to the rise of infectious diseases

Agriculture also influenced human biology. Lactase persistence emerged in Europe, west Africa, southwest Asia, and India, enabling adults to digest milk (see p.159), while lighter skin evolved in northern Asia and Europe potentially because it enhanced vitamin D absorption in low-sunlight regions. However, vitamin D can also be obtained from the diet, which may explain why Arctic foragers such as the Inuit – eating a marine diet rich in vitamin C, from fish and whale blubber – possessed darker skin.

EUROPE

ASIA

PACIFIC
OCEAN

AFRICA

China, 7000 BCE: Farmers grew rice in the south and millet in the northern river valleys, and domesticated water buffalo, pigs, and chickens

Western Africa, 2800–1500 BCE: Farmers grew rice, bulrush millet, cowpea, yam, and oil palm

Fertile Crescent, 9000 BCE: Early farming arose due to abundant domesticable plants (wheat, barley) and animals (cattle, goats, sheep)

INDIAN OCEAN

OCEANIA

Sub-Saharan Africa 3000–2000 BCE: Early crops were millet, sorghum, yam, cowpea, and oil palm

Polynesia, 1400 BCE–1100 CE: Farming began after settlers arrived, bringing taro, pigs, and chickens

Permanent homes and food surpluses reshaped social life, fostering inequality (see pp.296–99), patriarchy, and territoriality. Fortified settlements and mass graves suggest rising warfare (see pp.292–95), while farming – enabling the production of surpluses – supported the emergence of states and empires, including Sumer, Egypt, Nubia, the Indus Valley, Shang China, and later, the Maya. Agriculture also altered ecosystems through deforestation and soil depletion. Grazing and rice cultivation added gases to the atmosphere – marking the earliest detectable human contribution to climate change and the beginnings of the Anthropocene.

Despite its global reach, the Neolithic Revolution was never universal. Many communities continued foraging for millennia – and some still do today. But farming left lasting cultural and ecological legacies: monumental architecture, kinship tied to land ownership, and the rise of cities. By establishing towns and cities, farming set humanity on a settled trajectory, culminating in today's megacities, direct descendants of the first Neolithic communities.

Transition to farming resulted in a chain of events leading to states, empires, and megacities

Humans exhibit a remarkable capacity for cooperation. We donate blood to strangers, form alliances across cultures, and build nations. But while these behaviours may seem uniquely human, the roots of cooperation run deep in the animal world.

Cooperation

What sets humans apart is not that we cooperate, but how extensively and flexibly we do so – across large groups, with strangers, and in complex ways. This "puzzle of cooperation" has fascinated scientists for decades, revealing a powerful mix of biological and cultural strategies that have shaped human evolution and continue to influence modern societies.

FAMILY TIES
Some of our cooperative behaviour may be explained by kin selection – helping relatives who share our genes. Bees work collectively to support the hive, elephants care for extended family members, and humans often go to great lengths for family. Parents raise children, siblings support one another, and grandparents pass down resources. Helping relatives makes evolutionary sense: by supporting them, you indirectly support the survival of your own genes. However, this idea doesn't explain why humans also help strangers, support unrelated friends, or act kindly towards people they may never see again, even across vast distances or differing social groups.

RECIPROCAL ALTRUISM
One important explanation for this broader cooperation is reciprocal altruism – the idea that helping others can benefit us in return, especially when interactions are repeated. In nature, helping is a good strategy when there's a chance it will be repaid. Vampire bats, for example, share blood meals with others that have fed them before. Humans are especially skilled at such exchanges. Across cultures,

Reciprocal altruism means helping with the expectation of future help, strengthening cooperation through repeated exchanges

Mutual trust and cooperation brings abundance

Self-interest breaks cooperative balance

Solo hunters have limited rewards

COOPERATE

DEFECT

COOPERATE DEFECT

HUNTER'S DILEMMA
In this version of the Prisoner's Dilemma game, each hunter chooses to share the hunt or act selfishly. Cooperation yields a shared stag – optimal for both. But if one defects, they take all while the other gets nothing. Mutual distrust leads to rabbits: safer, but far less rewarding.

One player proposes how to split a shared reward, and the other can accept or reject. If rejected, neither receives anything. This game shows that people often refuse unfair offers, showing that fairness and justice can outweigh pure self-interest in decision-making.

Player 1

Player 1 decides how to split a fixed reward, often between a 50/50 split and a very unequal split

EVEN SPLIT

UNEVEN SPLIT

⊗
REJECT

Player 2 chooses to either accept or reject the offer

If the responder rejects, neither player receives any money

Player 2

If the responder accepts, the money is divided as proposed

✓
ACCEPT

favours, food, and assistance are shared with the expectation that generosity will eventually be returned in both small and significant ways.

But reciprocity only works if others also reciprocate. The success of cooperation depends not just on an individual's willingness to help, but on how others behave. This is where game theory becomes useful. It provides tools to explore how cooperation can evolve in strategic, often competitive environments. Even seemingly simple games, like the Prisoner's Dilemma, reveal important subtleties. In this game, betrayal brings a higher payoff in a single interaction. Yet, if enough individuals in a population behave cooperatively over time, cooperation becomes the more successful strategy. Reciprocal altruism, then, is not just a moral instinct – it's an adaptive response shaped by interaction patterns within groups.

Game theory shows reciprocity thrives when cooperation repeats, making mutual help a winning long-term strategy

Humans also take reciprocity even further. Studies show that people often help others even when there is no guaranteed return. This is known as generalized reciprocity, or "paying it forward". Helping one person makes it more likely others will help you or your group later. Over time, this kind of indirect helping builds trust and strengthens cooperation within communities.

Reputation also matters. People tend to behave more generously when others are watching. This connects to the evolutionary idea of costly signalling – performing costly acts to signal hidden traits. A peacock's extravagant tail, for example, is burdensome but shows strength and genetic fitness. In humans, generous actions like giving to charity or volunteering can signal trustworthiness, reliability, or social status. Even if costly in the short term, these behaviours can bring indirect benefits by attracting friends, allies, and opportunities. Experiments show that people tend to give more when their actions are visible, suggesting that our kindness is partly shaped by how others perceive us.

DEALING WITH TRAITORS

When people don't follow social rules, punishment helps to enforce cooperation. In both small and large groups, individuals who cheat, steal, or refuse to contribute are often punished. Sometimes this is formal – through laws and courts – but people also engage in costly punishment, where they punish wrongdoers even at a personal cost. Classic psychological games like the Ultimatum Game suggest an innate sense of fairness; participants

285

286

MONUMENTS OF COOPERATION
The standing stones at Calanais, erected over 5,000 years ago, are a powerful testament to humans working together to create something meaningful. Some have suggested that they are evidence of free cooperation and cultural unity. However, their origins are unknown, and they may have equally been erected using coercive means.

will often reject an unfair offer, even if it means getting nothing themselves, choosing to punish the selfish proposer. This signals commitment to group norms and deters future selfishness. Similarly, in the Public Goods Game with Punishment, players are willing to spend their own money to penalize free-riders, demonstrating how costly punishment maintains cooperation by discouraging selfish behaviour. While some debate remains over whether such punishment is truly selfless, it clearly helps stabilize cooperation by discouraging behaviours that harm the group.

COOPERATIVE CULTURE

Perhaps the most far-reaching explanation for human cooperation is the success of cooperative cultures. Humans don't just pass on genes – we pass on ideas, beliefs, and behaviours. Through teaching, imitation, storytelling, and shared rituals, people learn what is expected in their communities. Over generations, cooperative norms – such as fairness, generosity, and punishing cheaters – become part of a group's culture. Groups with strong internal cooperation often function better, compete more effectively, and survive longer than less united ones. Their cultural traits spread not because they benefit individuals directly, but because they help groups succeed.

Ultimately, human cooperation is not mysterious – it's an evolved, flexible strategy built on many layers: family ties, mutual exchange, reputation, punishment, and culture. These mechanisms reinforce one another, helping us build lasting partnerships and complex societies. Far from being the exception, cooperation is what makes human civilization possible.

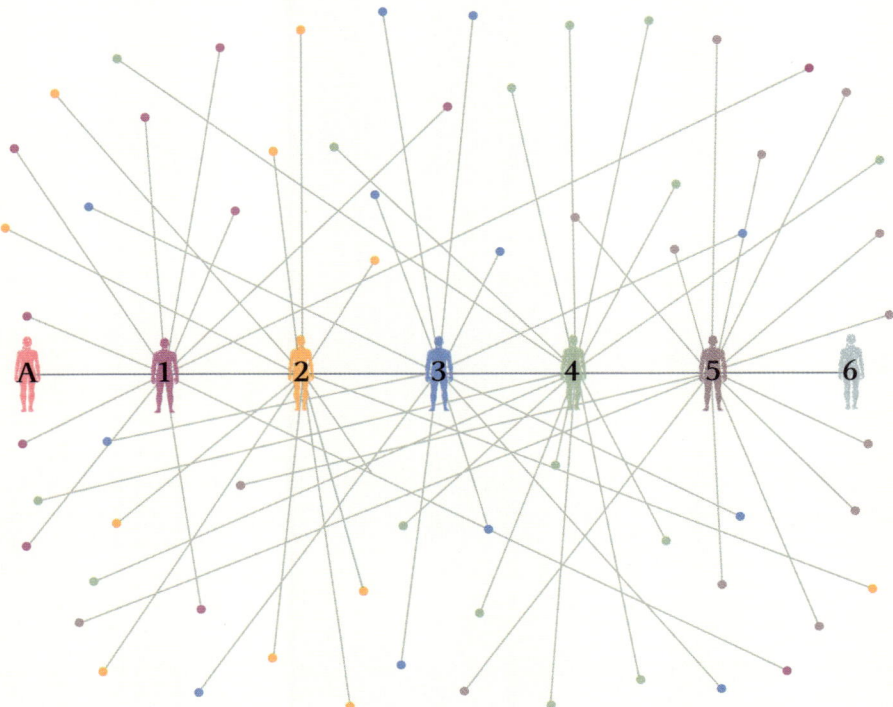

DEGREES OF SEPARATION
Mathematicians calculate that any two people on the planet are linked through a chain of just six relationships.

Social bonding

Many of our defining traits, such as our abilities to talk and to learn from one another, are linked to our propensity for social bonding and interaction – characteristics that extend back to our early ancestors and our primate heritage.

Researchers studied 30 billion electronic messages to confirm the six degrees concept

Many animals are gregarious, but humans take this to another level, cooperating with often unrelated individuals without the prospect of immediate benefit. Today, humans communicate across the globe, and mathematicians have calculated that any individual is linked to every other person on Earth by a chain of six acquaintances – the so-called six degrees of separation. Studying how human sociality evolved is difficult, because most social

behaviours leave no lasting evidence. However, by examining our archaeological prehistory and observing our closest ape relatives, we can make educated guesses.

OUR FAMILY AND OTHER SOCIAL ANIMALS
Our closest living cousins, chimpanzees and bonobos, are highly social, living in fluctuating groups of 10 to well over 100 individuals. Their social relationships are maintained through bonding activities, notably grooming one another's fur. Sessions may last for several hours. Like humans, these apes have friendships, rivalries, and power struggles. They learn from one another and may cooperate in defending territory and finding food. If our ancestors were anything like chimpanzees and bonobos, it is likely that social bonding runs in the family.

We humans are not regular social groomers. However, we do "hang out" and chat, using language not only as a tool for exchanging information, but also for bonding. Laughing, joking, and sharing anecdotes are activities that bring us closer together; some researchers have proposed that chatting has taken the place of grooming in the human lineage. British anthropologist Robin Dunbar argued that bonding through talk allowed us to expand our social group sizes and accommodate more close acquaintances than was possible in our ancestors. He suggested that there is a relationship between the relative size of the neocortex and primate group size. Dunbar extrapolated his theory to humans, and proposed that evolution permits us to foster around 150 meaningful relationships. Though influential in the 1990s, the statistics underlying Dunbar's number have been contested, and human group size and organisation is highly variable across cultures.

CENTRAL PLACES AND SOCIAL SPACES

Humans are "central-place foragers", returning to a home base at the end of each day. This base is where we prepare food, make tools, hang out, and sleep. Almost all people today (including hunter-gatherers) live in this way, but the origins of this habit are thought to predate our species. The first

solid evidence of hearths – where people sat together around the flames – dates to 780,000 years ago (see p.150), and evidence of wooden structures to almost half a million years ago. However, designated places for social interaction – where people gathered to process foods, eat, and make tools – appear 2.6 million years ago, and may have existed even earlier.

SOCIAL RITUALS

Many animals display ritual-like behaviour; elephants, for example, "bury" their dead by covering their bodies with leaves or soil. However, the scale and diversity of human rituals is unmatched. All human groups engage in collective rituals that may not be necessary for social bonding but can serve to foster social closeness and, perhaps, promote cooperation. The origins of ritual behaviour are hazy. The earliest artefacts with direct ties to modern-day

Early people may have gathered together at dedicated spaces more than 2.6 million ago

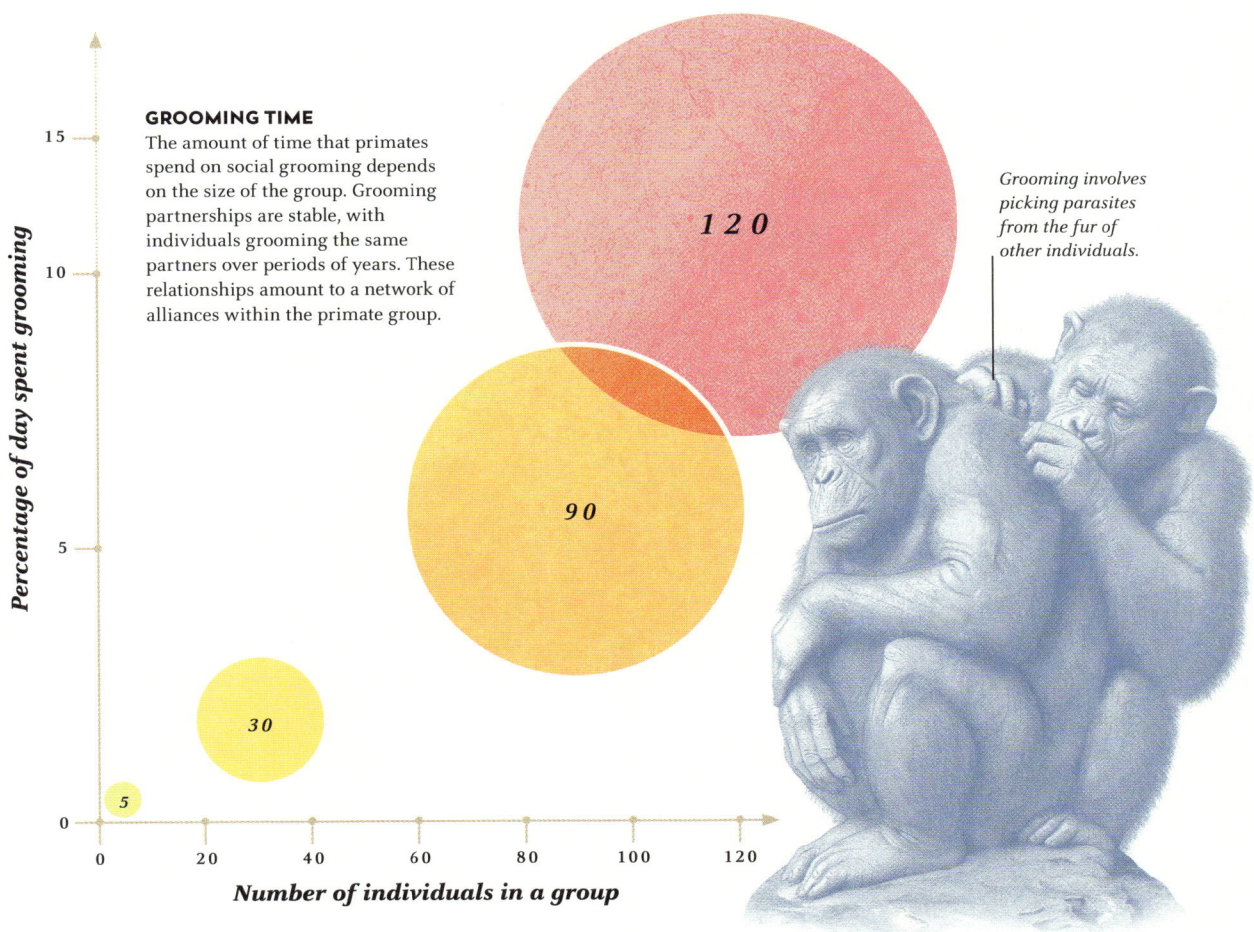

GROOMING TIME
The amount of time that primates spend on social grooming depends on the size of the group. Grooming partnerships are stable, with individuals grooming the same partners over periods of years. These relationships amount to a network of alliances within the primate group.

Grooming involves picking parasites from the fur of other individuals.

Percentage of day spent grooming

120

90

30

5

Number of individuals in a group

SOCIAL BONDING

Since we know most evidence decays, we should assume that many social behaviours – this includes language itself – could be much older than any surviving traces.

rituals are 12,000-year-old sticks found in a cave in southeastern Australia. These ritual sticks bear a striking similarity to the murrawan sticks used by Gunaikurnai medicine men and women in contemporary mulla-mulla rituals, which are performed to heal the sick and harm enemies. If the connection is genuine, it implies at least 10 millennia of uninterrupted social learning.

DEMOGRAPHY AND SOCIAL LEARNING

As human populations grew, so too did opportunities for innovation, just because more minds produced greater creativity and a wider sharing of ideas. This relationship is evidenced by a study of islands in Oceania from the time of early European contact. Islands with larger populations and connected communities developed more tools of greater sophistication than islands with smaller populations. Similar demographic processes may underlie the advent of technological and symbolic complexity over the last 100,000 years. As groups grow, so does a need for people to signal their identities and affiliations. For example, shell beads appear early in the archaeological record around 75,000

years ago at Blombos Cave and Border Cave, South Africa, and perhaps even earlier in north Africa and the Levant. Some archaeologists believe these were used as markers of social identity or group membership.

PRESERVATION AND UNCERTAINTY

Of course, most social bonding behaviours and artefacts do not stand the test of time. A recent study of contemporary African hunter-gatherers found that almost none of their ritual objects, instruments, or structures were made of enduring material. Most were made from wood, reed, and other materials that decay quickly, leaving no lasting evidence. We should assume that many social bonding behaviours – including language itself – could be considerably older than the existing evidence suggests. Anyone giving a precise date for the appearance of language, homes, hearths, music, or social rituals is concealing the fact that we do not know for sure when any of these appeared.

More people means greater collaboration in a group, generating more creativity

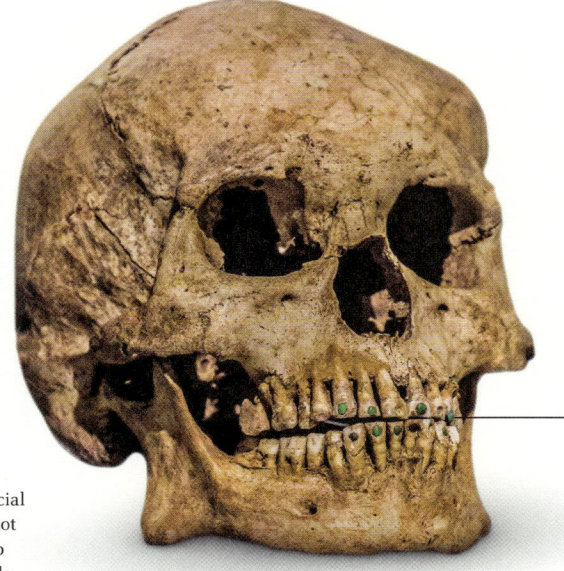

Teeth were drilled and inlaid with jade, which had spiritual meaning and may also have signified group identity and status

LEAVING A RECORD
One way to leave a record of social behaviour, although probably not the motivation in this case, is to decorate the body's most durable material – the teeth. This Mayan skull is more than 1,000 years old.

Mayan skull,
Classic Period (250–900 CE)

Competition and conflict

Wherever people live together, there is potential for conflict over space, status, and land that can trigger organized inter-group violence. Is such behaviour particular to modern humans or rooted more deeply in our ancestors?

Many group-living animals, from ants to meerkats, monkeys, and chimpanzees, may have coordinated territorial conflicts on the scale of large social groups. We would expect humans, as animals, to be no exception. However, researchers differ over how common or intense inter-group conflict has been in human prehistory. Were humans essentially peaceful before the origin of agriculture and settled life – or not?

PRESSURE FOR LAND

Human conflicts have been sparked by religion, ideology, and trade, but the principal cause of, or excuse for, war has always been competition for land. Land was always important, but a relatively recent invention made it extra valuable – farming. Agriculture requires fertile land. Not only that, farming boosts population growth, which in turn intensifies the need for more fertile land. Agriculture also permits the accumulation of surplus food, which can be used as a tool by elites to exert control over populations; there is a link between land and social status, or power.

Our species has been around for 300,000 years, but farming only really took off within the last 12,000, in a period called the Neolithic. Examples of inter-group violence and warfare from this time are unfortunately numerous. Mass graves in Talheim, Halberstadt, and Kilianstädten, Germany, dating to around 5000 BCE, contain skeletons of which more than 50 per cent show signs of injury; among the first gardeners of the Atacama Desert in Neolithic Chile, some 20 per cent of adult remains showed

WEAPONS OF WAR

Over time, hunting spear designs changed to reflect a new purpose – warfare. This is a replica of a spearhead found in Wales. First made in 1500 BCE, its shaft was 2.64 m (8 ft 8 in) long – too long and heavy to throw far in hunting. It was more likely made specifically for battle against humans or as a symbol of such martial potential and power.

Leaf-shaped blade made from bronze cast in a clay mould

Midrib divides the blade along its length

Loops allowed the wooden shaft to be secured and easily replaced

TERRITORIAL CONFLICT

This tracing of a rock painting in Castellón in the Spanish Pyrenees shows conflict between two rival groups. The date of around 5300 BCE is contested, so the painters may have been either early farmers or hunter-gatherers.

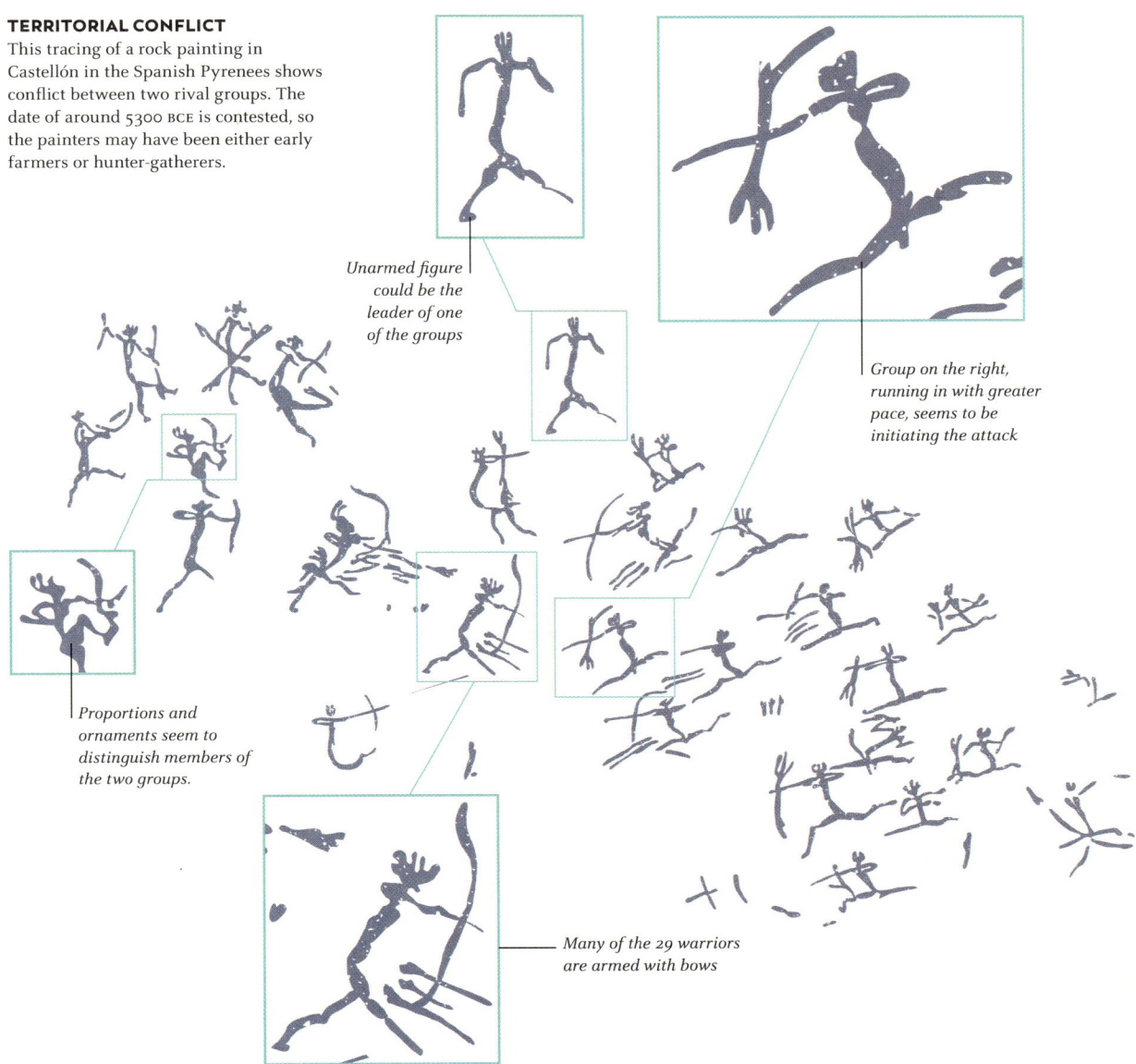

Unarmed figure could be the leader of one of the groups

Group on the right, running in with greater pace, seems to be initiating the attack

Proportions and ornaments seem to distinguish members of the two groups.

Many of the 29 warriors are armed with bows

Castellón battle scene

evidence of trauma; and in Honghe, China, 41 beheaded skeletons were buried together. There is indirect evidence of warfare from earlier in the Neolithic, too. The city of Jericho was bounded by walls, ditches, and towers by 8000 BCE. While these might have been built for flood protection, they probably also kept people out. After the Neolithic, and into the Bronze Age, conflicts only increased in scale and destruction. But is farmland the root of all conflict?

HUNTER-GATHERER CLASHES

Before the Neolithic, most humans lived as hunter-gatherers including during a period called the Mesolithic. While there is certainly less evidence of organized violence from this time, it is far from absent. Some of this evidence is indirect. Cave art in Castellón, Spain, appears to show a battle between Mesolithic archers, and fighting is pictured in stylized rock-art in Arnhem Land, Australia, some of it 10,000 years

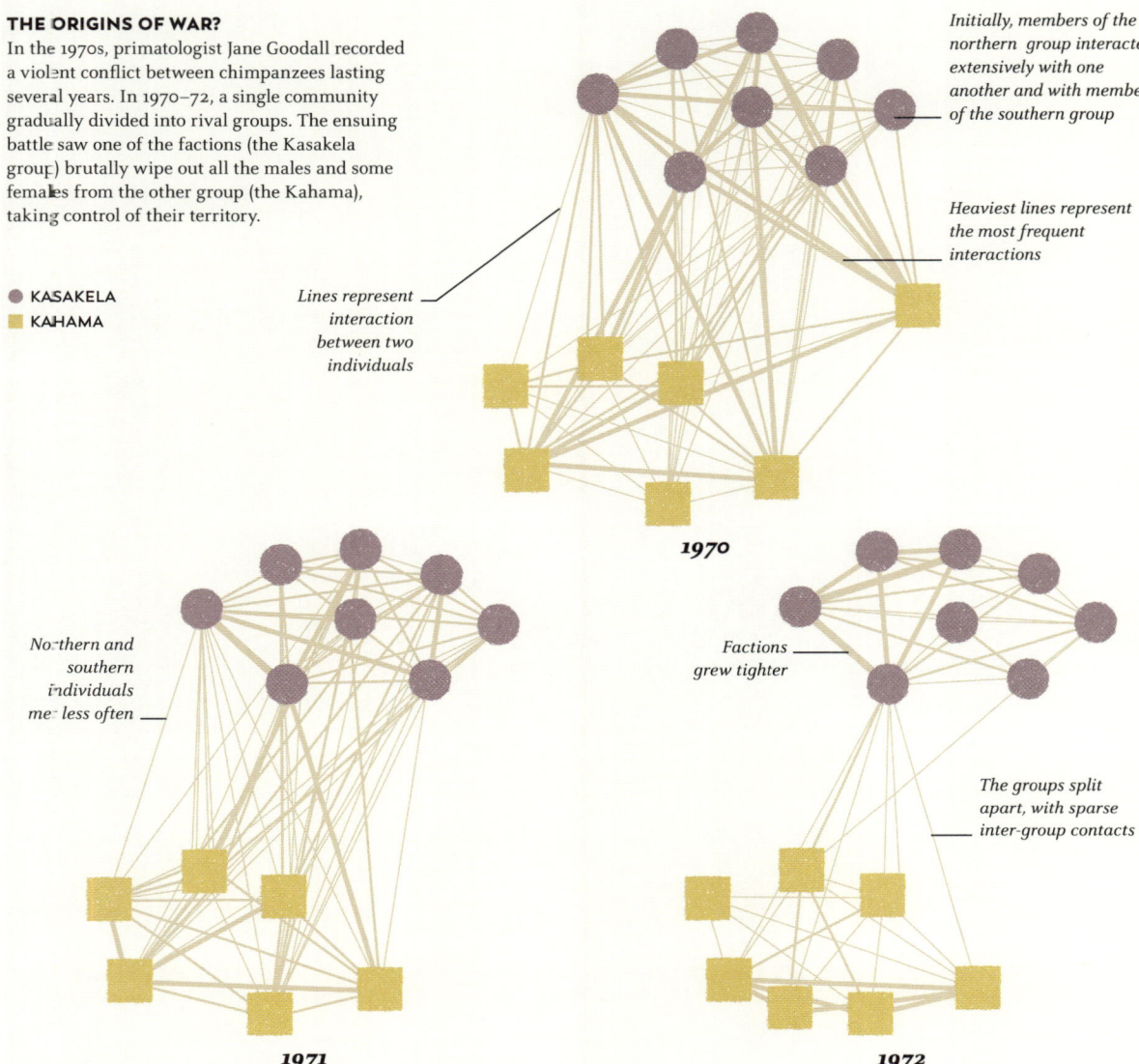

THE ORIGINS OF WAR?

In the 1970s, primatologist Jane Goodall recorded a violent conflict between chimpanzees lasting several years. In 1970–72, a single community gradually divided into rival groups. The ensuing battle saw one of the factions (the Kasakela group) brutally wipe out all the males and some females from the other group (the Kahama), taking control of their territory.

● KASAKELA
■ KAHAMA

Lines represent interaction between two individuals

Initially, members of the northern group interacted extensively with one another and with members of the southern group

Heaviest lines represent the most frequent interactions

1970

Northern and southern individuals met less often

1971

Factions grew tighter

The groups split apart, with sparse inter-group contacts

1972

old. Direct evidence of group conflict has been found in Mesolithic graveyards, including those in Vasylivka, Ukraine, where an unusually large proportion of skeletons shows signs of violent injury. And at Naratuk, in Kenya, 27 skeletons dating back 10,000 years display skull fractures and arrow wounds. These finds suggest that organized violence does predate farming. However, these ancient hunter-gatherers had settlements, resources, and stored food to fight over, and might be exceptions. It is safer to state that the prevalence of inter-group conflict in pre-agricultural societies remains contested.

CONFLICT BEFORE 10,000 BCE

Direct evidence for organized violence before about 10,000 BCE is in short supply. Since all of our ancestors lived as hunter-gatherers, some researchers have looked to modern hunter-gatherers for insights into past warfare. Many hunter-gatherers today display little coordinated violence and make few territorial claims over land. The Tanzanian Hadza, the Batek of Malaysia, and the Moriori of Chatham Island for example, move freely from place to place, making a living from resources that are highly dispersed throughout the landscape (and thus

Many modern hunter-gatherer groups travel widely and do not protect concentrated food sources or territories

difficult to protect or contest). Hadza oral tradition records only one historical conflict with neighbours, and they were not the aggressors. Some scholars argue that the peacefulness of contemporary hunter-gatherers is unnatural – a product of pacification by large nation-states – and point to people such as the Yolngu of Australia – who did have coordinated conflicts – as more indicative of prehistory.

Extending enquiry to the distant human past, around 5–7 million years ago, some researchers have looked to our closest cousins, the chimpanzees. They found that many chimpanzee groups engage in lethal raiding, and thinning the numbers of their rivals to expand their own territory. If chimpanzees do this, the logic goes, then early humans would have been able to do the same thing. Given the lack of evidence, these debates have been fraught and may remain unresolved. However, while the prehistory of coordinated conflicts is hard to characterize, there is plentiful ancient evidence of interpersonal strife.

INTERPERSONAL CONFLICT

Interpersonal violence certainly predates the evolution of our species. A 430,000-year-old skull, smashed by blunt force, was discovered in Atapuerca, Spain. Evidence of cannibalism has been found at the same site dated to at least 800,000 years ago. The development of hunting tools, including spears from 400,000 years ago, would have made lethal weapons available. Early humans had the opportunity of turning the tools they used for hunting on one another.

Throughout the animal kingdom, violence occurs between individuals. Social animals must compete within the group for food, territory, and mates. Some of our closest ape relatives are equipped with long canine teeth, due to this need to intimidate or defeat rivals. In the human lineage, the canines have shrunk, and some experts link this with a reduction in aggression as mating patterns changed in hominin societies (see p.186). Perhaps humans are not as aggressive as certain close relatives, but interpersonal conflict was common in the human past.

BLUNT FORCE INJURY
This pre-Neanderthal human died violently 430,000 years ago. Its injuries are consistent with powerful blunt-force trauma to the head.

Impact from a tool most likely caused the damage to the cranium

Skull from Atapuerca, Spain

Human history is one of serfdom and servitude, monarchs, and rulers, and the gap between the richest and poorest is wider today than ever. Inequality is embedded in human society but is it innate, and how and when did it originate?

Origins of inequality

Material wealth sustains the power structures that we have observed through history. Its allure, and its danger, is that it can be self-sustaining because it can be used to grow more wealth, by investing it or by paying others for their labour. By definition, material wealth must exist in a form that can be accumulated, hinting that wealth inequality may have emerged with the development of agriculture, crops being goods that could be stored, defended, and even taxed.

People who control resources and technologies are most likely to accumulate wealth.

STATUS IN DEATH

The clearest relics of ancient inequality – those that have survived long enough to be studied by archaeologists and anthropologists – are burial sites and grave goods, the assumption being that wealthy people were more likely to be buried at revered sites and with objects of value than the poor.

Some Neolithic sites dating back to early agricultural societies do indeed reveal signs of social inequality. The Great Barrows at Bougon (4800–3500 BCE), and the Cairn of Barnenez (4580–4200 BCE), in France, are among the earliest of all funerary monuments and would have required substantial manpower to build. Clearer examples of inequality appear later, in the copper and bronze ages. The Varna people who lived in northeast Bulgaria around 4500 BCE, were among the first to inter their elites together with golden jewellery – bangles, necklaces and pendants. And further east, burial sites at the Royal Cemetery at Ur, Mesopotamia, from

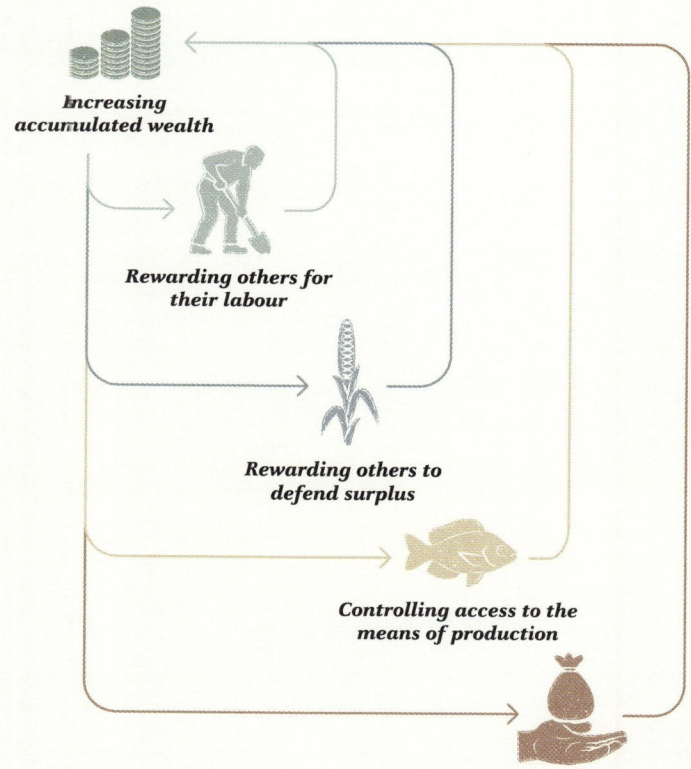

Increasing accumulated wealth

Rewarding others for their labour

Rewarding others to defend surplus

Controlling access to the means of production

Paying others to collect taxes

WEALTH FOR THE WEALTHY
Unlike inequality in physical strength, intelligence, or beauty, wealth inequality can be self-sustaining, creating a positive feedback loop that makes the wealthy ever wealthier.

The adornments found in burials at the Royal Cemetery at Ur reflected the power of Sumerian nobility.

Headdress of gold leaves adorned the forehead of a royal attendant in the so-called King's Grave

around 2600–2300 BCE contained not only gold and other treasures but also the bodies of royal attendants sacrificed to accompany their ruler to the afterlife.

THE PROPERTY PARADIGM

The key idea that only wealthy individuals were buried with valuable objects – known as the property paradigm – is supported by the analysis of bones at the Neolithic site at Osłonki in central Poland (c.4100 BCE). These studies show that individuals buried along with copper artefacts had more meat in their diets (and thus held more power and status) than those without. The trajectory of inequality was not, however, always a straight line. Analyses of the Neolithic settlement of Çatalhöyük (in modern-day Turkiye) from the 7th millennium BCE show that some burials had more grave goods than others, but that these inequalities appear to have decreased over the lifetime of the settlement.

ACCUMULATING INEQUALITY

Before the Neolithic, our ancestors lived as hunter-gatherers. Could their simpler lifestyle have supported a form of wealth accumulation

HIGH-STATUS BURIAL

The goods buried alongside the bodies of high-status individuals of the Varna culture include some of the oldest golden artefacts in the world.

capable of generating inequality in their societies? The evidence suggests that it could. This is because certain natural resources lend themselves to accumulation. Fish, such as salmon, which swim upriver annually to spawn, are a prime example. They suddenly appear in great numbers and can be caught, smoked, and stored. During salmon migrations, fishing weirs become prime real estate; whoever controls the weirs controls the food supply. The same is true of other wild resources; acorns, for example, were stored in ancient California and Japan (see p.170).

The use of fishing weirs began at least 5,000 years ago, and the resources accumulated through their use were leveraged to create inequality. Clear social distinctions existed in the hunter-gatherer-fisher groups of the Pacific Northwest of North America: the Chinook had hereditary castes and slavery; and the Tlingit and Kwakiutl chiefs held lavish ceremonies, called potlaches, to display wealth and prestige. Reaching back into history, the evidence for food storage is scant, and most people were

Inequality exists even in societies that do not farm the land.

assumed to be mobile rather than sedentary. This gap in our knowledge has prompted studies of living hunter-gatherers that allow us to make educated guesses about the hierarchies and inequalities that may have existed more than 16,000 years ago.

MOBILE HUNTER-GATHERERS

Today's hunter-gatherers are "residentially mobile", moving camp several times per year, so preventing resource storage of the type practiced by the Chinook. The societies of many mobile foragers are relatively egalitarian; hoarding is often frowned upon, and people will demand shares of each other's food. The group may actively prevent individuals getting too big for their boots: among the !Kung and the Mbendjele, braggarts are teased, while one anecdote from the Mbendjele tells of a man being ostracized for hunting too well. There are few formal positions of leadership and political power, and authority, where it exists, is exercised with humility and modesty.

But equality among contemporary hunter-gatherers is not a given. The Australian Yolngu, for example, have age-based status hierarchies grounded in secret ritual lore; the !Kung have hereditary land rights; and the Hadza reserve the best cuts of meat for men. Some societies are better at keeping the inequities in check, but a degree of inequality is typically present.

FISHING WEIR

The weirs used by ancient people may have been made of wood and reeds. They were used thousands of years ago not only in the Pacific Northwest, but also in Denmark, the Netherlands, Australia, and Southern Africa.

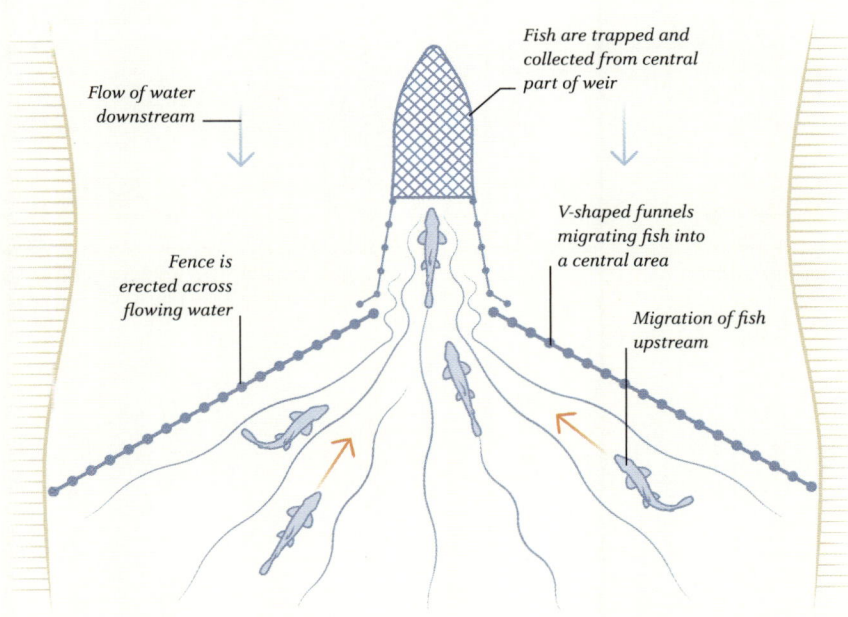

Flow of water downstream

Fish are trapped and collected from central part of weir

Fence is erected across flowing water

V-shaped funnels migrating fish into a central area

Migration of fish upstream

Individuals buried along with copper artefacts had more meat in their diets (and thus held more power and status) than those without.

ANCIENT COUNTING TOOL

The Lebombo bone is one of the oldest known mathematical artefacts. Featuring 29 distinct notches, it is believed to be a tally stick, used for counting, or a lunar calendar.

Notches intentionally cut into bone

Smooth, aged surface suggests frequent handling

Engraved dots may represent lunar phases, hence the days of a calendar

Bone about 7 cm (2.8 in) long

Engraved antler plaque

ANCIENT LUNAR CALENDAR

The Abri Blanchard plaque is around 32,000 years old and bears one of the earliest known examples of a notational system – possibly a lunar calendar. Its markings suggest that Palaeolithic hunter-gatherers tracked and recorded time and seasons.

Side and top views

Humans have long felt compelled to record their world and their thoughts, seeking to preserve memory and make sense of the life around them. This is a need that lies at the heart of how we understand and share our experiences.

Recording and writing

The human impulse to encode and convey meaning beyond speech can be traced back tens of thousands of years. During the Ice Age, we see evidence of this in the geometric symbols that often accompany paintings of animals. Across numerous European caves, certain motifs – including dots, lines, chevrons, and hand stencils – appear repeatedly over long periods. Their consistency suggests these symbols were used deliberately for communication, rather than decoration. Researchers have proposed that these markings indicated group identity, marked events, or reflected associations with animals and places. What's more, the repeated symbols may have functioned as records of crucial seasonal events, such as the numbers and timing of animal migration or breeding cycles, which was critical knowledge for survival.

TRACKING TIME AND QUANTITY

Building on the foundation of symbolic communication, early humans also developed specific ways to systematically record quantities and time. Evidence of numerical awareness appears in tally sticks and incised bones, such as the Lebombo bone, which is around 42,000 years old and was found in a cave on the border of South Africa and Eswatini. As humans transitioned to farming and settled life, pressures to monitor crops, herds, and property grew. Monumental structures such as stone circles and passage tombs, often aligned with the Sun and stars during solstices, suggest systematic observation of time, as marked out precisely by lights in the sky. Together, these developments suggest an evolution towards organized measurement of quantity and time.

Early farming societies tracked seasons through monuments aligned with the Sun and stars

FIRST WRITING SYSTEMS

As human societies became larger and more complex, the need for consistent record-keeping grew. Early farmers and traders developed

DAWN OF WRITING

These clay tablets found in Uruk, Mesopotamia, date to about 3200 BCE. They feature some of the earliest Sumerian pictograms, representing a crucial step in the development of the cuneiform writing system and the beginning of recorded history.

Pictogram relates to records of grain distribution

Circles, wedges, and slashes denote numbers

Administrative tablet

Impression pushed into wet clay before firing

Possible list of slave names

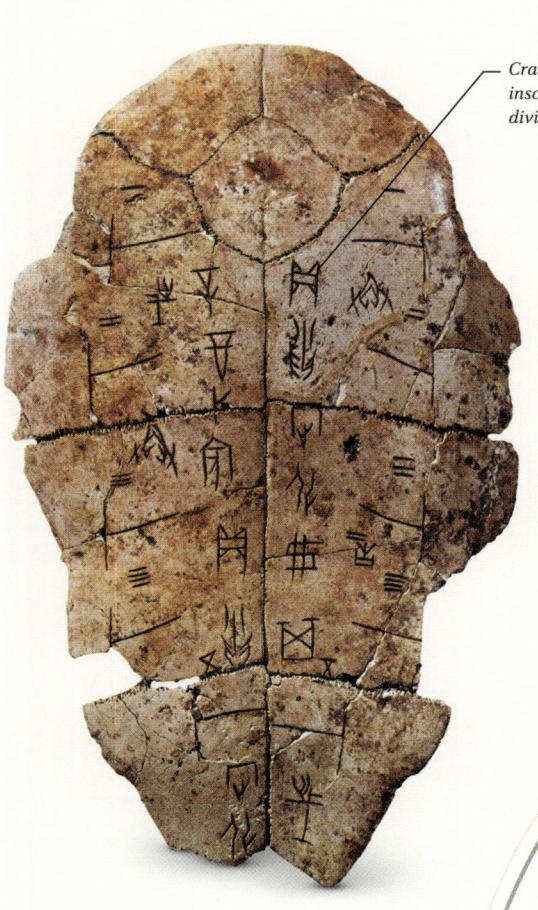

Cracks and carved inscriptions record divinations

RECORDED HISTORY

From prehistoric tally sticks and clay tokens to early scripts across Mesopotamia, Egypt, China, and the Americas, these artefacts trace humanity's development of counting, recording, and writing, revealing the foundations of communication and administration.

c.8,600–8,200 YA
Chinese bones and shells bear Jiahu symbols – incised marks predating writing

c.7,300–6,500 YA
Vinča symbols – early southeastern European markings on pottery, tablets

c.10,000–5,500 YA
Mesopotamian clay counting tokens shaped as spheres, crescents, and disks

c.8,000–6,000 YA
Neolithic pottery, southwest Asia and Europe, marked with geometric patterns

10,000 YA

EARLY CHINESE SCRIPT

This Shang dynasty tortoise shell, dating to around 1200 BCE, was used in divination rituals to communicate with ancestors and deities, providing the earliest known examples of a fully-developed Chinese writing system.

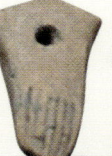

Clay counting tokens

c.30,000–10,000 YA
European incised stones and engraved pebbles possibly bear mnemonic marks

c.25,000 YA
Ishango bone, DR Congo, bears engraved marks, likely used as a counting device

c.42–17,000 YA
Symbols in cave art suggest a system of recording numbers or calendar information

50,000 YA

c.44–42,000 YEARS AGO (YA)
Lebombo bone, South Africa, features notches, likely used for counting or tracking lunar cycles

systems to manage goods, property, and labour. In Mesopotamia, clay tokens used for counting evolved into impressed symbols on tablets, which developed gradually into cuneiform writing in 3200–2600 BCE. At first it was used to record transactions, deliveries, inventories, and labour lists, forming the basis of commerce and administration. Eventually, symbols called logographs emerged that encoded words and ideas, and an entire language could be written.

In Egypt, hieroglyphs served both sacred and administrative purposes. Like cuneiform, the symbols were primarily logographic, although they incorporated many phonetic elements that represented sounds. This phonetic aspect was used by people in the Sinai region of Egypt to

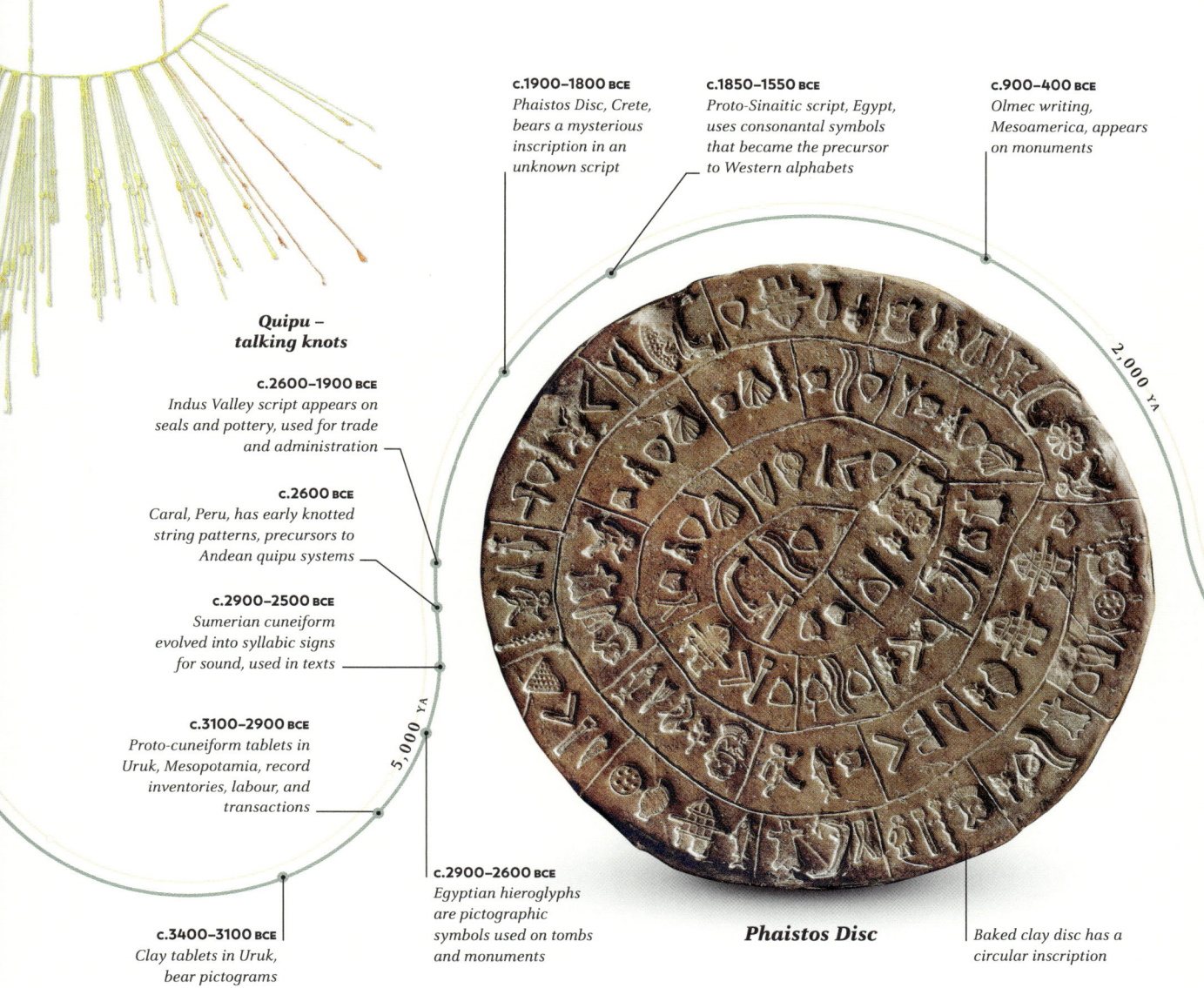

c.1900–1800 BCE
Phaistos Disc, Crete, bears a mysterious inscription in an unknown script

c.1850–1550 BCE
Proto-Sinaitic script, Egypt, uses consonantal symbols that became the precursor to Western alphabets

c.900–400 BCE
Olmec writing, Mesoamerica, appears on monuments

Quipu – talking knots

c.2600–1900 BCE
Indus Valley script appears on seals and pottery, used for trade and administration

c.2600 BCE
Caral, Peru, has early knotted string patterns, precursors to Andean quipu systems

c.2900–2500 BCE
Sumerian cuneiform evolved into syllabic signs for sound, used in texts

c.3100–2900 BCE
Proto-cuneiform tablets in Uruk, Mesopotamia, record inventories, labour, and transactions

2,000 YA

5,000 YA

c.3400–3100 BCE
Clay tablets in Uruk, bear pictograms

c.2900–2600 BCE
Egyptian hieroglyphs are pictographic symbols used on tombs and monuments

Phaistos Disc

Baked clay disc has a circular inscription

write their own Semitic language. Their phonetic system was later developed by the Phoenicians, Hebrews, and Greeks, and led to many alphabetic systems familiar today. Meanwhile in China, writing appeared on oracle bones and bronze vessels around 1200 BCE, using logographs to link language with divination and political power. In Mesoamerica, the Maya developed complex glyphs that similarly blended logographic and phonetic elements, recording dynastic histories, rituals, and astronomical observations.

Across these early civilizations, writing was a tool for control. It allowed rulers, priests, and administrators to govern increasingly organized societies, creating the first bureaucracies and formal histories. Over time, these utilitarian systems expanded into instruments for literature, culture, and self-expression, and advanced humans' drive to preserve knowledge.

CULTURAL AND EVOLUTIONARY IMPACT

Recording and writing represent key revolutions in human cultural evolution. Writing externalized memory, allowing ideas to be stored and shared across great distance and time, and enabled the rise of history, law, and science. Once limited to scribes and priests, literacy eventually became a universal skill, transforming how humans communicate, learn, and build knowledge.

Writing recorded memory, enabling knowledge sharing across generations

The present
and future

*For millennia, human evolution was governed by
the blind forces of nature and chance. Today,
culture and technology have become the dominant
selective drivers, dramatically reshaping the
pressures that determine survival and reproduction.*

Following the transition from foraging to
farming, humans entered a period of accelerating
change. Agriculture created food surpluses and
set the stage for population expansion. From the
15th century onwards, European expansionism
connected distant regions. The resulting gene
flow – the mixing of populations due to migration
– has been a dominant force in changing human
genetics over the last few millennia. These
forces were later amplified by industrialization,
driving mass urbanization and rapid innovation.
Together, these shifts prompt an enduring
question: as selection weakens, how do chance,
movement, and mutation increasingly shape
human evolution?

Population growth remained gradual for millennia, only accelerating sharply with industrialization. Global numbers rose from roughly 1 billion in 1800 to over 8 billion today. This massive population effectively wraps around the globe, meaning there is almost no geographical isolation left. Consequently, gene flow acts to homogenize populations, while the sheer size of the human gene pool buffers against rapid shifts caused by random chance.

Post-WWII innovations triggered huge global growth in population

NATURAL WORLD IMPACT

This explosive growth defines the Anthropocene, an era in which human domination of the Earth's systems has introduced entirely novel

MAJOR MEGACITY

From above, Tokyo sprawls like a living organism – humanity's grand experiment in cooperation, innovation, and adaptation. This vast metropolis embodies our species' evolution from tribal wanderers to architects of planetary-scale civilizations.

evolutionary pressures. This dominance is clear in biomass: 30 per cent of all land mammals are humans, 67 per cent are livestock, and only 3 per cent are wild animals. We now produce over 300 million tonnes of plastic annually, and have created more than 170,000 synthetic mineral-like substances. These contaminants,

THE PRESENT AND FUTURE

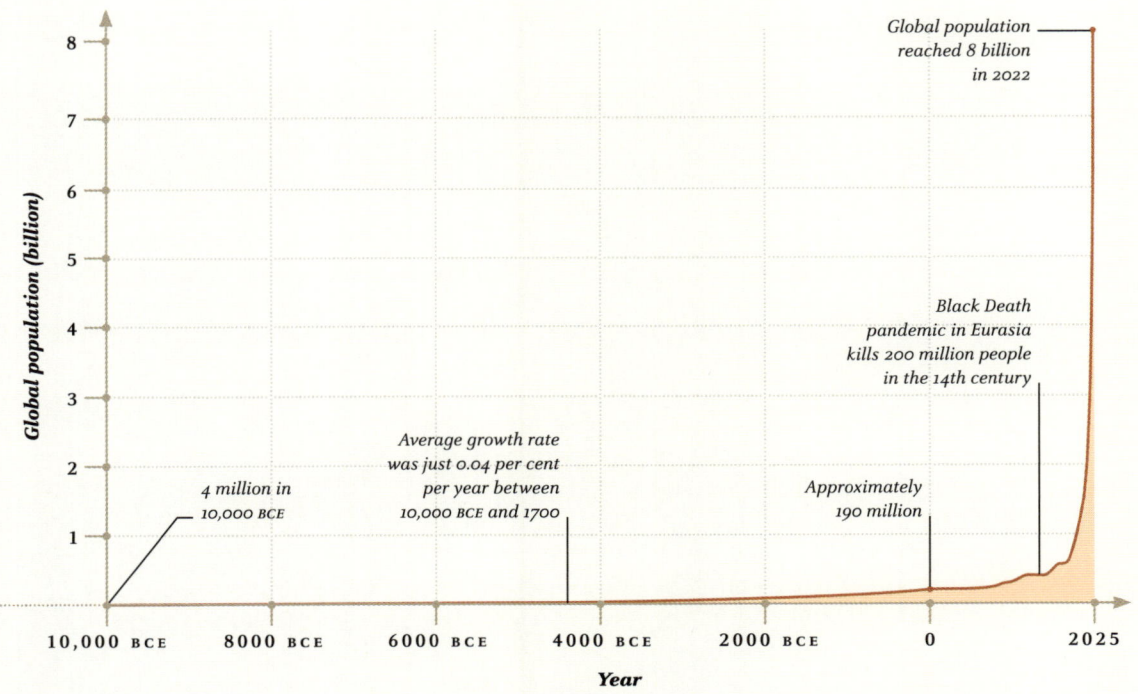

POPULATION EXPLOSION
A steep, unrelenting curve traces humanity's rapid growth, highlighting both our expanding presence and the mounting pressures on the planet's finite resources.

particularly microplastics now found in human blood, breast milk, and major organs, introduce novel physiological and potential genetically damaging stressors that human biology has never encountered.

Climate change adds another layer of impact. Since the late 19th century, global temperatures have risen by over 1.5 °C (2.7 °F), and sea levels by over 25 cm (10 in), reshaping coastlines and increasing exposure to extreme weather and pollutants. These human-driven shifts ensure that evolution continues in a world increasingly shaped by our own activity.

RELAXED EVOLUTIONARY PRESSURES

In prehistory, survival depended on traits such as strength, disease resistance, and hunting skill. In developed regions, modern healthcare, vaccination, sanitation, and global food systems now buffer many of these pressures, creating

Modern advances lessen survival pressures on humans, but billions continue to experience them

"relaxed selection" for traits once essential. This relief is highly uneven, however: billions still lack safe sanitation, clean water, adequate food, or access to medical care so traditional survival pressures persist for much of the world. The easing of old pressures can also generate new ones. The "hygiene hypothesis", for instance, suggests that reduced microbial exposure in some societies may contribute to immune-related conditions such as allergies and asthma.

MODERN STRESSES

Modern life introduces intense new evolutionary forces: lifestyle diseases, mental health stressors, and pervasive chemical exposure. However, in many developed regions these conditions alter selection in subtle and indirect ways rather than acting as consistently lethal pressures. By contrast, fast-evolving pathogens remain a potent driver of human evolution. Microbes adapt on timescales

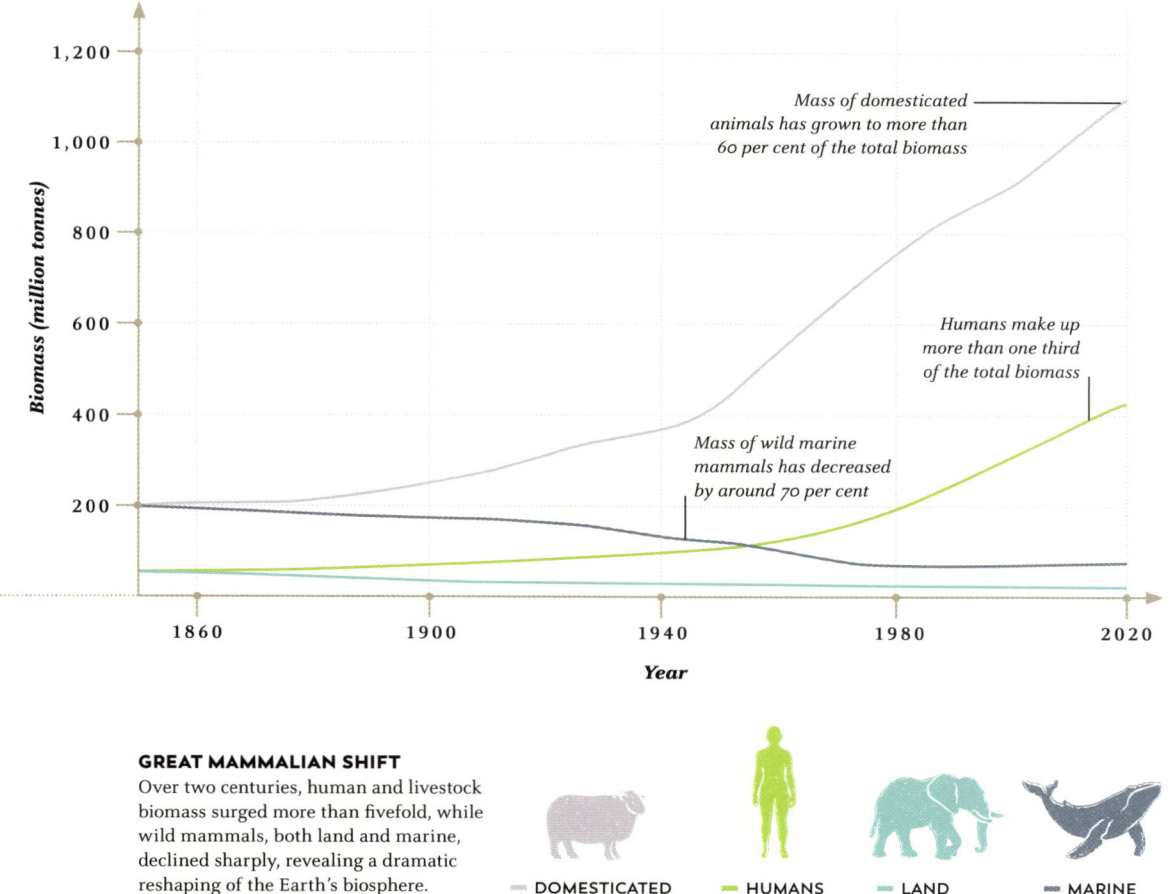

GREAT MAMMALIAN SHIFT
Over two centuries, human and livestock biomass surged more than fivefold, while wild mammals, both land and marine, declined sharply, revealing a dramatic reshaping of the Earth's biosphere.

■ DOMESTICATED　　■ HUMANS　　■ LAND　　■ MARINE

Chart labels:

Mass of domesticated animals has grown to more than 60 per cent of the total biomass

Humans make up more than one third of the total biomass

Mass of wild marine mammals has decreased by around 70 per cent

Y-axis: **Biomass (million tonnes)** — 1,200 / 1,000 / 800 / 600 / 400 / 200

X-axis: **Year** — 1860 / 1900 / 1940 / 1980 / 2020

far shorter than human generations, and outbreaks such as COVID-19 demonstrate how infectious diseases can still influence survival, including among people of reproductive age.

Globally, the rise of chronic conditions highlights "mismatch diseases", in which our ancient physiology is poorly adapted to sedentary lives and processed diets. Additionally, air pollution kills millions annually, and cancer rates among under-50s have risen 79 per cent in 30 years, underscoring the potency of novel environmental stressors alongside persistent pathogen threats.

Despite these challenges, humans continue to evolve in response to their environments. Modern populations show shifts in gut and skin microbiomes as diets, hygiene, and antibiotic use reshape our resident microbes, alongside immune-system changes influenced by widespread vaccination and early-life microbial exposure.

Humans continue to evolve socially and biologically in response to enduring challenges

LOOKING FORWARDS
The future of human evolution will be defined by a profound blend of natural and human-guided processes. Technological advancements in genetic engineering, assisted reproduction, and emerging biotechnologies allow for "directed evolution", enabling the conscious shaping of traits. Environmental pressures – particularly those related to climate and pollution – will continue to shape human biology alongside these interventions. Evolution is increasingly becoming a biological-cultural coevolution, amplified by technology and cross-generational knowledge transfer. Ultimately, the story of human evolution may be defined as much by conscious, collective choice as by the indifferent forces of nature.

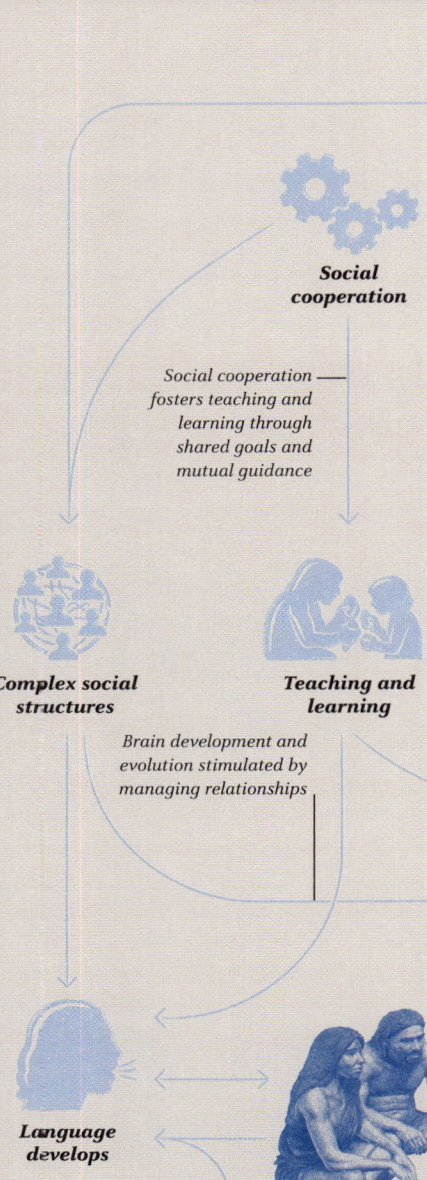

Social cooperation

Social cooperation fosters teaching and learning through shared goals and mutual guidance

Complex social structures

Brain development and evolution stimulated by managing relationships

Teaching and learning

Permanent settlements

Social care and social bonding

Extended childhood deepens social bonding via dependence and shared play

Extended childhood

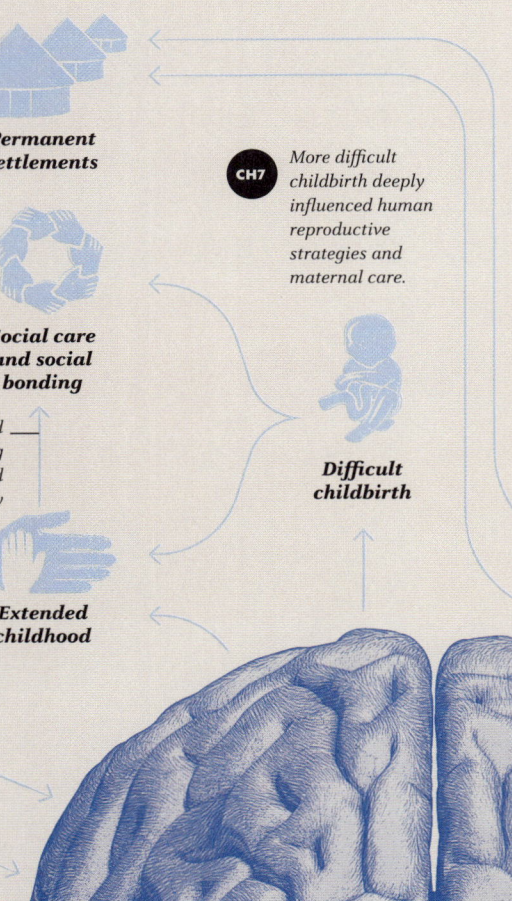

CH7 More difficult childbirth deeply influenced human reproductive strategies and maternal care.

Difficult childbirth

Brain development and evolution stimulated by cultural exchange

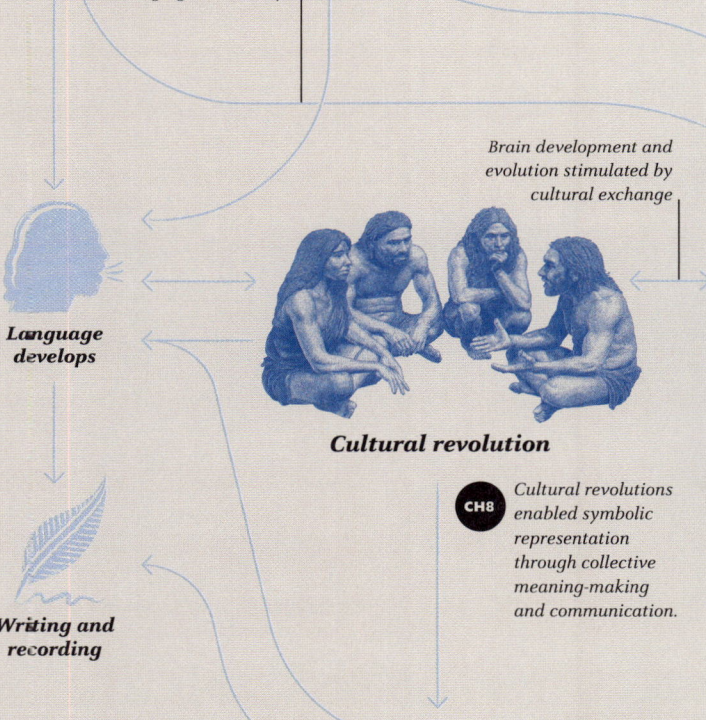

Cultural revolution

CH8 Cultural revolutions enabled symbolic representation through collective meaning-making and communication.

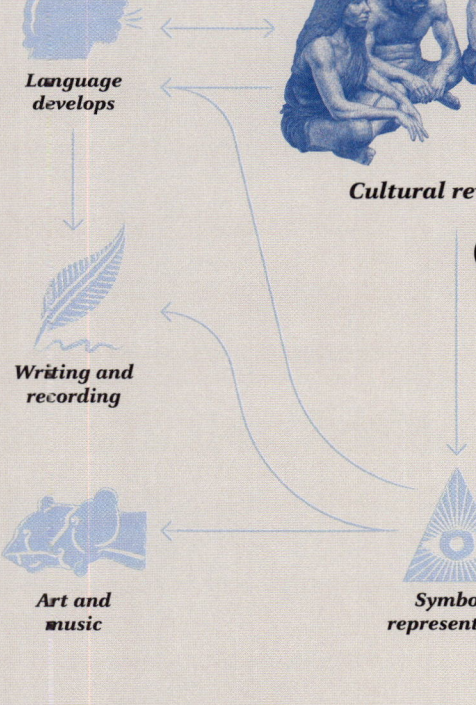

Language develops

Writing and recording

Art and music

Symbolic representation

MIND MAP
This network connects the many threads of evolution, from the numerous factors that may have contributed to the increased size and complexity of the human brain, to the far reaching consequences and feedback loops that resulted.

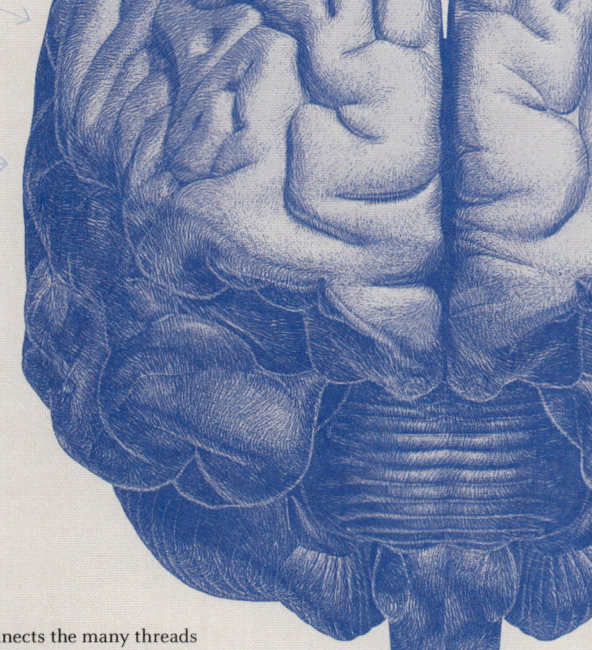

Inputs and outputs

Brain development knits threads of human evolution, but what mixture enlarged it? Dexterity, technology, teaching, and social interaction likely drove growth, yielding language, art, music, global dispersal, and eventual transformation of Earth's ecosystem.

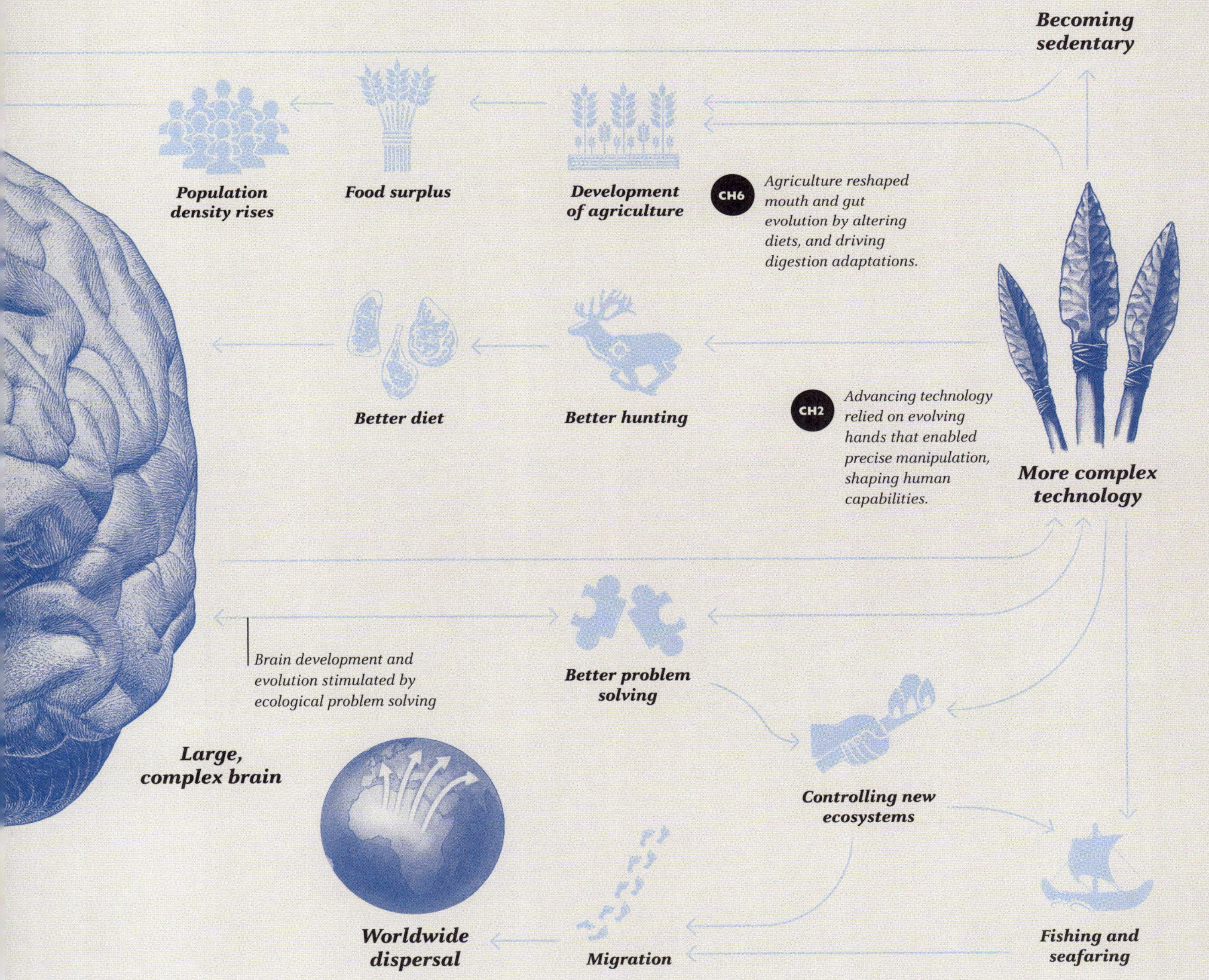

Becoming sedentary

Population density rises

Food surplus

Development of agriculture

CH6 Agriculture reshaped mouth and gut evolution by altering diets, and driving digestion adaptations.

Better diet

Better hunting

CH2 Advancing technology relied on evolving hands that enabled precise manipulation, shaping human capabilities.

More complex technology

Brain development and evolution stimulated by ecological problem solving

Better problem solving

Large, complex brain

Controlling new ecosystems

Worldwide dispersal

Migration

Fishing and seafaring

Index

Acknowledgments

DK would like to thank picture researchers Aditya Kaytal and Samrajkumar S and cartographer Simon Mumford. DK would also like to thank Steve Crozier for image retouching, Sarah Carpenter for proofreading, Helen Peters for creating the index, and fact-checkers Bharti Bedi, Katie Cederborg, Katie John, and Priyanka Kharbanda. Editorial support by Becky Gee.

The publisher would like to thank the following for their kind permission to reproduce their photographs:
(Key: a-above; b-below/bottom; c-centre; f-far; l-left; r-right; t-top)

6 Science Photo Library: Paul D Stewart (br). Shutterstock.com: Sipa (bl). 7 Stephen Smith: (t) 25 Tim White: Illustration based on mCT imagery- C. Owen Lovejoy et al. ,Careful Climbing in the Miocene: The Forelimbs of Ardipithecus ramidus and Humans Are Primitive.Science326,70, 70e1-70e8(2009).DOI:10.1126 / science.1175827 for Ardipithecus ramidus hand (c). 29 Wim Lustenhouwer / Vrije Universiteit University Amsterdam: Adapted Figure on the photo-Joordens, J., dErrico, F., Wesselingh, F. et al. Homo erectus at Trinil on Java used shells for tool production and engraving. Nature 518, 228–231 (2015). https://doi.org/10.1038/nature13962 (b). 34 Dreamstime.com: Dmitrii Moroz (c). Museum of Stone Tools-https:// stonetoolsmuseum.com/, https://creativecommons.org/licenses/by/4.0/: Michael Curry (crb); Emma Watt (crb/Point). 35 Alamy Stock Photo: Granger - Historical Picture Archive (bl). Science Photo Library: Philippe Psaila (br). 36-37 Shutterstock. com: Yes058 Montree Nanta (c). 37 Sonia Harmand MPK/WTAP: (tl). 38 Upper Galilee Museum of Prehistory: (l). 38-39 The Metropolitan Museum of Art: Purchase, Arthur Ochs Sulzberger and Friends of Arms and Armor Gifts, Arthur Ochs Sulzberger Bequest, and funds from various donors, 2018 (t). 41 AncientCraftUK: Dr. James Dilley & Emma Jones (bl, c). 43 AncientCraftUK: Dr. James Dilley & Emma Jones (t). 44 AncientCraftUK: Dr. James Dilley & Emma Jones (br). 44-45 AncientCraftUK: Dr. James Dilley & Emma Jones (b). 45 AncientCraftUK: Dr. James Dilley & Emma Jones (b). 64-65 Tim White: Illustration based on mCT imagery courtesy of G. Suwa, U. Tokyo and the Middle Awash research project; see Science 2009 C. Owen Lovejoy et al. ,Combining Prehension and Propulsion: The Foot of Ardipithecus ramidus.Science326,72, 72e1-72e8(2009).DOI:10.1126 / science.1175832 for description of the Ardipithecus foot. 67 Science Photo Library: John Reader. 92 Elsevier: Infographic based on-Willian da Silva, Juan R. Godoy-López, Álvaro Sosa Machado, Andressa Lemes Lemos, Carlos Sendra-Pérez, Manuel Gallango Brejano, Felipe P. Carpes, Jose Ignacio Priego-Quesada, (bl). 96 Francesco d'Errico: (crb). 97 Francesco d'Errico: (clb). Institute of Archaeology and Ethnography of the Siberian Branch of the Russian Academy of Sciences, Novosibirsk, Russia: (cb). Science Photo Library: Eht Zurich (cra). 98 Alamy Stock Photo: Penta Springs Limited / Artokoloro (bl). Don Hitchcock, donsmaps.com: (tc). PNAS: M. Soressi, S.P. McPherron, M. Lenoir, T. Dogandži, P. Goldberg, Z. Jacobs, Y. Maigrot, N.L. Martisius, C.E. Miller, W. Rendu, M. Richards, M.M. Skinner, T.E. Steele, S. Talamo, & J. Texier, Neandertals made the first specialized bone tools in Europe, Proc. Natl. Acad. Sci. U.S.A. 110 (35) 14186-14190, https://doi.org/10.1073/pnas.1302730110 (2013). (br). 99 Photo Scala, Florence: AGF (r). 101 The Metropolitan Museum of Art: The Charles and Valerie Diker Collection of Native American Art, Gift of Charles and Valerie Diker, 2019 (tl). 102 Reproduced with permission from Renée Rust and Charles Helm. 103 Alamy Stock Photo: Heritage Image Partnership Ltd / © Fine Art Images (br). Qasigiannguit Museum, Greenland: (tl). 104 AncientCraftUK: Dr. James Dilley & Emma Jones (t). 105 AncientCraftUK: Dr. James Dilley & Emma Jones (tl, tr). Aaron Deter-Wolf: After Deter-Wolf et al. 2021, "Ancient Native American bone tattooing tools and pigments: Evidence from central Tennessee (Fig. 3) (clb). 128 Science Photo Library: Zephyr. 142 Dikika Research Project: (tr). 143 Niedersächsisches Landesamt für Denkmalpflege (NLD): Volker Minkus, © NLD (t). Smithsonian Institution: Human Origins Program, NMNH, Smithsonian Institution (b). 146 Smithsonian Institution: Human Origins Program, NMNH, Smithsonian Institution

(cla). 146-147 Photo Scala, Florence: RMN-Grand Palais / Gérard Blot / Dist. Foto SCALA, Florence. 147 Photo Scala, Florence: RMN-Grand Palais / Gérard Blot / Dist. Photo SCALA, Florence (cr). 148 AncientCraftUK: Dr. James Dilley & Emma Jones (bl). 149 AncientCraftUK: Dr. James Dilley & Emma Jones (c, bc). 150 AncientCraftUK: Dr. James Dilley & Emma Jones (tl). 152 Alamy Stock Photo: Science History Images (tc). 154 Alamy Stock Photo: TT News Agency / Johan Nilsson. 160-161 AncientCraftUK: Dr. James Dilley & Emma Jones (t). 166 Science Photo Library: Dennis Kunkel Microscopy. 171 Alamy Stock Photo: V. Dorosz. 175 Alamy Stock Photo: Penta Springs Limited / Artokoloro. 176 Alamy Stock Photo: Ed Buziak (br). Getty Images / iStock: Ivan-96 (bl, bc). 177 Science Photo Library: Science Museum Group (br, t). 204 Elissa L. Newport, Ph.D.: Infographic based on-O.A. Olulade, A. Seydell-Greenwald, C.E. Chambers, P.E. Turkeltaub, A.W. Dromerick, M.M. Berl, W.D. Gaillard, & E.L. Newport, The neural basis of language development: Changes in lateralization over age, Proc. Natl. Acad. Sci. U.S.A. 117 (38) 23477-23483, https://doi. org/10.1073/pnas.1905590117 (2020). (c). 226 Katja Heuer & Roberto Toro: Figure adapted from Heuer, K., Traut, N., Aristide L. et al. Principles of neocortical organisation and behaviour in primates. bioRxiv (2025). https://doi. org/10.1101/2025.07.17.665410. 234-235 Alamy Stock Photo: Robert J Preston (t). 236 AncientCraftUK: Dr. James Dilley & Emma Jones (x3). 237 AncientCraftUK: Dr. James Dilley & Emma Jones (tr). 238 AncientCraftUK: Dr. James Dilley & Emma Jones (c, r). 239 Science Photo Library: Philippe Psaila (tr). 245 Alamy Stock Photo: Adam Ján Fige (crb). Bridgeman Images: © Kenneth Garrett / GEO Image Collection (t). 253 Adam Brumm: Maxime Aubert (tr). 254 Alamy Stock Photo: Robertharding / Michael Nolan. 256 Alamy Stock Photo: Album (c). 257 Alamy Stock Photo: Ann Ronan Picture Library / Photo12 (t). 258 Pierre-Jean Texier. 259 Getty Images: The Image Bank / Alcibbum Photograph / Fadil Aziz. 260-261 Getty Images: Patrick Aventurier (c). 262 Alamy Stock Photo: Album (cl); The Natural History Museum, London (cr). 263 Alamy Stock Photo: Historic Images (cl). SuperStock: Fine-Art-Images / A. Burkatovski (cr). 264 Alamy Stock Photo: Dinodia Photos RF (tl). 264-265 Getty Images: Stockbyte / Paul Souders (t). 266-267 University of Tübingen: Hilde Jensen (c). 267 Alamy Stock Photo: Maxppp / Frederic Charmeux (br). Don Hitchcock, donsmaps.com: (bl). Science Photo Library: MSF / Javier Trueba (bc). 269 Dreamstime.com: Zhasminaivanova (tr). 274 PNAS: M. Fujita, S. Yamasaki, C. Katagiri, I. Oshiro, K. Sano, T. Kurozumi, H. Sugawara, D. Kunikita, H. Matsuzaki, A. Kano, T. Okumura, T. Sone, H. Fujita, S. Kobayashi, T. Naruse, M. Kondo, S. Matsuura, G. Suwa, & Y. Kaifu, Advanced maritime adaptation in the western Pacific coastal region extends back to 35,000–30,000 years before present, Proc. Natl. Acad. Sci. U.S.A. 113 (40) 11184-11189, https://doi.org/10.1073/pnas.1607857113 (2016). (c). 275 Alamy Stock Photo: Dietmar Rauscher (c). 277 Alamy Stock Photo: Rolf Richardson. 280 Getty Images / iStock: HilalOnac. 281 Bridgeman Images: © 2025 Museum of Fine Arts, Boston. All rights reserved. (cra); © NPL - DeA Picture Library (tl); © Ashmolean Museum (tr); © 2025 Museum of Fine Arts, Boston. All rights reserved. / E. Rhodes and Leona B. Carpenter Foundation Grant and Edwin E. Jack Fund (cr). 286-287 Alamy Stock Photo: Unknown1861. 291 Alamy Stock Photo: Imagebroker. com / Raimund Franken. 292 AncientCraftUK: Dr. James Dilley & Emma Jones (r). 293 Nature Publishing Group (www.nature.com): Figure Adapted from-Alt, K.W., Tejedor Rodríguez, C., Nicklisch, N. et al. A massacre of early Neolithic farmers in the high Pyrenees at Els Trocs, Spain. Sci Rep 10, 2131 (2020). https://doi.org/10.1038/ s41598-020-58483-9. http: / / creativecommons.org / licenses / by / 4.0 / (t). 295 Science Photo Library: MSF / Javier Trueba. 297 Alamy Stock Photo: Edwin Baker (br); MET / BOT (t). 300 Francesco d'Errico: (c). Shutterstock.com: Yes058 Montree Nanta (clb). 301 Alamy Stock Photo: BibleLandPictures / Zev Radovan (cb); Zev Radovan (bc). 302 Alamy Stock Photo: Granger - Historical Picture Archive (tl); The History Collection (cr). 303 Getty Images: Corbis Historical / Leemage (ca). 304-305 Getty Images / iStock: E+ / FotoVoyager (t)

Cover images: Back: Getty Images: David Levenson bl

DK LONDON

Senior Editor Rob Houston
Senior Art Editor Jessica Tapolcai
Editors Sarah Bailey, Abigail Ellis, Rebecca Fry,
David Summers, Marek Walisiewicz
Project Art Editor Francis Wong
Designers Noor Ali, Steve Bere
Illustrators Arran Lewis,
Phil Gamble, Mesa Schumacher
Managing Editor Angeles Gavira
Managing Art Editor Michael Duffy
Production Editor Andy Hilliard
Production Controller Rebecca Parton
Jacket Designer Luke Bird
Publishing Director Georgina Dee
Art Director Maxine Pedliham

DK DELHI

Senior Jacket Designer Suhita Dharamjit
Senior DTP Designer Tanya Mehrotra
Senior Jackets Coordinator
Priyanka Sharma Saddi

First published in Great Britain in 2026 by
Dorling Kindersley Limited
20 Vauxhall Bridge Road,
London SW1V 2SA

The authorized representative in the EEA is
Dorling Kindersley Verlag GmbH. Arnulfstr. 124,
80636 Munich, Germany

Copyright © 2026 Dorling Kindersley Limited
A Penguin Random House Company
10 9 8 7 6 5 4 3 2 1
001–341849–06/2026

A CIP catalogue record for this book
is available from the British Library.
ISBN: 978-0-2416-8274-6

Printed and bound in China

www.dk.com